Intellectual Property Issues in Biotechnology

Intellectual Property Issues in Biotechnology

Edited by

Harikesh B. Singh
Banaras Hindu University, India

Alok Jha
International Livestock Research Institute, India

Chetan Keswani
Banaras Hindu University, India

CABI

CABI is a trading name of CAB International

CABI
Nosworthy Way
Wallingford
Oxfordshire OX10 8DE
UK

Tel: +44 (0)1491 832111
Fax: +44 (0)1491 833508
E-mail: info@cabi.org
Website: www.cabi.org

CABI
745 Atlantic Avenue
8th Floor
Boston, MA 02111
USA

Tel: +1 617 682 9015
E-mail: cabi-nao@cabi.org

A catalogue record for this book is available from the British Library, London, UK.

Library of Congress Cataloging-in-Publication Data

Names: Singh, H. B., Dr., editor. | Jha, Alok, 1966- editor. | Keswani, Chetan, editor. | C.A.B. International, issuing body.
Title: Intellectual property issues in biotechnology / edited by Harikesh B. Singh, Alok Jha, Chetan Keswani.
Description: Wallingford, Oxfordshire ; Boston, MA : CABI, [2016] | Includes bibliographical references and index.
Identifiers: LCCN 2016022764 (print) | LCCN 2016023846 (ebook) | ISBN 9781780646534 (hbk : alk. paper) | ISBN 9781780646541 (ePDF) | ISBN 9781780646558 (ePub)
Subjects: LCSH: Biotechnology--Patents. | Biotechnology--Social aspects. | Biotechnology--Moral and ethical aspects. | Intellectual property. | MESH: Intellectual Property | Biotechnology | Developing Countries
Classification: LCC TP248.175 .I58 2016 (print) | LCC TP248.175 (ebook) | NLM TP 248.175 | DDC 660.6--dc23
LC record available at https://lccn.loc.gov/2016022764

ISBN-13: 978 1 78064 653 4

Commissioning editor: David Hemming
Editorial assistant: Emma McCann
Production editor: Shankari Wilford

Typeset by SPi, Pondicherry, India
Printed and bound in the UK by CPI Group (UK) Ltd, Croydon, CR0 4YY

Contents

Contributors

Elisabeth Abergel, Department of Sociology, Institut des Sciences de l'Environnement, Université du Québec à Montréal (UQAM), 405 Rue Sainte-Catherine Est, Montréal, QC H2L 2C4, Canada

Monica Alandete-Saez, The Public Intellectual Property Resource for Agriculture, University of California Davis, Davis, California, USA

Adelaide Antunes, School of Chemistry, Federal University of Rio de Janeiro (EQ/UFRJ); National Institute of Industrial Property (INPI), Rio de Janeiro, Brazil

Avinash Arivazhahan, Department of Pharmacology, Kasturba Medical College, Manipal-576104, India

Tanoy Bandyopadhyay, Department of Biotechnology, Tocklai Tea Research Institute, Tea Research Association, Jorhat-785008, Assam, India

Tomas G. Bas, Institute of Innovation based in Science, Vice-rectorate for Innovation and Technology Transfer, University of Talca, Av. Lircay S/N, Talca, Chile

Nitai Basak, Department of Biotechnology, National Institute of Technology, Jalandhar-144011, Punjab, India

Alan B. Bennett, The Public Intellectual Property Resource for Agriculture, University of California Davis, Davis, California, USA

Kartikay Bisen, Department of Mycology and Plant Pathology, Institute of Agricultural Sciences, Banaras Hindu University, Varanasi-221005, India

Michael Blakeney, Faculty of Law, University of Western Australia, 35, Stirling Highway, Crawley, Perth, WA 6009, Australia

Ashim Borah, Department of Biotechnology, National Institute of Technology, Jalandhar-144011, Punjab, India

Sangeeta Borchetia, Department of Biotechnology, Tocklai Tea Research Institute, Tea Research Association, Jorhat-785008, Assam, India

Dhanay M. Cadillo Chandler, Faculty of Law, University of Helsinki, Helsinki, Finland

Ananda M. Chakrabarty, Department of Microbiology & Immunology, University of Illinois College of Medicine, Chicago, IL 60612, USA

Puneet S. Chauhan, CSIR-National Botanical Research Institute, Rana Pratap Marg, Lucknow-226001, India

Pritom Chowdhury, Department of Biotechnology, Tocklai Tea Research Institute, Tea Research Association, Jorhat-785008, Assam, India

Sunil K. Dhar, Microbial Culture Collection, National Centre for Cell Science, Pune-411021, India

Graham Dutfield, School of Law, Liberty Building, University of Leeds, Leeds LS2 9JT, UK

Chien-Te Fan, Institute of Law for Science & Technology, National Tsing Hua University, Taiwan

Kathleen L. Hefferon, Department of Food Sciences, Cornell University, Ithaca, NY 14850, USA

Akira Homma, Bio-Manguinhos, Oswaldo Cruz Foundation (FIOCRUZ), Rio de Janeiro, Brazil

Tzu-Hsun Hung, Attorney-at-Law, Taiwan

Asim K. Jana, Department of Biotechnology, National Institute of Technology, Jalandhar-144011, Punjab, India

David J. Jefferson, The Public Intellectual Property Resource for Agriculture, University of California Davis, Davis, California, USA

Alok Jha, International Livestock Research Institute, New Delhi-110008, India

Chetan Keswani, Department of Mycology and Plant Pathology, Institute of Agricultural Sciences, Banaras Hindu University, Varanasi-221005, India

Ashwani Kumar, UP Council of Agricultural Research, Lucknow-226010, Uttar Pradesh, India

Rajendra Kumar, UP Council of Agricultural Research, Lucknow-226010, Uttar Pradesh, India

Ravindra Kumar, ICAR – Indian Agricultural Research Institute, Regional Station, Karnal-132001, Haryana, India

Jiang Li, Kenneth Wang School of Law, Soochow University, Shi Zi Jie 1 Hao, Su Zhou, 215006, China

Jui-Chu Lin, College of Intellectual Property Studies, National Taiwan University of Science & Technology, Taiwan

Reinaldo M. Martins, Bio-Manguinhos, Oswaldo Cruz Foundation (FIOCRUZ), Rio de Janeiro, Brazil
Flavia M.L. Mendes, School of Chemistry, Federal University of Rio de Janeiro (EQ/UFRJ), Brazil
Aradhana Mishra, CSIR-National Botanical Research Institute, Rana Pratap Marg, Lucknow-226001, India
Sandhya Mishra, Department of Mycology and Plant Pathology, Institute of Agricultural Sciences, Banaras Hindu University, Varanasi-221005, India
Sankalp Misra, CSIR-National Botanical Research Institute, Rana Pratap Marg, Lucknow-226001, India
Liliana Mitkova, Institut de Recherche en Gestion, University of Paris Est Marne-la-Vallée (UPEM), Bât. BDE, Rue Gallilée, 77420 Champs-sur-Marne, France
Chandra S. Nautiyal, CSIR-National Botanical Research Institute, Rana Pratap Marg, Lucknow-226001, India
Rakesh Pandey, Department of Microbial Technology and Nematology, CSIR-Central Institute of Medicinal and Aromatic Plants, Lucknow-226015, India
Aabheejeet Pansare, Microbial Culture Collection, National Centre for Cell Science, Pune-411021, India
Cristina Possas, Bio-Manguinhos, Oswaldo Cruz Foundation (FIOCRUZ), Rio de Janeiro; Evandro Chagas National Institute of Infectious Diseases, Oswaldo Cruz Foundation, Rio de Janeiro, Brazil
Om Prakash, Maharana Pratap Govt P.G. College, Hardoi-241001, Uttar Pradesh, India
Om Prakash, Microbial Culture Collection, National Centre for Cell Science, Pune-411021, India
Gerard M. Raj, Department of Pharmacology, Jawaharlal Institute of Postgraduate Medical Education and Research, Pondicherry-605006, India
Birichi K. Sarma, Department of Mycology and Plant Pathology, Institute of Agricultural Sciences, Banaras Hindu University, Varanasi-221005, India
Kanikaram Satyanarayana, Intellectual Property Rights Unit, Indian Council of Medical Research, Department of Health Research, Ansari Nagar, New Delhi-110029; Biosafety Unit, Department of Biotechnology, New Delhi-110003
Swati Sharma, CSIR-National Botanical Research Institute, Rana Pratap Marg, Lucknow-226001, India
Akanksha Singh, Department of Microbial Technology and Nematology, CSIR-Central Institute of Medicinal and Aromatic Plants, Lucknow-226015, India
Harikesh B. Singh, Department of Mycology and Plant Pathology, Institute of Agricultural Sciences, Banaras Hindu University, Varanasi-221005, India
Poonam C. Singh, CSIR-National Botanical Research Institute, Lucknow-226001, Uttar Pradesh, India
Vinit P. Singh, School of Agriculture, Lovely Professional University, Phagwara-144411, Punjab, India
Asha Sinha, Department of Mycology & Plant Pathology, Institute of Agricultural Sciences, B.H.U., Varanasi-221005, Uttar Pradesh, India
Sadhana Srivastava, Intellectual Property Rights Unit, Indian Council of Medical Research, Department of Health Research, Ansari Nagar, New Delhi-110029, India
Seweta Srivastava, UP Council of Agricultural Research, Lucknow-226010, Uttar Pradesh, India
Vivek Srivastava, CSIR-National Botanical Research Institute, Rana Pratap Marg, Lucknow-226001, India
Xi Wang, Central University of Finance and Economics, 39 South College Road, Haidian District, 100081, Beijing, China
Sujit K. Yadav, UP Council of Agricultural Research, Lucknow-226010, Uttar Pradesh, India

Foreword

The rapidly evolving international scenario of the biotechnology landscape can provide a leading economic edge to any nation if legitimately protected through intellectual property (IP) rights. Innovations and developments in the biotechnological sector have been directly linked to foreign investments and economic development. The significance of effective protection of intellectual property in biotechnological innovations in agriculture, healthcare and industries is poorly understood. Taking into consideration that IP protection has a major impact on national progress, there is a catastrophic need to emphasize the role of IP issues in economic development in a way that it is readily adapted by industry, academia and research institutes.

Recent breakthroughs in the area of genome editing, synthetic biology, life sciences, crop improvement and pharmaceutics have created vast opportunities but simultaneously posed new legal questions and challenges. Agreement on Trade Related Aspects on Intellectual Property Rights (TRIPS Agreement) has extended IP protection to inventions in all fields of technology. However, there are countries still fighting to understand the TRIPS agreement and struggling to incorporate it within their legal frameworks.

In this context, this volume includes contributions from vastly experienced, global experts of IP-related issues in agriculture, pharmaceutics, healthcare and industrial biotechnology, in a comprehensive and influential manner, with most up-to-date facts and extended case studies. This volume addresses the most recent and vital issues in various sectors of biotechnology including agriculture, pharmaceutics, healthcare and industry. This book presents IP issues in biotechnology in a manner that will prove to be an invaluable tool for academicians, scientists, legal practitioners and policy makers.

I congratulate the editors for synchronizing with global authorities on the subject to underline the upcoming challenges and present the most viable options for translating commercially viable ideas into easily affordable products and technologies.

S. Ayyappan

Preface

The term Intellectual Property Rights (IPR) describes the set of rules and regulations imposed to protect the application of ideas of commercial value in a geo-temporal space. IPR, particularly patents, occupy a prominent position in the innovation systems in life sciences; but to what extent they support or hinder innovation is widely disputed. Biotechnology is a rapidly booming field including pharmaceuticals, food, agriculture, etc. However, despite broad applications of biotechnology, profits on investments are surprisingly low. Thus, it is vitally important for universities, institutions, public research organizations and private enterprises to protect their innovation. At present, IP protection for biotechnological inventions is in a state of uncertainty. The concurrence of biotechnology and IPR has caused adverse challenges before the IP laws and regulations. There are vast differences in opinions on patentability of living organisms, which are largely barred from patentability. However, mindsets are rapidly shifting and IP issues in biotechnology are receiving great attention. In order to compete in global biotechnology-based markets, developing countries are amending their patent laws in order to encourage investment in biotechnology. Recent advances in biotechnological innovation have resulted in remarkable transformations in the fields of genetics and molecular biology and have opened new vistas in the healthcare, pharmaceutical and agricultural sectors.

India is the third largest biotech industry in the Asia-Pacific region and amongst the 12 largest biotech destinations in the world. Increasing government expenses augment the growth of this sector and the Government of India has planned to spend US$3.7 billion on biotechnology from 2012 to 2017 during the 'Make in India' campaign. It is estimated that the biotech industry in India will grow at an average rate of around 30% and reach US$100 billion by 2025. Growing demand for healthcare services, advanced R&D activities and strong government initiatives have resulted in an increase in the market size of the biotech sector, which is expected to reach US$11.6 billion by 2017. In India, the Biotechnology Industry Research Assistance Council (BIRAC) supports the biotech industry through mentoring, financial aid and infrastructure development. The Indian biotech sector is primarily led by pharmaceuticals, which has accounted for the largest share in the biotech industry followed by bio-services, agri-biotech, industrial biotech and bioinformatics.

This volume addresses the most recent and vital issues in various sectors of biotechnology including agriculture, pharmaceutics, healthcare and industry. The book is divided into four parts, each representing a different area of biotechnology. Part 1 comprises five chapters and discusses the: applications of biotechnology in agriculture, industry and medicine; status of biotechnology in developing countries; patent eligibility issues in life science; the Indian Patents Act (chapters 1–5). Part 2 includes six chapters (chapters 6–11), which address the IP issues in agricultural biotechnology: IP in agricultural biotechnology; bioprospecting for soil health and crop production; genetically modified crops and the future of Canadian agriculture; recent innovations in agricultural biotechnology; Chinese innovation system and knowledge sharing; and IPR in agricultural biotechnology in India. Part 3 (chapters 12–15) emphasizes the role of IP issues in industrial biotechnology including: DNA patenting; genetic patents in China and Taiwan; IP issues in bioprospecting microbial diversity; and IP disputes in CRISPR/Cas9 genome editing technology. Part four discusses IP issues in pharmaceutical biotechnology and comprises seven chapters (chapters 16–22). These address: public interest and the patent system in healthcare innovations; patentability of human embryo stem cells in the USA and Europe; a Brazilian perspective of innovations and IP issues in vaccines development; regulatory barriers and health access through biogenerics in India; Indian status in IPR protection in drug research; and IP issues in drug development.

The present book highlights emerging legal, social and regulatory issues pertaining to biotechnology and IPRs in agriculture, food, pharmaceutics, healthcare and industry. It presents expert opinions by premier academicians, scientists and industrialists to counter the critical questions such as whether the existing IP frameworks at both national and international levels are durable to withstand the challenges posed by contemporary biotechnology. It also offers an inclusive outline of biotechnology and IPR and connects various aspects of the topic in an integrated manner. It is aimed at providing comprehensive knowledge to a wide range of audiences, including students, scholars, researchers, legal practitioners, policy makers and others interested in IP issues in biotechnology.

Harikesh B. Singh
Alok Jha
Chetan Keswani

1 Biotechnology in Agriculture, Medicine and Industry: An Overview

Harikesh B. Singh,[1]* Alok Jha[2] and Chetan Keswani[1]

[1]Institute of Agricultural Sciences, Banaras Hindu University, Varanasi, India;
[2]International Livestock Research Institute, New Delhi, India

1.1 Introduction

Biotechnology is a rapidly booming field influencing pharmaceutical, food, agriculture and related industries. Biotechnological based interventions are vital components in virtually any industry employing microorganisms, cells and tissues for the production of biologicals. The global market for biotechnology is estimated to grow at an average compound annual growth rate (CAGR) of 11.6% from 2012 to 2017 and to reach a value of US$414.5 billion by the end of 2017. The global market was valued at approximately US$216.5 billion in 2011 (PR Newswire, 2014). India has the third largest biotech industry in the Asia-Pacific region and is amongst the 12 largest biotech destinations in the world. Increasing government expenses augment the growth of this sector and the Government of India has planned to spend US$3.7 billion on biotechnology from 2012 to 2017 during the 'Make in India' campaign. It is estimated that the biotech industry in India will grow at an average growth rate of around 30% and reach US$100 billion by 2025 (DIPP, 2014). Growing demand for healthcare services, advanced R&D activities and strong government initiatives have resulted in an increase in the market size of the biotech sector, which is expected to reach US$11.6 billion by 2017. In India, a special organization including the Biotechnology Industry Research Assistance Council (BIRAC) supports the biotech industry through mentoring, financial aid and infrastructure development. The Indian biotech sector is primarily led by pharmaceuticals, which accounted for the largest share in the biotech industry followed by bio-services, agri-biotech, industrial biotech and bioinformatics. The effect of biotechnology can be observed predominantly in major domains including agriculture, healthcare and industry.

In the area of human welfare and medicine, monoclonal antibodies, vaccines, lifesaving drugs, human insulin, disease diagnostic kits and gene therapy are some of the notable achievements of red biotechnology (pharmaceutical biotechnology). Microorganisms have formed a major role in the industrial production of biochemicals, including ethanol, enzymes, antibiotics etc. They are also being employed in the food processing industry for enhancing industrial processes leading to sustainable commercial production (industrial biotechnology). In agriculture, rapid production of disease-free planting material through tissue culture, development of novel crop varieties, somaclonal variations and transgenic crops with valuable genes for biotic and abiotic resistance through genetic engineering have ensured the enhanced crop production and better livelihood of growers (agricultural biotechnology).

1.2 Biotechnology in Agriculture

Cultivation of crops is as old as the history of humans, with the selection of better planting materials with superior traits through unintentional breeding being a predominant feature. Conventional plant breeding over the last 200 years has offered various superior crop varieties resulting in enhanced crop production and better livelihoods. In recent decades, modern agriculture has gone through significant changes, including the introduction

*Corresponding author, e-mail: hbs1@rediffmail.com

of new crop varieties through mutation breeding and transgenics. Intervention of biotechnological applications has revolutionized the current scenario of food production.

The term 'biotechnology' has been used to refer to various biological processes in the production of beneficial products (Coombs, 1992; Zaid *et al.*, 1999); however, in the present scenario it covers a broader area (McCouch, 2001). Earlier methods involved in modification of the genotype of plants and animals, and still used alone or in combination with various DNA-based methods, are referred to as 'conventional' plant breeding. Plants with alien genes transferred from unrelated organisms by genetic engineering are referred to as genetically modified (GM).

In 2014, the area under biotech crops was growing at the rate of 3–4% (6.3 million ha/year), reaching 181.5 million ha (Mha) and considered as the most widely adapted technology in modern agriculture (James, 2014). The most popular transgenics in 2014 included Bt cotton (25.1 Mha), herbicide-tolerant soybean (90.7 Mha), canola (9 Mha) and maize (55.2 Mha) (James, 2014). GM crops were valued at US$15.7 billion (2014) of which 72% market share was from developed countries and the remaining 28% was contributed by developing countries (James, 2014)

1.3 Biotechnology in Industry

Microbes have been associated with human health and welfare for a long time. The ability to manipulate the genotype of microbes has revolutionized life science and laid the foundation for innovations in industrial applications. Biotechnological applications in industry allow novel manufacturing processes that are environmentally safe and economically sustainable. Industrial biotechnology, commonly referred to as white biotechnology, is the application of modern biotechnology for sustainable production and processing of materials, chemicals, fuels, textiles, food and feeds, electronics and bio-energy using living microorganism and their enzymes (Tang and Zhao, 2009). Demand for sustainable production of various materials and chemicals has led to the advancement of industrial biotechnology, where microbe-based fermentation is used to produce biofuels and biochemicals (Nielsen and Jewett, 2008; Tang and Zhao, 2009; Otero and Nielsen, 2010; Du *et al.*, 2011; Sauer and Mattanovich, 2012).

The industrial biotechnology sector in India recorded an 11.2% growth in 2011–2012 in terms of revenue, with total sales of Rs6970 million. This growth is credited to increased enzyme consumption in the food, detergent, pharmaceutical and energy sectors. Multinational companies have contributed about 65% of the total market whereas local companies have shared about 35%. The enzyme-based biotechnology sector recorded an 11.47% growth during the period 2002–2012. For 2011–2012, enzyme consumption in India reached Rs5270 million and exports attained Rs1700 million. For 2012–2013, estimated growth was 11.17% (Parmeshwaram *et al.*, 2013). Taking into consideration that biotechnology is frequently connected with reduction in energy expenditure, lower greenhouse gas emissions and low waste production, much attention has been focused on this field. A key factor driving the growth of industrial applications of biotechnology is guaranteed economic profit, as biotechnology assures efficient production at lower operating costs. Rising demands for sustainability and eco-friendly industrial production systems associated with continuous exhaustion of natural crude oil reserves will continue to boost this trend (Soetaert and Vandamme, 2006). Expanding R&D sectors in biotechnology, encouraging government policies and other positive responses have aided the execution of industrial applications of biotechnology. Moreover, the ground-breaking innovations in enzyme engineering and technology, metabolic engineering, '-omics' science and computational systems biology are also anticipated to boost this sector.

1.4 Biotechnology in Medicine

Biotechnology encompasses processes used since biblical times, such as wine fermentation, and more recently discovered genetic engineering (Salomaa *et al.*, 1995; Lubiniecki, 1997; Walsh, 2000). Biotechnology in drug development and pharmacology is generally referred to as pharmaceutical biotechnology. Biopharmaceuticals are various proteins with therapeutic activity and nucleic acid for application in gene therapy. The pharmaceutical biotechnology industry is the most dynamically growing sector over the last 30 years. The success of insulin and sulfonamides gave new heights to the pharma industry (Walsh, 2000; Weng and DeLisi, 2000). Biotechnological processes used most regularly in the pharmaceutics industry include recombined DNA technology, monoclonal antibodies, directed mutagenesis, biocatalysts, metabolic engineering, vaccine technology and gene therapy.

Table 1.1. Drugs produced by recombinant technology (Stryjewska *et al.*, 2013).

Drug name	Active molecule	Disease
Humulin R, Humalog, NovoRapid, Gensulin R, Actrapid HM	Insulin	Diabetes
Factor VIII	Factor VIII	Haemophilia A in male
Factor IX	Factor IX	Haemophilia B
Human growth hormone (HGH)	Serostin, genotropin, humatrope	Growth hormone deficiency
Eprex, Epogen	Erythropoietin (EPO)	Anaemia
Rebif, Avonex, Betaseron	Interferons	Hepatitis B
Aldurazyme	α-L-iduronidase (rhIDU; laronidase)	Mucopolysaccharidosis type I non-neurological symptoms
Naglazyme	N-acetylgalactosamine-4-sulfatase	Mucopolysaccharidosis type VI
Neupogen	Granulocyte colony-stimulating factor (G-CSF)	Neutrophil production
Engerix B	Hepatitis B surface antigen (HBsAg)	Hepatitis B
Pulmozyme	Dornase alfa	Cystic fibrosis
Ceredase	Glucocerebrosidase	Gaucher's disease type 1
C1INH	C1 inhibitor	Hereditary angioedema
Activase	Tissue plasminogen activator	Dissolving blood clots
Pulmozyme	Dornase alfa	Cystic fibrosis

Pharmaceutical biotechnology was first exemplified by discovery of penicillin and streptomycin antibiotics in the 1940s. The revolution in pharmaceutical biotechnology redirected the research and development of the drugs industry. Recombinant DNA technologies have been extensively used in the production of a wide range of drugs/hormones/enzymes etc. (Table 1.1) (Bristow, 1993; Drews, 1993; Weng and DeLisi, 2000; Ausubel *et al.*, 2002; Demkow *et al.*, 2012). In 2014 the global biopharmaceuticals market was estimated to be US$162 billion and projected to swell to US$278 billion by 2020 growing at a CAGR of 9.4% (Persistence Market Research, 2015). Meanwhile, the global vaccine market is growing at CAGR of 12.65% and is estimated to reach US$43.5 billion by the year 2018 (PR Newswire, 2015).

1.5 Conclusion

Recent advances in biotechnological innovation have resulted in a remarkable transformation in the fields of genetics and molecular biology and have opened new vistas in the healthcare, pharmaceutical and agricultural sectors. However, despite broad applications of biotechnology, profits on investments are surprisingly slow. Intellectual property rights (IPRs), particularly patents, occupy a prominent position in the innovation systems in life sciences; but to what extent they support or hinder innovation is widely disputed. Thus, it is vitally important for universities, institutions, public research organizations and private enterprises to protect their innovation. At present, IP protection for biotechnological inventions is in a state of uncertainty. Modern biotechnological innovations are facing severe challenges under the existing IP framework. There are vast differences in opinions on the patentability of living organisms which are largely barred from patentability. However, mindsets are rapidly shifting and IP issues in biotechnology are receiving massive attention. To compete in global biotechnology-based markets developing countries are amending their patent laws in order to encourage foreign investment. Thus synchronization of international policy frameworks to provide unified guidelines for the patenting of life forms will be indispensable for the overall progress of life sciences in general and biotechnology in particular.

References

Ausubel, F.M., Brent, R., Kingston, R.E., Moore, D.D., Seidman, J.G., Smith, J.A. and Struhl, K. (2002) *Short Protocols in Molecular Biology* (5th edn). Wiley, New York.

Bristow, A.F. (1993) Recombinant-DNA-derived insulin analogues as potentially useful therapeutic agents. *Trends in Biotechnology* 11, 301–305.

Coombs, J.M. (1992) *Macmillan Dictionary of Biotechnology*. Macmillan, Basingstoke, UK.

Demkow, U., Winklewski, P., Ciepiela, O., Popko, K., Lipińska, A., Kucharska, A. and Wąsik, M.M. (2012) Modulatory effect of insulin on T cell receptor mediated calcium signaling is blunted in long lasting type 1 diabetes mellitus. *Pharmacology Reports* 64, 50–156.

Department of Industrial Policy & Promotion (DIPP), Government of India (2014) Make in India – Biotechnology. Available at: http://www.makeinindia.com/sector/biotechnology (accessed 14 December 2015).

Drews, J. (1993) Into the twenty-first century. Biotechnology and the pharmaceutical industry in the next 10 years. *Biotechnology* 11, 516–520.

Du, J., Shao, Z. and Zhao, H. (2011) Engineering microbial factories for synthesis of value-added products. *Journal of Industrial Microbiology and Biotechnology* 38, 873–890.

James, C. (2014) Global Status of Commercialized Biotech/GM Crops: 2014. ISAAA Brief No. 49. ISAAA, Ithaca, New York.

Lubiniecki, A. (1997) Potential influence of international harmonization of pharmaceutical regulation on biopharmaceutical development. *Current Opinion in Biotechnology* 8, 350–356.

McCouch, S.R. (2001) Is biotechnology an answer? In: Wiebe, K., Ballenger, N. and Pinstrup-Andersen, P. (eds) *Who Will Be Fed in the 21st Century? Challenges for Science and Policy*. International Food Policy Research Institute, Washington, DC, pp. 29–40.

Nielsen, J. and Jewett, M.C. (2008) Impact of systems biology on metabolic engineering of *Saccharomyces cerevisiae. FEMS Yeast Research* 8, 122–131.

Otero, J.M. and Nielsen, J. (2010) Industrial systems biology. *Biotechnology and Bioengineering* 105, 439–460.

Parmeshwaram, B., Palkhiwala, P., Gaikaiwari, R., Nampoothri, K.M., Duggal, A. and Pandey, A. (2013) Industrial enzyme: present status and future prospective for India. *Journal of Scientific & Industrial Research* 72, 271–286.

Persistence Market Research (2015) Global Biopharmaceuticals Market. Available at: http://www.persistencemarketresearch.com/mediarelease/biopharmaceutical-market.asp (accessed 14 December 2015).

PR Newswire (2014) Global Biotechnology Market to Value USD 414.5 Billion 2017: Transparency Market Research. Available at: http://www.prnewswire.com/news-releases/global-biotechnology-market-to-value-usd-4145-billion-2017-transparency-market-research-276915571.html (accessed 14 December 2015).

PR Newswire (2015) Global Human Vaccine Technologies Market 2015 – Forecasts to 2018. Available at: http://www.prnewswire.com/news-releases/global-human-vaccine-technologies-market-2015---forecasts-to-2018-with-aduro-biotech-antigen-express-sanofi-pfizer-gsk-novartis--merck--co-dominating-300077735.html (accessed 14 December 2015).

Salomaa, V., Stinson, V. and Kark, J.D. (1995) Association of fibrinolytic parameters with early atherosclerosis. *The ARIC Study Circulation* 91, 284–290.

Sauer, M. and Mattanovich, D. (2012) Construction of microbial cell factories for industrial bioprocesses. *Journal of Chemical Technology and Biotechnology* 87, 445–450.

Soetaert, W. and Vandamme, E. (2006) The impact of industrial biotechnology. *Biotechnology Journal* 1, 756–769.

Stryjewska, A., Kiepura, K., Librowski, T. and Lochyński, S. (2013) Biotechnology and genetic engineering in the new drug development. Part I. DNA technology and recombinant proteins. *Pharmacological Reports* 65, 1075–1085.

Tang, W.L. and Zhao, H. (2009) Industrial biotechnology: tools and applications. *Biotechnology Journal* 4, 1725–1739.

Walsh, G. (2000) Biopharmaceutical benchmarks. *Nature Biotechnology* 18, 831–833.

Weng, Z. and DeLisi, C. (2000) Protein therapeutics: promises and challenges of the twenty-first century. *Trends in Biotechnology* 20, 29–36.

Zaid, A., Hughes, H.G., Porceddu, E. and Nicholas, F.W. (1999) *Glossary of Biotechnology and Genetic Engineering*. FAO, Rome.

2 Biotechnology and its Development in Developing Countries: Can Intellectual Property Rights Foster Innovation in the Field?

Dhanay M. Cadillo Chandler*

Faculty of Law, University of Helsinki, Finland

2.1 Introduction

Intellectual property rights (IPR) in principle and according to the Agreement on Trade Related Aspects on Intellectual Property Rights (TRIPS Agreement) have extended IP protection to inventions in all fields of technology. However, there are countries still struggling to cope with the implementation and interpretation of TRIPS within their national frameworks. Developments within the biotechnological field have been correlated to inflow of foreign direct investments and national development as well. Understanding each country's industrial capacity and needs are also key factors when developing both an industry and a regulatory framework.

Among BRICS, China and India are known to have a strong biotechnological industry catering pharmaceuticals to an important share of the world market. The same economies could be considered key players within the agriculture-biotechnological field. Therefore, this chapter intends to analyse how the international framework on IP and biotechnology is fostering developments within the field. To accomplish this goal the chapter will be structured in four parts. The first section will address the concept of biotechnology in general. The second section will present an overview of the relevant international framework to biotechnology and biosafety, and the interplay with TRIPS. This is followed by an overview of the patent, biotechnology and biodiversity legal frameworks in China and India in order to understand each country's position and interpretation of said norms. The final section aims to present a comparison between the countries to determine if the current framework has had an impact within their national development.

2.2 The Concept of Biotechnology and the Legal Implications

By 1991, 'modern' biotechnology had been defined in various ways. For instance, the US Office of Technology Assessment provided that biotechnology 'includes any technique that uses living organisms (or part of organisms) to make or modify products to improve plants, animals or to develop micro-organisms for specific uses' (Wiegele, 1991). Today, biotechnology is to be understood in a much broader sense than the aforementioned definition, which has been voided. Even when significant parts of the 'old' definition are encompassed within the current one – 'comprehends techniques applied to living organisms and parts thereof to produce, identify or design substances or to modify organisms for specific applications' (Usha and Annadurai, 2010) – this broader sense foresees or recognizes the multidisciplinary character of biotechnology in itself as a science involving several disciplines, i.e. biochemistry, cell biology and embryology among others.

*Corresponding author e-mail: cadillo@hanken.fi

 Intellectual Property Issues in Biotechnology (Eds H.B. Singh, A. Jha and C. Keswani)

The OECD working paper (Devlin, 2003) on biotechnology statistics highlights that members had agreed to a single definition of biotechnology, and also complement it with a non-exhaustive list of biotechnologies to serve as a guideline to interpret such definition when aiming to regulate it. Thereafter, biotechnology was understood as:

> the application of Science and Technology to living organisms as well as parts, products and models thereof, to alter living or non-living materials for the production of knowledge, goods and services. Including, DNA (the coding): genomics, pharmacogenetics, gene probes, DNA sequencing/synthesis/amplification, genetic engineering; proteins and molecules (the functional blocks): protein/peptide sequencing/synthesis, lipid/protein glyco-engineering, proteomics, hormones, and growth factors, cell receptors/signalling/pheromones; cell and tissue culture and engineering: cell/tissue culture, tissue engineering, hybridization, cellular fusion, vaccine/immune stimulants, embryo manipulation; process biotechnologies: bioreactors, fermentation, bioprocessing, bioleaching, bio-pulping, bio-bleaching, bio desulphurization, bioremediation, and bio filtration; and sub-cellular organisms: gene therapy, viral vectors.
>
> Devlin, 2003

It is important to point out that even when there is a certain level of agreement in relation to the definition per se, there is no single concept used unanimously in each and every agreement or convention setting the framework for biotechnology. One of the challenges related to the field arises from the need to set a framework efficient enough as to deal with both the opportunities and risks brought by biotechnological developments. Thus, not having a common definition poses a challenge in terms of the scope and extent of protection granted to inventions derived from biotechnology.

Admittedly, the field has revolutionized the world as we know it with its scientific discoveries, which has not only brought expectations to address fundamental food and health needs globally by fostering R&D in the field of healthcare, agriculture, food production and environmental protection (Horng, 2009), but also the need to provide a legal framework.

Agreeably, biotechnology has become 'the amalgamation of several technologies ranging from the field of biomedical research and synthetic biology and further transcending to engineering, nanotechnology, and bioinformatics' (Ganguly *et al.*, 2014). Given the pressing need for further innovation addressed for instance to alleviate suffering from illnesses, to ensure sustainability and agricultural development, innovations within the fields of biomedical research and biotechnological agriculture are of particular importance to society. Thus, its foreseeable impact relates to access to information, treatment, affordability and sustainability.

In the case of biomedicine and genetics the concerns regarding the impact on public healthcare policy and development were already palpable when the Human Genome Project (HGP) began in the early 1990s given the implications and expectations on the dramatic advancements for the healthcare sector, e.g. preventive medicine (Taylor, 1999). The HGP mapped and sequenced DNA and the location of all genes, which later on translated into the expectation to have 'personalized medicine' or 'preventive medicine' as scientists gather further understanding about the genetic basis of illnesses (Taylor, 1999).

In principle, personalized medicine aims at using each individual's unique genome to determine which is the right medicine and treatment, thereafter voiding the 'one size fits all' medicine concept that we currently know (Annas, 2014). In light of these developments, as early as 1999 scholars worried and called upon international organizations to provide further and adequate framework/guidelines to govern biotechnology, i.e. HGP (Taylor, 1999).

To treat patients or even gather information associated to genomic research, some countries have established biobanks to collect human biological data (Chen and Pang, 2015) and precisely the use of these data could be challenged by e.g. property rights and privacy laws, etc. In the field of agriculture, property rights are very important as plant breeders have the right to 'own' their creations/innovations before commercializing them, and countries also have the right to protect the use or access to their genetic resources. As depicted above, biotechnology covers many aspects, which cannot and have not been able to be compiled or regulated under a single comprehensive legal instrument. None the less, there are in place a number of international agreements, conventions and protocols to serve biotechnology with a legal framework.

2.3 Institutions, Legal Framework and Intellectual Property Law

When regulating biotechnology there are a number of international organizations working together with country members to provide the standards necessary to protect inventions in the field and also to protect the environment from possible negative effects derived from the use of modern biotechnology, for example the World Health Organization (WHO), the World Trade Organization (WTO), the United Nations Food and Agriculture Organization (FAO), the United Nations Environmental Programme (UNEP), the United Nations Educational, Scientific and Cultural Organization (UNESCO), the United Nations Industrial Development Organization (UNIDO), etc. Under the umbrella of these organizations a few treaties and conventions have been ratified.

In the context of bioethics – concerning i.e. biomedical research – UNESCO has claimed that only the Universal Declaration on the Human Genome and Human Rights and the International Declaration on Human Genetic Data are to be considered to be the relevant international framework of reference in terms of bioethics. These two declarations are particularly important since both acknowledge the primacy of the human being and the importance of informed consent when researching on genetic data (Cordeiro, 2014).

Biotechnology has become an important tool for the development of agriculture and food production as already established above. Therefore, the three main aspects of biotechnology also relevant for agricultural development regulated under the scope of international treaties are the following:

> (1) access to genetic resources – a key input for agricultural research – and sharing of the benefits from its utilization, such as the Convention on Biological Diversity (CBD) and the International Treaty on Plant Genetic Resources for Food and Agriculture (ITPGRFA); (2) intellectual property rights governing the use of proprietary technologies, such as the International Union for the Protection of New Varieties of Plants (UPOV) Conventions and the Agreement on Trade-Related Aspects of Intellectual Property Rights (TRIPs) under the World Trade Organization; and (3) biosafety, including environmental and food and feed safety.
>
> Komen, 2012

2.4 Treaties, Conventions and Declarations in Brief

The international or emerging international framework, as some scholars have pointed out, is not only broad enough but also targets specific aspects of biotechnology developments, which need to be complemented with national policies 'reflecting the country's objectives for the development of biotechnology capacity, and society's desires on how technology will affect it' (Komen, 2012).

2.4.1 Declaration on the Human Genome and Human Rights

This was adopted in UNESCO's 29th Session of the General Conference by Resolution 29 C/I7 on 11 November 1997.[1] The United Nations (UN) endorsed the Declaration in its 53rd Session of the General Assembly by its Resolution AIRES/53/152 on 9 December 1998. This declaration recognized the importance and need to regulate the human genome as it does not only belong to humankind but research should also be carried out with respect to human dignity, freedoms and human rights. In order to safeguard respect for human rights, fundamental freedoms and human dignity and to protect public health, Article 15 of the declaration stresses the need for states to provide the adequate framework for the free exercise of research on the human genome while at the same time bearing in mind the principles set out within the declaration, and demanding to ensure the lawful use of the research results.

2.4.2 International Declaration on Human Genetic Data

This was adopted unanimously in the context of UNESCO's 32nd General Conference on 16 October 2003.[2] This declaration recognizes that genetic information belongs to the overall spectrum of medical data and as such its content is not only highly contextual but dependent on the particular circumstances. Another important aspect of this declaration is the elevation of the status quo of human genetic data, which is highly sensitive thus requiring further considerations due to the cultural significance for individuals or groups. Furthermore, Article 1 clearly defines the aim and scope of the declaration, which is to protect human dignity throughout the processes of collection, processing, use and storage of the information.

2.4.3 The International Convention for the New Plant Varieties

The International Union for the Protection of New Varieties of Plants (UPOV) was established at the International Convention for the Protection of New Varieties of Plants in 1961,[3] which has also been revised in 1972, 1978 and 1991. With the adoption of this convention the system for the protection for plant breeders' rights was established. Article 2(a) of the convention grants breeders' rights protection both through patents and a special title of protection for one or the same botanical genus or species. If the member state of the union foresees protection under both patent and special title, then only one type of protection should be granted.

Each revision incorporated new terms and conditions that reflected the evolution within the field. The revision in 1978 (1978 International Convention for the Protection of New Varieties of Plants, also known as the 1978 Act) differed from the previous one by focusing on the subject matter and the system of protection around it, i.e. eligibility requirements, exclusive rights, term of protection and limitation to exclusive rights (Papadopoulou, 2014). An interesting distinction between the revisions from 1978 and 1991 has to do with farmers' right to save and exchange seeds without this constituting an infringement. Within the 1978 revision saving seeds and exchanging them between farmers did not constitute an infringement, whereas the 1991 revision limited the scope of the exceptions settled in the previous text by limiting the exchange and the seeds that could be saved for use during the following season.

2.4.4 Convention on Biological Diversity

The Convention on Biological Diversity (CBD) was opened for signature on 5 June 1992 at the UN Conference on Environment and Development, also known as the 'Rio Earth Summit'. The CBD is said to be an expression of the world community's increasing commitment to sustainable development, conservation of biological diversity and the sustainable use of its components, and the equitable sharing of the benefits emerging from the use of genetic resources (CBD, 2015). An interesting fact about the CBD is that this convention was initially envisaged as an environmental law convention (Morgera, 2011) instead of a convention conceived for the purpose of promoting and conserving biological diversity. The convention's preamble is particularly important for a few reasons. For instance, it is not only acknowledged that human activity reduces biological diversity but it also recognizes each country's sovereignty over their own biological resources, and calls upon each state to take measures consistent with both conserving biological diversity and ensuring sustainability when making use of their own biological resources (UNTS 79, 1992).[4] Article 8(j) from the CBD re-conceptualizes the subject matter since it introduces the notion of protection of knowledge, innovation and practices of indigenous and local communities; thus also establishing collaboration practices and benefit sharing between the holders of the information – indigenous communities – and the third party by requiring the right holder's involvement and approval. Given the scope of the CBD – traditional knowledge and biosafety – two other protocols have been drafted to complement and guide further the implementation of the convention, namely the Nagoya Protocol and the Cartagena Protocol.

2.4.5 Cartagena Protocol on Biosafety

This is a complementary agreement to the CBD adopted in Decision EM-I/3 at the Conference of the Parties to the Convention on Biological Diversity from January 2000. This protocol intends to protect biological diversity from the potential risks posed by living modified organisms resulting from modern biotechnology that may have adverse effects on the conservation and sustainable use of biological diversity taking also into account risks to human health, and specifically focusing on transboundary movements (UNTS 208, 2000).[5] The main focus of this convention seems to be trade, given the concerns that arise from the commercialization of genetically modified seeds 'intended for planting the field, or for direct use as food and feed or for processing' (Komen, 2012). The main objective of this protocol is to provide an adequate framework for protection in the field of the safe transfer, handling and use of living modified organisms resulting from modern biotechnology that may have adverse effects on biological diversity.

2.4.6 The Nagoya Protocol on Access and Benefits Sharing

The Nagoya Protocol is one of the most recent agreements to enter into force – October 2014 – regulating or framing the use of biotechnological inventions. This protocol is important not only because it seeks to set out the framework to share the benefits arising from the utilization of genetic resources in a fair and equitable way, but also because it settles the ABS Clearing-House system to achieve its objective. Scholars have pointed out the binding effect of this protocol for member states, since it creates the obligation for these to take appropriate measures at a national level to protect traditional knowledge and genetic resources accessed inside their jurisdictions according to prior-informed consent regulations as established within Articles 15 and 16 from the protocol (Papadopoulou, 2014).

The ABS Clearing-House as set out in Article 14 constitutes a platform to exchange information on access and benefit sharing, thus aims to increase transparency on the procedures for access, benefit sharing and to monitor the use and value of genetic resources. The protocol also highlights the protection of local and indigenous communities' know-how by requiring mutual agreements between the communities and the third parties delimiting the scope of the 'collaboration' and the benefits to be received for the use and access of that knowledge (UNEP, 2010).[6]

2.5 Biotechnology and Intellectual Property Rights as set out within TRIPS

Among the descriptions above a few concepts emerged, namely innovation, patents, traditional knowledge, genetic resources, trade, etc., signalling the existence of a dependence between law and science to reassure the balance between rights and obligations among the parties involved. Intellectual property rights (IPRs) have played a significant role within the field of innovation. In an era when standardization seeks to establish clearer terms, conditions and guidelines to implement levels of protection among parties – member states – which are not always at the same level of development, it is important to understand not only about the basics of a system construed to foster innovation but also the challenges faced by the legislation in itself when regulating rapidly evolving scientific developments, e.g. biotechnology.

Prior to the TRIPS, a few conventions were in place to provide biotechnological developments with a 'non-TRIPS' system of protection. With the creation of the WTO also came the mandatory adherence to TRIPS; even though the purpose of this chapter is not to analyse in detail the creation and background of the agreement, it is important to remember the agreement's preamble where its purpose is highlighted: to reduce distortions and impediments to international trade and taking into account the need to promote effective and adequate protection of IPRs (WTO, 1994).

On the one hand, TRIPS supports the claim that IPR provides an incentive to innovate, but on the other hand, by extending patent right to all fields of technology also brought several concerns. TRIPS Article 27 foresees patents in all fields of technology, whether products or processes as long as these inventions meet with the patentability requirements set out in the agreement, i.e. are new, involve inventive step and are capable of industrial application. In terms of biotechnology TRIPS is not as straightforward as with other patentable subject matters, and it is here where the challenge began. Biotechnology is addressed within the scope of non-patentable subject matters settled within Article 27(3), however it is part 3(b) that provides that a country member may exclude from patentability 'plants and animals other than *micro-organisms, and essentially biological processes* for the production of plants or animals other than non-biological and microbiological processes'.[7] In principle, essentially biological processes are not patentable, however microorganisms and microbiological processes are. This difference is an important one given that TRIPS adds to the challenge the fact that a country member cannot deny patent protection on the aforementioned processes because its exploitation is prohibited by law (Alam, 1997).

The language used within the agreement has been regarded to be vague, which allows member states to interpret these provisions when tailoring the national system to protect plant varieties as mandated either through patents or by a *sui generis* system or a combination thereof.[8] For each country member it is crucial to understand or bear in mind its own interests, development objectives and industrial capabilities before effectively tailoring the national legislation after the minimum standards of protection within TRIPS, and especially within the fields of biotechnology and biodiversity.

Within TRIPS there is an array of circumstances when patents are not to be granted or cases when it is possible to make use of flexibilities. Accordingly, Article 27(2) envisages the protection of '*ordre public* or morality, including to protect human, animal or plant life or health or to avoid *serious prejudice to the environment*'[9] as valid grounds to make use of certain flexibilities, e.g. compulsory licences, as long as such use does not unreasonably prejudice with the legitimate interests of the patent owner'. Hence, exercising TRIPS flexibilities need to follow a set of terms and conditions also settled within the agreement, e.g. Article 31 TRIPS.

Generally, patents are both an incentive and a safeguard for inventors and investors. Under the scope of patent protection, the invention is protected against third parties making use of it in any way – unless licensing agreements are in place or other forms of consent. Hence, use and ownership rest with the right holder for a period of 20 years commonly from the application filing date, which is also regarded as a monopoly right. Patents allow investors to recoup their investments, but these are also perceived as a possible impediment to dissemination of information thus hindering progress (Hing and Back, 2009).

Dissemination of information – disclosure of country or origin – together with benefit sharing are two of the concerns brought to the TRIPS revision by developing countries in the contexts of biotechnology. China and India have been particularly active together with other developing countries in lobbying towards the creation of better guidelines and provisions within the international framework. More precisely, together with public health there were a series of issues dealt with in relation to the implementation of TRIPS at the Doha round in 2001. The Doha WTO Ministerial 2001: Ministerial Declaration WT/MIN(01)/DEC/1, 20 November 2001,[10] adopted an extensive work programme where paragraphs 17, 18 and 19 aimed at clarifying further both the patent title within the agreement and the interface with other rights affected or not sufficiently addressed within the agreement, e.g. public health and the CBD.

The Doha Declaration was a milestone for developing and least developed countries in terms of public health, given that the concerns related to this issue were not only successfully acknowledged by the international community, but also TRIPS was amended accordingly. The full extent of the declaration falls outside the scope of this chapter, thus the focus is on the working programme paragraph 19 where the Council for TRIPS is instructed to review Article 27.3(b) and its relationship between TRIPS and the CBD (WTO, 2001).[11] Prior to Doha in 1999 when the review of Article 27.3(b) began, the topics raised in the TRIPS Council's discussions in terms of biodiversity sought to clarify the use of the flexibilities, i.e. to continue allowing farmers to save and exchange seeds harvested by them, and to define whether or not TRIPS collided with the CBD.

In principle, the agreement had to be supportive of the convention and vice versa, however the CBD's primary objectives are the conservation of genetic resources and their sustainable use (Papadopoulou, 2014), whereas TRIPS's objective is to ensure the protection and enforcement of IPR, since these should contribute with member countries' development.

The real core of the dilemma brought upon the revision as mandated by Doha, relies on the fact that TRIPS does not call for disclosure on the origin of the genetic resource and traditional knowledge used in the invention within the patent application. Such disclosure is said to grant some degree of certainty in terms of compliance with the prior informed consent requirement settled within the CBD. In 2008 a group of WTO members – among them China and India – requested a procedural decision on IPR related to disclosure, geographical indications and traditional knowledge. The document proposed the following:[12]

- Members agree to amend the TRIPS Agreement to include a mandatory requirement for the disclosure of the country providing/source of genetic resources, and/or associated traditional knowledge for which a definition will be agreed, in patent applications. Patent applications will not be processed without completion of the disclosure requirement.
- Members agree to define the nature and extent of a reference to Prior Informed Consent (PIC) and Access and Benefit Sharing.
- Text-based negotiations shall be undertaken, in Special Sessions of the TRIPS Council, and as an integral part of the Single Undertaking, to implement the above. Additional elements contained in members' proposals, such as PIC and ABS as an integral part of the disclosure requirement and post grant sanctions, may also be raised and shall be considered in these negotiations.

China and India are two important players within the pharma-biotechnological fields. For instance, already in 2004 China alone was the second largest producer in the world of pharmaceutical active components (Grace, 2004). India's biotechnological industry, on the other hand, has been portrayed to be one of the fastest growing knowledge base sectors within the country, accounting for a

US$2.5 billion turnover during 2008–2009 (Usha and Annadurai, 2010). Given the importance of China and India for both the international supply and their efforts to clarify the position between TRIPS and the CBD, the following section of the chapter will overview each country's approach towards IPR and biotechnology.

2.6 China and India Revisited: Intellectual Property Rights and Biotechnology

> The ownership and exploitation of intellectual property rights are the key factors in determining success of any technological innovation introduced in the market that provide the means for technological progress to continue, to be made and thereby support competitiveness of industry of the country.
>
> (Usha and Annadurai, 2010)

India has become a leader in the supply and manufacture of generics, supplying its internal market and also aggressively competing to earn a share of other countries' markets. Biopharmaceuticals, in 2005, accounted for 60% of the biotechnological market in India. China on the other hand, has a longer tradition within the pharmaceutical sector than India. However, both countries have a similarly long history in the use and manufacture of traditional medicines that account for an important share within the pharmaceutical market in each country (Rao, 2008). The importance of traditional medicine for both countries explains their active role within the TRIPS review process in terms of paragraph 19 from the Doha Declaration as mentioned above.

China and India have struggled to implement TRIPS minimum standards of protection, but beyond this struggle both countries share a few other similarities that are by far more important, for instance, their push to protect each country's national industry by implementing a regulatory framework consistent with their development goals. In this respect, it is necessary to overview the national regulatory frameworks in terms of biotechnology to have a better picture of each country's direction for the years to come.

2.6.1 China

Patent regulation in China dates back to 1898 when the Emperor Guanxu of the Qing Dynasty provided protection for inventions and creations of technology through the Reward Technical Arts Laws and Regulations (Sanqiang, 2012). Later on in 1911 the Department of Trade and Industry of the Republic of China promulgated a modernized patent law version, namely 'Provisional Regulations on Crafts Award', where many of the principles of patent law that we know today were settled, e.g. first to file principle, assignment of rights, and legal responsibility (Sanqiang, 2012), but it was not until 1982 that China actually enacted the 'Patent Law of the People's Republic of China'. Three amendments took place before the patent law could be considered in compliance with the TRIPS Agreement. China seems to be keen on promoting development and having an innovation-oriented society throughout the implementation and use of IPR (Kshetri, 2009).

Strengthening the protection of IPR in China is believed to be the rationale behind the increasing foreign direct investments within the country (Awokuse and Yin, 2010). Even when the patent system has been reformed to comply with TRIPS, it has been highlighted that obtaining patent protection for pharma and biotechnology inventions is less predictable than it used to be, as stringent patentability requirements are in place. Patentability requirements are similar to anywhere else, however the 'support requirement' implemented by the Guidelines for Patent Examination[13] (the Guidelines) enacted by the Patent Office of the People's Republic of China (SIPO) is not a universal requirement, meaning it is not a requirement imposed to the rest of the patent offices around the world. The support requirement within the Guidelines seeks to determine whether or not the invention is supported by the specification, thus 'answering if the man skill in the art can really expect the invention to work' (Mak, 2014).

Another difference between the patent law in China and other Western jurisdictions relates to the patent scope of protection of the subject matter, and the application's supportive description of the claim (Weping, 2014). To determine whether or not the invention falls within the scope of patentable subject matter, the invention must be 'a new technical solution relating to a product, a process or improvement thereof' according to Article 2 of the Chinese Patent Act (Weping, 2014). This highlights that the 'technical nature' of the invention is the key element to obtain patent protection. Even though the purpose of this chapter is not to make an extensive analysis of the patent system in either

China or India, it is necessary to draw attention to the provisions relating to biotechnology since this is the focus of the chapter after all.

The Chinese Patent Act in Article 25(1) excludes from patent protection plant and animal varieties, however these are in practice interpreted literally, meaning a plant or an animal cannot be protected under patent protection (Weping, 2014). Accordingly, the processes to produce plants or animals are in principle excluded from patent protection in Article 25(2). Nevertheless, the Guidelines for Patent Examination 2010 clarifies in Part II, Chapter 10, Section 4.4 that processes used to produce animal or plant varieties may be patentable as long as these processes are non-biological processes and do not include essentially biological processes used for the production of plants and animals. The key factor determining whether a process can be regarded as essentially biological depends on the level of human technical involvement in said process, thus if the human technical involvement is paramount to achieve the result then the process is definitely not regarded as essentially biological.[14] In terms of plant varieties protection, the Guidelines expressly revert to the Regulations on the Protection of New Varieties of Plants as an avenue to obtain protection.

The Ministry of Agriculture of the People's Republic of China (MOA, 2015) is a component of the States Council in charge of agriculture and rural economic development. Among its functions is to:

> draw up technical standards for various agricultural industries and organize their implementation thereof; to organize the implementation of quality supervision and certification of various agricultural products and green food products and the protection of new varieties of agricultural plants; to organize and coordinate the monitoring and defining of quality of agricultural inputs such as seeds, chemicals ... and of related law supervision and executions; to organize the registration of domestic manufactured as well as imported seeds, chemicals, vet drugs and related fertilizers, etc. and the safety supervision of agricultural machines.

Both the protection of plant varieties and biosafety are important for China's agricultural development.

Towards the end of the 1990s, China's biotechnological industry grew at a rapid pace, calling for the country to take measures in the field of biosafety. Thereafter, in 2001 China's State Council promulgated by Decree No. 304 of the State Council of the People's Republic of China of 2001 on 23 May 2001 the Regulations on Administration of Agricultural Genetically Modified Organisms Safety. MOA has been referred to as China's key player in terms of biosafety management of agriculture. Hence, the organization provided further regulations to ensure that genetically modified organisms (GMOs) are dealt with responsibly due to their plausible harmful effects to the environment if not treated with care.

China aims at becoming a world reference on biotechnological issues, hence, decisively developing its industry and the IP regime since 1980. It is essential for China to invest in biosafety management capacity and policy implementation given the pressing need to create improvements within the healthcare sector and to provide the industry with sustainable development (Huang and Wang, 2002).

2.6.2 India

The IP system in India was established as early as 1856 with the adoption of the Patent Act. Several reforms have taken place throughout the years, illustrating India's commitments to both international partners and also their national development goals. The patent legislation in force by 1911 is said to have contained standards of protection at the same level as developed countries, however, during 1970 the legislation was amended 'to encourage economic development through reverse engineering of western technologies' (Graff, 2007). Like most developing countries at that point, India did not grant patent protection either for pharmaceutical products or processes. The situation shifted with India's membership of WTO when ratifying TRIPS was among the mandatory package. Thus, the Patent Act was amended in 1999, 2002 and 2005 to comply with the minimum standards of protection as set out in TRIPS.

During the 1970s, India's pharmaceutical industry flourished to the extent that India has been reported to supply up to 77% of its internal market and 22% of the world market (Rao, 2008). Moreover, the pharmaceutical industry in India is considered among the leaders in manufacturing of generics, and within the biotechnological field the country is also developing a strong industry, placing them at the same level of many multinational corporations competing in the field (Connor and Corrado, 2007).

Recent foreign direct investment (FDI) has also played a significant role in developing the industry in India, specifically the human resources available in the country have been regarded partially responsible for the foreign direct investment inflow into the country (Verma and Rao, 2007). Besides skilled human recourses India seems to offer several other advantages justifying FDI, for instance strong chemistry-based knowledge, sophisticated R&D facilities and cost effectiveness (Usha and Annadurai, 2010). Thus, Indian companies are focusing on collaborative research, clinical trials and production of generics as a strategy to foster development by creating further knowledge that will allow them to shift from pure generic producers into innovators of new drugs (Rao, 2008).

Implementing the TRIPS Agreement in India has brought concerns for both the country's pharmaceutical industry and foreign investors. On one hand (Usha and Annadurai, 2010) drew attention to challenges brought upon Article 27.3(b) in terms of the patentability of GMOs, and the interplay between TRIPS and the CBD. On the other hand, two aspects of India's 'calibrated' approach towards TRIPS implementation are causes of concern for trading partners, primarily the use of undisclosed or sensitive information to grant marketing approvals for new pharmaceutical products, and second the patent standards in place within the system that restricts patents for second medical uses – also known as derivative pharmaceuticals (Connor and Corrado, 2007).

Article 27.3.(b) may be challenged in itself by India's Patent Act – 1970 definition of invention. Chapter I.2.(j) defines patentable inventions as a new product or process involving an inventive step and capable of industrial application. These definitions' use of the term 'industrial application' has been highlighted as an obstacle for biotechnology-based patents since it is intertwined with the concept of a mere discovery. In other words, the patent grant seems to rely solely on the examiner's point of view with regards to the 'invention' using something that freely occurs in nature – discovery – as the claim for patent rights on a living form (Singh, 2009). Moreover, the basis within the legislation to prevent the evergreening of the patent system relies on the provision in Section 3(b) where these new uses are defined.

> *(d)* the mere discovery of a new form of a known substance which does not result in the enhancement of the known efficacy of that substance or the mere discovery of any new property or new use for a known substance or of the mere use of a known process, machine or apparatus unless such known process results in a new product or employs at least one new reactant.
>
> *Explanation.* – For the purposes of this clause, salts, esters, ethers, polymorphs, metabolites, pure form, particle size, isomers, mixtures of isomers, complexes, combinations and other derivatives of known substance shall be considered to be the same substance, *unless they differ significantly in properties with regard to efficacy*
>
> Ministry of Justice, 2015.

The provision above provides within the explanation the key factor determining whether the 'improvement' could be regarded as a new use, namely efficacy. Connor and Corrado (2007) explained that a new use under this provision will only be considered 'new' if proven to be considerably more efficacious than the known substance. This provision has already brought significant challenges for foreign parties to protect inventions in India. Recent court cases[15] demonstrated the use of the provision is meant to protect public health to a certain extent by preventing patents for uses that are not so 'new'. Even though the cases do not directly discuss biotechnology issues, it is relevant to highlight the importance of implementing international agreements in a manner that is consistent with the country's (i.e. India) interests.

In terms of biodiversity and biotechnology it is relevant to point out that India did not ratify UPOV, but nevertheless opted for a *sui generis* system of protection for plant varieties – Plant Varieties Protection and Farmers Right Act, 2001 – and hence also establishing the Protection of Plant Varieties and Farmers' Rights Authority of India:

> to encourage the development of new varieties of plants it has been considered necessary to recognize and protect the rights of the farmers in respect of their contribution made at any time in conserving, improving and making available plant genetic resources for the development of the new plant varieties. Moreover to accelerate agricultural development, it is necessary to protect plants breeders' rights to stimulate investment for research and development for the development of new plant varieties (PPV and FR, 2015).

Given the vagueness invoked within TRIPS in terms of biotechnology, India has put in place a few regulations to address the terms to manufacture, import and store genetically modified organisms. The Ministry of Environment and Forests having in mind the need to protect the environment, nature and health, enacted in 1989 'the Manufacture, Use, Import, Export and Storage of Hazardous Micro-organisms/Genetically Modified Organisms or Cell Rules, 1989'. However, India does have a regulatory authority for biotechnological inventions. Nevertheless, the creation of one has been suggested in the Draft National Biotechnology Regulatory Bill from 2008. Currently several other reports from the Ministry of Agriculture, Ministry of Environment and Forests and the Planning Commission of India have provided the field of biotechnology in India with a framework of guidelines to address pressing issues related to it. India ratified the Cartagena Protocol on Biosafety of the Convention on Biological Diversity on 17 January 2003, this entering into force on 11 September 2003.

2.7 Concluding and Comparative Remarks

The differences portrayed at the beginning of the chapter in terms of the biotechnology concept brought to the spotlight the need to have at least a general agreement on the basic meaning of concepts before the legal framework can be settled. The concept in itself provided a starting point to prove the broadness of the term and the various applications existent and available under the umbrella of 'biotechnological developments'. The author recognizes the fact that having a unanimous definition on a field so fast evolving might not be easy or even advisable since over- or under legislating the field may simply prevent its development. Nevertheless, it is definitely advisable to have at the very least a common ground of understanding over the implications of such a definition. Having such an understanding may work in favour of member states to implement further rules clarifying the use of certain terminologies within their legislation. Such is a first comparison between China and India's understanding of the protection of 'plant and animal varieties' within their respective patent law.

On the one hand, China's guidelines made express reference to the literal interpretation given to the provision on non-patentable subject matters. However, the same guidelines (Mitkova and Wang, Chapter 10, this volume) clarify that processes used to produce animals or plant varieties may be patentable as long as these processes are non-biological processes and do not include essentially biological processes used for the production of plants and animals. The key element rests on determining whether or not the invention relies on an essentially biological process. And on the other hand, India's interpretation of the same exclusion within TRIPS suggested that animal and plant varieties could not be protected if already protected under a *sui generis* system, and if processes related to these were to be granted protection the key element determining the non-essentially biological character of the process was given by the level of technical and human involvement in achieving the said process.

To the extent of the comparison China seems to be a step ahead of India in terms of a biosafety and biotechnology legal framework. China has been harshly criticized for rapidly developing a framework on biotechnology but at the same time having little regard for biosafety measures, and more recently the country was criticized again for implementing labelling requirements as part of their new biosafety measures that at the time appeared to be stringent. However, India's efforts in regulating biotechnology cannot be disregarded either.

Another important issue within the chapter for both countries is the UPOV Convention as a means to protect and foster developments within the national agricultural industry. India is the first country to adopt a *sui generis* system of protection that recognizes both breeders' and farmers' rights. Most importantly, this system allows farmers to save and share seeds, which in principle is limited under other legal frameworks, whereas China expressly reverts to UPOV as an avenue of protection since China became a member after 1999.

To the extent of the commonalities between China and India within the context of this short comparison, is the fact that both countries have implemented TRIPS in a more or less consistent manner with their national needs as mandated or allowed by TRIPS and the Doha Declaration. This has not been free of concern to other trading partners, since in the case of China, patentability requirements for chemical, pharmaceutical and biotechnological inventions have been referred to as more stringent or restrictive due to the use of support material which, ultimately verses the interpretation or expectation from the human skill in the art over the invention. On the other hand, India has implemented a provision restricting the protection for new uses unless these are proven considerably efficacious over the known substance. This provision preventing patent evergreening is said to continue allowing generic competition and thus access to medicines in India. The other provision calling for 'concern' refers to data protection, which has been pushed under the frame of TRIPS plus agreements. Unless there is an established linkage between the Patent Office and Health Institute or the authority responsible for granting marketing approvals it is highly unlikely that the agencies will communicate to prevent a product from an already known product to obtain marketing approval. In theory the authority should not rely upon the submitted data to grant protection to a similar product for a period of time, but this is something not defined yet within India's Patent Act or TRIPS.

Both countries seem to be aware of the benefits derived from modern biotechnology and its uses in both health and agriculture. Perhaps this is the rationale behind their 'cautious' approach towards implementing TRIPS, UPOV and CBD. Admittedly there is still a lot of work to carry out within the regulatory sector, but this task needs to be done in a consistent manner not only efficaciously but also carefully, since each countries' development goals and capacities need to be taken into consideration if the goal nevertheless is to foster innovation.

Notes

All the views in this chapter belong to the author and in no way implies an official position from the organization represented by the author.

1 UNESCO (1997) *Universal Declaration on the Human Genome and Human Rights, 29th Session General Conference, Resolution 29 C/I7.*
2 UNESCO (2003) *International Declaration on Human Genetic Data, 32nd General Conference, 32 C/ Resolution 15.*
3 UPOV (1961) *International Convention For New Plant Varieties, Diplomatic Conference, Act of 1961.*
4 UNTS 79 (1992) *Convention on Biological Diversity, UNTS 79; 31 ILM 818 (1992).*
5 UNTS 208 (2000) *Cartagena Protocol on Biosafety to the Convention on Biological Diversity, 2226 U.N.T.S. 208; 39 ILM 1027 (2000); UN Doc. UNEP/CBD/ ExCOP/1/3, at 42 (2000).*
6 UNEP (2010) *8.b Nagoya Protocol on Access to Genetic Resources and the Fair and Equitable Sharing of Benefits Arising from their Utilization to the Convention on Biological Diversity, UNEP/CBD/COP/ DEC/X/1 of 29 October 2010.*
7 WTO (1994) *TRIPS: Agreement on Trade-Related Aspects of Intellectual Property Rights, Marrakesh Agreement Establishing the World Trade Organization, Annex 1C, 1869 UNTS 299, 33 I.L.M. 1197.*
8 WTO (1994)
9 WTO (1994)
10 WTO (2001) *Doha WTO Ministerial 2001: Ministerial Declaration, WT/MIN(01)/DEC/1.*
11 WTO (2001)
12 TNC (2008) *Draft Modalities for TRIPS-related Issues*, TN/C/W/52, from 19 July 2008. Communication from Albania, Brazil, China, Colombia, Ecuador, the European Communities, Iceland, India, Indonesia, the Kyrgyz Republic, Liechtenstein, the Former Yugoslav Republic of Macedonia, Pakistan, Peru, Sri Lanka, Switzerland, Thailand, Turkey, the ACP Group and the African Group, TN/C/W/52.
13 SIPO (2010) *Guidelines for Patent Examination* (promulgated by Order No. 55 of the State Intellectual Property Office (SIPO)). Available at: http://www.wipo.int/edocs/lexdocs/laws/en/cn/cn079en.pdf (accessed 18 July 2015).
14 SIPO (2010)
15 Supreme Court of India, Civil Appeal No. 2706–2716 of 2013. *Novartis AG* v. *Union of India (UoI) and Others*; *NatcoPharma Ltd.* v. *UoI and Others*; *M/S Cancer Patients Aid Association* v. *UoI and Others*

References

Alam, G. (1997) *TRIPS, Biotechnology and Global Competition*. Vol. 9709, CUTS Centre for International Trade, Economics & Environment, Jaipur, India.

Annas, G.J. (2014) Personalized medicine or public health? Bioethics, human rights, and choice. *Revista Portuguesa de saúde pública* 32(2), 158–163.

Awokuse, T. and Yin, H. (2010) Intellectual property rights protection and the surge in FDI in China. *Journal of Comparative Economics* 38, 217–224.

CBD (2015) History of the Convention. Available at: http://www.cbd.int/history (accessed 10 July 2015).

Chen, H. and Pang, T. (2015) A call for global governance of biobanks. *Bulletin of the World Health Organization* 93, 113–117.

Connor, K. and Corrado, N. (2007) A 'Calibrated Approach': pharmaceutical FDI and the evolution of Indian patent law. *Journal of International Commerce and Economies*, August.

Cordeiro, J. (2014) Ethical and legal challenges of personalized medicine: paradigmatic examples of research, prevention, diagnosis and treatment. *Revista Portuguesa de Saúde Pública* 32(2), 164–180.

Devlin, A. (2003) An overview of biotechnology statistics in selected countries. *OECD Science, Technology and Industry Working Papers*, No. 2003/13. OECD Publishing, Paris.

Ganguly, N.K., Croft, S., Singh, L., Sinha, S. and Balganesh, T. (2014) Biomedicine and biotechnology: public health impact. *BioMed Research International* 2014, 524785.

Grace, C. (2004) The effect of changing intellectual property on pharmaceutical industry prospects in India and China: considerations for access to medicines. *Issue Paper – Access to Medicines*. DFID, London.

Graff, G.D. (2007) Echoes of Bayh-Dole? A survey of IP and technology transfer policies in emerging and developing economies. In: Krattiger, A., Mahoney, R.T. and Nelsen, L. (eds) *Intellectual Property Management in Health and Agricultural Innovation: A Handbook of Best Practices*. MIHR, Oxford, UK.

Hing, C.B. and Back, D.L. (2009) A review of intellectual property rights and biotechnology. *The Surgeon: Journal of the Royal Colleges of Surgeons of Edinburgh and Ireland* 7(4), 228–231.

Horng, D.C. (2009) International Law on Biotechnology. In: UNESCO-EOLSS Joint Commitee (eds) *Encyclopedia of Life Science Systems*. Encyclopedia of Life Support Systems (EOLSS), UNESCO–EOLSS Publishers, Oxford, UK, 27 pp.

Huang, J. and Wang, Q. (2002) Agricultural biotecnology development and policy in China. *Agricultural Biotechnological Development and Policy in China* 5(4), 122–135.

Komen, J. (2012) The emerging international regulatory framework for biotechnology. *GM Crops & Food: Biotechnology in Agriculture and the Food Chain* 3(1), 78–84.

Kshetri, N. (2009) Institutionalization of intellectual property right in China. *European Management Journal* 27, 155–164.

Mak, T. (2014) Biotechnological, chemical and pharmaceutical inventions II: support and experimental data. In: Luginbuehl, S. and Ganea, P. (eds) *Patent Law in Greater China*. Edward Elgar Publishing, Cheltenham, UK.

Ministry of Justice (2015) *The Patents (Amendment) Act, 2005, N 15 of 2005*. Ministry of Law and Justice, New Delhi.

MOA (2015) About MOA/Main functions of the Ministry of Agriculture. Available at: http://english.agri.gov.cn/aboutmoa/mandates (accessed 20 July 2015).

Morgera, E. (2011) Faraway, so lose: a legal analysis of the increasing interactions between the convention on biological diversity and climate change law. *University of Edinburgh School of Law Working Paper No. 2011/05*.

Papadopoulou, F. (2014) *Opening Pandora's Box: Exploring Flexibilities and Alternatives for Protecting Traditional Knowledge and Genetic Resources under the Intellectual Property Framework*. Acta Universitatis Stockhomlmiensis, Stockholm.

PPV and FR (2015) About the PPV&FR Authority. Available at: http://www.plantauthority.gov.in/about-authority.htm (accessed 18 July 2015).

Rao, P.M. (2008) The emergence of the pharmaceutical industry in the developming world and its implications for multinational enterprise strategies. *International Journal of Pharmaceutical and Healthcare Marketing* 2(2), 103–116.

Sanqiang, Q. (2012) Overview of patent law. In: *Intellectual Property Law in China*. Kluwer Law International, Alphen Aan Den Rijn, the Netherlands, pp. 243–263.

Singh, A. (2009) Critical issues relating to intellectual property rights (IPR) in biotechnology: developing countries' perspective and India. *National Seminar on Limits and Scope of Intellectual Property Rights*, Gorakhpur, India.

Taylor, A. (1999) Globalization and biotechnology: UNESCO and an international strategy to advance human rights and public health. *American Journal of Law & Medicine* 25, 479–541.

Usha, S. and Annadurai, C. (2010) The intellectual property rights and pharmaceutical biotechnology in India. *Journal of Pharmacy Research* 3(10), 2466–2469.

Verma, S.K. and Rao, M.N.C. (2007) Impact of the intellectual property system on economic growth: fact-finding surveys and analysis in the Asian region. *Country Report – India, WIPO–UNU Joint Research Project.*

Weping, C. (2014) Biotechnological, chemical and pharmaceutical inventions I: general. In: Luginbuehl, S. and Ganea, P. (eds) *Patent Law in Greater China.* Edward Elgar Publishing, Cheltenham, UK.

Wiegele, T.C. (1991) *Biotechnology and International Relations.* University of Florida Press, Gainesville, Florida.

3 Patent Eligibility Issues in Life Science Innovations: Contentious Court Cases

ANANDA M. CHAKRABARTY*

University of Illinois College of Medicine, Chicago, Illinois

3.1 Introduction

While industrial biotechnology primarily involves chemicals, drugs and pharmaceuticals, the recent advancements in biotechnology in the area of genome manipulation, synthetic biology, stem cells and improved plant breeding have created many opportunities but also new legal questions and challenges. Intellectual property (IP) generation, particularly having patentable inventions in industrial biotechnology and other life science areas, is prerequisite for both industrial and economic development. The patent laws in the USA date back from 1790 to 1793 to cover, in the words of Thomas Jefferson, one of the framers of the US Constitution, 'any new and useful art, machine, manufacture, or composition of matter, or any new or useful improvement thereof'. The US Congress in 1952 replaced the word 'art' with the word 'process'. In subsequent decisions by the courts, the predominant theme of the US patent laws has been to emphasize Thomas Jefferson's philosophy that 'ingenuity should receive a liberal encouragement' and the 1980 Supreme Court verdict that 'anything under the sun that is made by man' is patent eligible, including genetically engineered life forms, the only exclusions from patentability being 'laws of nature, natural phenomena and abstract ideas'. However, the 2013 US Supreme Court verdict that products of nature such as human genes *BRCA1* and *BRCA2* are not patent eligible has thrown considerable uncertainties with regard to the established patents involving natural products such as antibiotics, drugs from medicinal plants, bacterial anticancer proteins/peptides as well as human embryonic stem cells. This chapter deals with some of the contentious court cases involving patent eligibility issues in life science areas in several countries, including the European Union (EU) and India. In India, the emerging patent laws are designed not so much to encourage innovation but to prevent the so-called 'evergreening' of issued patents held by foreign multinational corporations.

3.2 Genetic Technology: The Beginning

The emergence of the recombinant DNA technology in the early- to mid-1970s triggered a plethora of new innovations encompassing the crossing of the evolutionary barrier and leading to such early interesting products as fermentative production of human insulin by the bacterium *Escherichia coli*. However, in the 1970s, there was a great deal of uncertainty whether newly developed recombinant life forms, their cellular constituents or products derived from them, could be patent eligible to allow them to reach the market place and bolster the economy of the USA. Thus the 5–4 narrow decision in June 1980 of the US Supreme Court in the *Diamond* v. *Chakrabarty* case[1] to allow modified life forms to be patent eligible played a major role in helping biotechnology innovations to take root and contribute to economic gains (Chakrabarty, 2002, 2003a). However, with the advent of new technologies in life sciences as well as in computer-related information technology, opening of new vistas also led to controversies regarding the patent

*Corresponding author, e-mail: pseudomo@uic.edu

eligibility and IP considerations of many emerging innovations and technologies, both in life sciences and beyond. Below is a very limited discussion on some of the more recent patent-related controversies that have dogged the lower courts and ultimately the US Supreme Court with interesting, and sometimes unpredictable, scientific and technological questions and consequences. Such court cases are not limited to the USA, however, and I discuss a Patent Court, UK case, going forward at the Court of Justice of the EU and in UK, encompassing the issue of the definition of a human embryo.

3.3 Patent Eligibility of Diagnostic Procedures and Drug Dosing: The Case of *Mayo Collaborative Services* v. *Prometheus Laboratories*

Prometheus Laboratories had two patents covering the use of thiopurine drugs to treat autoimmune diseases. The claims in the patents covered the low and high levels of the thiopurine drug metabolites in the patients' blood and how these different amounts produced either ineffectual effects or toxic symptoms. The claims then suggested how to arrive at a dose that would be optimally effective. Mayo Collaborative Services and Mayo Clinic initially used such diagnostic tests based on Prometheus patents but later came up with their own tests, whereupon Prometheus sued Mayo alleging patent infringement. The District Court agreed that Mayo's tests did infringe Prometheus' patents but granted Mayo summary judgment due to subject matter ineligibility, ruling that the patents claimed natural phenomena and laws of nature. On appeal, the US Court of Appeals for the Federal Circuit (CAFC) reversed the District Court ruling, and reaffirmed, on remand from the Supreme Court with regard to its previous *Bilski* v. *Kappos* decision, that the claims were pertinent to the machine or transformation test and were patent eligible. On appeal from Mayo, in March 2012, a unanimous Supreme Court reversed the CAFC ruling, holding that claims directed to methods of drug dosage optimization are basically a manifestation of the laws of nature and are therefore ineligible for patenting under 35 USC Section 101.

A more recent case involves the diagnostic method for prenatal testing of Down's syndrome in the case *PerkinElmer, Inc.* v. *Intema, Ltd.* The diagnostic test involves measuring known biomarkers from the first trimester and one or more biomarkers from the second trimester and comparing the levels with observed frequency of marker levels in normal and Down's syndrome pregnancies. The District Court found the tests to be patent eligible since the data gathering steps satisfied the machine or transformation test. On appeal, however, the CAFC panel reversed the District Court ruling based in part on the Supreme Court's ruling on *Prometheus* v. *Mayo* that the patent claims had no inventive steps and were basically based on routine and conventional activities in the field.

3.4 Patenting Genetically Engineered Plant Seeds: The Doctrine of Patent Exhaustion and Self-Replicating Genetic Entities

An interesting question in genetic technology is the patent protection until patent expiry of plant seeds where the genome has been modified for a useful purpose. The pertinent question then is: do such genetically modified seeds that can replicate for many generations and where patents have been obtained for such genetic modification, allow the patent holder the rights to the patent beyond the first generation? This question is pertinent because of the doctrine of exhaustion of a patent after the first sale or licensing. This controversy has been addressed by the US Supreme Court in the case *Bowman* v. *Monsanto Co.*, which involved Monsanto's Roundup Ready soybean seeds. These seeds have been genetically modified by a bacterial gene that encodes an enzyme capable of inactivating a weed killer called glyphosate (also called Roundup). Thus the soybean plants with the modified genome can grow freely in presence of the weed killer Roundup, and such plants/seeds are called Roundup Ready. The Roundup Ready soybean plants can grow in the presence of Roundup while the weeds cannot. Monsanto Company held patent rights for these modified plants/seeds and farmers, who use such seeds, sign an agreement with Monsanto not to use the engineered seeds after the first harvest unless they pay royalty to Monsanto for continued use of the seeds for many generations.

An Indiana farmer, Vernon Hugh Bowman, initially signed such a contract with Monsanto but later bought mixtures of seeds from a grain elevator where the seeds also contained the Roundup Ready soybean seeds. Mr Bowman did not pay any royalty to Monsanto, reasoning that he bought a mixture of seeds from another source and such seeds are obtained after first generation products so that the doctrine of patent exhaustion will apply even if the seed mixture happens to contain any Monsanto patented seeds. Upon learning the use of such seeds for growing soybean plants in presence of the weed killer Roundup, Monsanto sued Mr Bowman for patent infringement. A District Court in Indiana affirmed infringement of Monsanto's patents and awarded the company US$84,456 as compensation. On appeal, the CAFC agreed with the District Court verdict, rejecting the doctrine of the patent exhaustion defence of Mr Bowman's lawyers. On further appeal to the Supreme Court by Mr Bowman, the Supreme Court in May 2013 affirmed the judgment of the CAFC and upheld the validity of the Monsanto patents on the self-replicating Roundup Ready seeds for use beyond the first harvest. This is thus an interesting example of the validity of patents on inventions that are self-replicating so that the patents can be enforced for many generations of the plant growth.

This Supreme Court victory for Monsanto is in contrast to another patent-enforcement related controversy of the same Roundup Ready soybean plant in Europe, and this decision points to the unique role of DNA and genes as patentable material. The European case involves patent eligibility claims of the Roundup resistance gene present in cooked soybean. Unlike the Bowman case where Mr Bowman was growing the Roundup Ready soybean plants in presence of the glyphosate herbicide, where the gene was functionally active to help the plant become glyphosate-resistant, the Monsanto lawsuit in Europe involved the presence of the DNA of the gene in a non-functional form in soymeals exported by an Argentinean company to a Dutch group in the Netherlands. The Roundup Ready soybean meal containing the patented bacterial glyphosate-resistance gene was, however, not functioning in its role in promoting growth of the soybean plants in presence of the herbicide. Thus while Monsanto demonstrated the presence of the glyphosate-resistant gene in the soymeal and sued for patent protection, the European Court of Justice cited article 9 of the Directive 98/44/EC of the EU Patent Law, which mandates that the patented DNA must perform its function in the material where it is present. The European Court of Justice, therefore, ruled against Monsanto because of the absence of the intended function of the Roundup Ready gene present in the soymeal, thus providing an interesting perspective of the patent eligibility of a gene, only when tied to its functionality.

3.5 Beyond Genetic Technology: Patent Eligibility Issues on Business Methods and Computer- Related Inventions

A contentious patent eligibility issue in the US patent laws is the patent eligibility of business methods, particularly those involving the use of computers. A case in point is the *Bilski* v. *Kappos* case involving a business method like hedging commodity-associated risks for energy commodities such as natural gas or electricity. The patent application covered claims on hedging the risks associated with consumption of a commodity such as natural gas used for home heating during winter months, which in turn involved both natural gas suppliers and consumers. The Claim 1 in the patent covered a series of steps on hedging the risks on price fluctuations and the rest of the claims involved mathematical formulae to allow energy buyers and sellers to reduce risks due to market price fluctuations. The patent examiner rejected the *Bilski* application as encompassing abstract ideas and mental steps without any machine involvement or physical transformation of materials. On appeal to the Board of Patent Appeals and Interferences (BPAI), this rejection was upheld on the ground of lack of any type of physical transformation of materials and the fact that the claims did not produce any useful, concrete and tangible results. *Bilski* appealed this decision to the CAFC and the case was heard by the entire panel at CAFC.

In a 9–3 majority opinion in October 2008, the *en banc* CAFC set forth the criteria for subject matter eligibility for patentable process claims, i.e. that such claims must be tied to a machine or apparatus, and/or it is involved in the transformation of a material to a different state or thing. Since the *Bilski* claims did not address either, the CAFC decided against the patent eligibility of the *Bilski* claims. On appeal, the Supreme Court granted *certiorari* and agreed to review the CAFC decision.

In June 2010, the Supreme Court affirmed the CAFC decision on the patent ineligibility of the *Bilski* claims as representing abstract ideas which are not patent eligible. The court, however, held that the machine or transformation test, that the CAFC applied to determine the patent eligibility of the *Bilski* claim, was not the sole test for determining patent eligibility and there is no categorical exclusion of business method or process patents, leaving the business method patents to a somewhat unpredictable and uncertain future.

The patent eligibility criteria for computer-associated business methods have not fared any better. In fact, the patent eligibility of mathematical formula or algorithm has an interesting history. In *Parker* v. *Flook*,[1] the respondent's patent application involved a method of alarm limits during catalytic conversion processes under operating conditions such as temperature, pressure and flow rates and their measurements. Thus an alarm signals the presence of an abnormal condition and the patent application involved a step using an algorithm to calculate alarm-limit value and a set of computerized calculations producing automatic adjustments in alarm settings. Both the patent examiner and the Board of Appeals of the PTO rejected the application based on the fact that mathematical formula is a form of mental exercise ineligible for patent protection. This decision was reversed by the Court of Customs and Patent Appeals (CCPA) and the case finally moved to the Supreme Court, which reversed the CCPA decision in a divided opinion. However, the Supreme Court in *Diamond* v. *Diehr*[2] did allow the patent eligibility of mathematical calculation, in this case the Arrhenius equation, as an integral part of a useful process such as curing of rubber. The patent application involved a process for moulding raw, uncured synthetic rubber into cured precision products using a mould as well as heat, pressure and temperature measurements by a computer. The computer then signals a device to open a press for cured rubber. The patent examiner and the PTO Board of Appeal rejected the claims as part of a non-statutory subject matter under *Gottschalk* v. *Benson*.[3] This decision was reversed by the CCPA and the case ultimately went to the Supreme Court. The Supreme Court recognized the fact that the respondents were not seeking to patent the computer-generated mathematical formula but simply using the formula for a useful process of curing synthetic rubber. The court thus ruled the patent application as eligible for protection under 35 USC Section 101.

To determine the computer's role in the patenting process, the US President's Commission on the Patent System in 1965 recommended exclusion of computer programs from the coverage of patenting. This recommendation was finally modified by the CCPA in 1968 leading to decisions such as *Gottschalk* v. *Benson* in 1972 and *Parker* v. *Flook* in 1978. The uncertainty about the patenting of computer programming continues as of today, as illustrated by the case *CLS Bank Int'l* v. *Alice Corp. Private Ltd.* Alice Corp. holds several patents covering the risk minimization for foreign currency exchange and other monetary exchanges. Alice brought a lawsuit in the US District Court for the District of Columbia against CLS Bank International for infringing several patents on the use of computer storage media and software for currency exchanges. However, in 2011, the District Court issued a Summary Judgment holding these patents as invalid because the patent claims were viewed as abstract ideas and therefore ineligible for patenting under the Supreme Court verdict in *Diamond* v. *Diehr*[4] that laws of nature, natural phenomena and abstract ideas are not patent eligible under 35 USC Section 101. On appeal, a panel at CAFC in a divided judgment reversed this decision. The majority argued that the patent has a focus on using computer systems for performing exchanges, and is not simply a compilation of business methods that have been ruled to be patent ineligible because of their abstract ideas, as decided by the minority.

On petition from the CLS Bank, the CAFC granted a rehearing *en banc*, basically to address two important questions: (i) what test should the court adopt to determine if the involvement of a computer program or its use represents only an abstract idea and therefore makes it patent ineligible; and (ii) for patent eligibility determination, should the criterion rest on whether the claims cover a method, a program or a storage medium? The CAFC then issued six rather confusing and split decisions, holding Alice's patents as invalid, but providing no clear guidance with respect to patent eligibility of computer methods, program or storage medium. With split decisions and numerous divided opinions, the CAFC ruling only provided uncertainties on patent eligibility issues on computer-related inventions. In June 2014, the US Supreme Court issued a unanimous verdict holding that the software patent claims were drawn to a patent-ineligible abstract idea and were thus patent ineligible.

Another interesting case that the court ruled in a 7–2 opinion in June 2011 is known as *Stanford* v. *Roche* on the patent right ownership by academic institutions (or small businesses) on inventions that are federally funded under the University and Small Business Patent Procedures Act of 1980, commonly known as the Bayh-Dole Act. In this decision, the Supreme Court sided on the part of Roche Molecular Systems, who was sued by

Stanford University for infringing its patents on the use of the PCR (polymerase chain reaction) technique in the detection and quantitation of HIV viral DNA in infected patients. The Supreme Court affirmed the ruling by the CAFC that the patent rights under the 1790 US Patent Laws belong to the inventor and not to an entity where the inventor was employed simply because the entity was funded under the Bayh-Dole Act and was entitled to any IP created under such federal funding. In this case, the Stanford researcher signed a separate agreement with a company, Cetus, where the Stanford employee went to learn how to use PCR for screening of HIV viral DNA and gave up the right of ownership to Cetus, which was later acquired by Roche.

3.6 Patent Eligibility of Human Genes and Mutations: *The Association for Molecular Pathology* v. *US PTO/Myriad Genetics*

Many human diseases such as cystic fibrosis, sickle cell anaemia and many others are due to mutations in genes (Couzin-Frankel, 2010). Cancers are a clear example, where sequencing of more than 100 tumour genomes has demonstrated the presence of about 50 mutations for pancreatic cancer, 100 or more mutations in lung and melanoma, and a varying numbers for others. Thus cancer is basically a disease of old age where accumulation of such mutations, particularly in some key genes where the mutations are known as driver mutations while the others are called passenger mutations, can lead to cancer. For example, for the skin cancer melanoma, noted for its metastatic (movement to other organs) propensity, mutations in a gene known as *PREX2* are more frequently found, making *PREX2* appear like a driver mutation (Berger *et al.*, 2012). Besides old age, certain chemicals or radiation, such as nicotine in tobacco smoke or UV radiations in sunlight can trigger cancer. Thus mutations in genes that encode enzymes responsible for the metabolism and excretion of environmental xenobiotic compounds and carcinogens, such as N-acetyl-transferase (NAT1 and NAT2), glutathione-S-transferase (GST M1, M3, P1, T1 etc.) or cytochrome P50 (CYP1A1, CYP2D6, etc.), have been implicated in the causation of bladder, colon or lung cancers (Chakrabarty, 2003b). Besides old age and exposure to environmental toxicants, a third way to cancer susceptibility is genetic inheritance from parents of certain mutations that can in turn confer cancer susceptibility. Thus knowing the presence of such inherited genetic mutations in one's genome can allow a potentially susceptible person to take precautionary measures, as prenatal or postnatal genetic screenings have shown (Chakrabarty, 2003a, b). Often, the developers of such diagnostic testing procedures patent their methods to prevent others from carrying out the screening of the mutations for commercial gain, and then are alleged to charge high fees or limit the testing by others, thereby fostering frustrations among clinicians, patients and advocacy groups. This was the case when the Canavan Foundation and parents of children suffering from Canavan disease, a central nervous system disorder due to mutations in chromosome 17 in a gene encoding an enzyme aspartoacylase, brought a lawsuit in 2000 in the District Court of Chicago against Miami Children's Hospital alleging high costs of testing because of limitations put on other clinicians for carrying out the patented testing. Such cases have created confusions and apprehensions in the general public about the value and ethics of the patenting of human genes and mutations (Chakrabarty, 2003a, b).

Another example of a similar lawsuit because of alleged high cost of genetic testing and the limits of such testing by other clinicians is the case known as *AMP et al.* v. *US PTO et al.* (Fialho and Chakrabarty, 2012). Around 1994/1995, the University of Utah Research Foundation and a company Myriad Genetics in Salt Lake City filed several patent applications to cover the role of two genes known as *BRCA1* and *BRCA2* where certain frameshift and deletion/genetic rearrangement mutations led to a very high (about 80%) incidence of breast or ovarian cancers in women where the normal incidence is about 10% or less (Fialho and Chakrabarty, 2012; Chakrabarty, 2013). The claims in these patents related to isolation, purification and sequence determination of a gene *BRCA1* present on chromosome 17 and a second gene *BRCA2* on chromosome 13, where certain mutations were determined to foster susceptibility to breast, or with a lower frequency to ovarian, cancers. Seven such patents were issued between 1997 and 2000 on *BRCA1* and *BRCA2* genes, whereupon Myriad Genetics was alleged to charge high fees for testing and prevent other clinicians and medical groups from conducting the genetic tests under an acceptable licensing agreement with Myriad Genetics. Myriad Genetics was also alleged not to allow women with a family history of breast or ovarian cancer, who were diagnosed by Myriad Genetics to harbour the BRCA mutations, to have the tests conducted by a second testing group to have an independent evaluation.

Frustrated with Myriad Genetics' strict enforcement of its patent rights on *BRCA1* and *BRCA2* genes, a group of patients, clinicians, the Association for Molecular Pathology (AMP), American Civil Liberties Union (ACLU) and the Public Patent Foundation brought a lawsuit in May 2009 in the District Court of the Southern District of New York against the US Patent & Trademark Office (US PTO), Myriad Genetics and the University of Utah Research Foundation. They argued that the US PTO should not have issued the *BRCA* gene patents to the patentees since these human genes are products of nature, common to mankind and should not be the products of commerce. In March 2010, Judge Robert Sweet of the District Court in Manhattan granted a summary judgment in favour of the plaintiffs, revoking the seven patents issued to Myriad Genetics.

In June 2010, Myriad Genetics appealed the District Court ruling to the CAFC for reversal, arguing that the isolated and purified *BRCA* genes were not the same as they occurred in the human genome and required considerable human ingenuity and intervention. The isolated DNA had free 5′ and 3′ ends that were not free when the genes were part of the genome. The three-judge CAFC panel, in July 2011, then reversed the New York District Court ruling, upholding the patent eligibility of isolated and purified human genes under 35 USC Section 101. The panel unanimously held the c-DNA forms to be patent eligible, while isolated and purified DNA fragments led to a split 2–1 decision with one judge finding the scope of the structural changes insufficient for patent eligibility. The CAFC held a negative view of the patentability of the sequence comparison for defining mutations in the *BRCA* genes, however, asserting that such sequence comparisons are mental exercises that do not fall under the patent eligibility laws.

After the issuance of the CAFC decision, the parties approached the Supreme Court for a final resolution, submitting a petition for writ of *certiorari* in December 2011, which was granted. However, in March 2012, the Supreme Court vacated and remanded the *AMP* v. *Myriad Genetics* case, No. 11-725, back to the CAFC to reconsider its decision based on the Supreme Court's ruling on *Mayo Collaborative Services* v. *Prometheus Laboratories* where the Supreme Court held the Prometheus diagnostic tests to be patent-ineligible, as mentioned earlier. In August 2012, however, the three-judge panel of the CAFC reaffirmed its earlier decision on the patent eligibility of the isolated and purified *BRCA1* and *BRCA2* genes as distinct chemical entities, as opposed to the genes on the chromosome, thus leading the plaintiffs to appeal to the Supreme Court for a final resolution of the case. On 13 June 2013, in a unanimous decision, the Supreme Court held that isolated and purified *BRCA1* and *BRCA2* genes are naturally occurring DNA segments without any modifications and are not patent eligible. The Court ruled that any cDNA that is different from the naturally occurring intron-containing DNA will be patent eligible provided it satisfies the statutory requirements of patentability. This decision thus reversed the 2–1 majority decision by the CAFC that Myriad's claims on isolated and purified *BRCA* genes were patent eligible. The Supreme Court acknowledged that Myriad found an important and useful gene(s), but separating that gene from its surrounding genetic material was not an act of invention, contending that ground breaking, innovative or even a brilliant discovery does not necessarily satisfy the Section 101 inquiry.

3.7 Why Is It Important to Allow Patenting of *BRCA* Gene Mutations?

Nearly 10% of breast cancer positive women harbour *BRCA* mutations. Since about 8–10% of all women will likely get breast cancer in their lifetime, about 1% or slightly less of all women will develop breast cancer due to *BRCA* mutations. This is indeed a huge number that requires urgent remedy. The first remedy is, of course, to determine who are the most vulnerable persons, particularly with a family history of breast or ovarian cancers. This requires the genetic screening for the presence of the *BRCA* mutations, which has led to the patent-related monopoly problem. We have previously argued (Fialho and Chakrabarty, 2012) that the US PTO, as dictated by their revised guidelines of 2001 requiring specific, substantial and credible utility of a gene to be patent eligible under 35 USC Section 101 and 112, should not have granted patents on the non-mutated *BRCA1* and *BRCA2* genes used as reference genes to determine the nature and number of *BRCA* gene mutations/deletions/genetic rearrangements in the genomes of vulnerable women. However, we have also argued (Fialho and Chakrabarty, 2012) that determination of the mutations in the *BRCA1* and *BRCA2* genes, even though involving nucleotide sequence comparison-related mental exercises, deserves patent protection based on the Supreme Court precedent *Diamond* v. *Diehr*.[5] This Supreme Court decision allowed patentability of mental exercises when such exercises are tied to a useful invention (optimum time of curing of rubber in this instance). Given the enormous importance of the screening of *BRCA1* and *BRCA2* mutations in vulnerable women to allow them to take preventive measures, we argued that the screening for mutations, but not the reference non-mutated genes, should be patent eligible. A simple example will illustrate this. A company comes up with a more colourful and more fragrant genetically modified rose. When demonstrating the improved nature of the genetically engineered rose for patent protection, the company makes a comparative study of the brilliance of colour and more fragrant nature of the genetically modified rose with a common garden-variety rose. The company then asks for patent protection of both the reference garden-variety rose and the genetically modified rose, thus having the patent-assured limited monopoly on all roses, natural and genetically modified (Chakrabarty, 2013).

A second and equally, if not more, important consideration is what to do if or when a woman, particularly when a teenage girl, with a family history of breast or ovarian cancer is tested positive for the presence of the mutated *BRCA* genes. One choice is not to take any particular preventive measure but to remain vigilant with the hope that the cancer never makes an appearance, or if detected early, can be surgically removed without any after effect. One concern is the emergence of ovarian cancer, which usually becomes detectable only after its metastatic migration to other body parts, leading to the appearance of symptoms. However, it should be noted that there are emerging protein-based diagnostic tests for ovarian cancer such as OVA1 approved by the US Food and Drug Administration (FDA). This test uses a set of biomarkers and appropriate software to predict the probability of having ovarian cancer, which could be useful for early detection. A less desirable and traumatic approach, used sometimes, is the surgical removal of the breasts and the ovary to prevent the onset of the cancers. This is particularly devastating for young women of child-bearing age, as recently reported in many national newspapers. Ideally, the *BRCA* mutation-positive women should have access to drugs that are non-toxic for long term use with both cancer therapeutic and preventive activities. While no such drug currently exists, there appears to be on the horizon potential candidate drugs, such as p28, with some of these desirable properties (Fialho and Chakrabarty, 2012). P28, a peptide of 28 amino acids, has recently been shown in phase I clinical trials in 15 stage IV cancer patients with solid metastatic tumours to have no significant toxicity even at the highest dose but considerable beneficial effect including partial and sometimes complete regression of the tumours that were resistant to all conventional drugs (Warso *et al.*, 2013). These patients had an average life expectancy of less than 6 months, but after the termination of the trials, three patients were alive beyond 2 and 3 years with their tumours regressed partially or completely (Warso *et al.*, 2013). It is not known if the tumours that were uniquely susceptible to p28 harboured any particular type of driver or passenger mutations that conferred such susceptibility. P28 is also undergoing a second successful phase I trial in paediatric brain tumour patients in 11 hospitals in the USA (http://clinicaltrials.gov/ct2/show/NCT01975116). While this therapeutic activity of p28 with no toxicity but some efficacy against the drug-resistant metastatic tumours is encouraging, more important is the fact that its therapeutic activity aside, p28 has been shown to have cancer preventive activity as well (Fialho and Chakrabarty, 2012). As we suggested earlier (Fialho and Chakrabarty, 2012), this combined therapeutic and preventive activity without any demonstration of toxicity can be tested in mouse models. Introduction of activated oncogenes such as H-RAS, AKT and Tp53 through Cre-loxP-controlled lentiviral vectors can trigger brain tumour glioblastoma in all the mice (Marumoto *et al.*, 2009), while the introduction of activated oncogenes K-RAS and Tp53 can trigger adenocarcinomas of the lungs in mice. Oral and oesophageal cancers can also be induced in male CBA mice when given a carcinogen 4-nitroquinoline-1-oxide in drinking water for several weeks (Hasina *et al.*, 2013). P28 at various doses can be given to these mice through intravenous (i.v.) injections, or through nasal spray for lung cancers, to examine its potency in preventing the induction of cancers in such susceptible mice. If p28 can be shown to be functional in preventing the emergence of cancer in mice models, it can then be tested in *BRCA1/BRCA2* mutation-harbouring women in long term experiments to determine if p28 can be effective in significantly reducing the incidence of breast or ovarian cancers in women taking i.v. injections of p28 once a week, as opposed to women not taking p28. If the initial results look encouraging, further modifications of p28 can be made using polymers such as polyethyleneglycol to produce an oral variety of p28 for long term consumption. Ultimately, if all these potential experiments show some degree of success, that will demonstrate the importance of the genetic screening of *BRCA* mutations in vulnerable women for them to take non-surgical preventive measures (Chakrabarty, 2012).

3.8 India's IP Scenario

India's patent laws were enacted in 1856 under the British rule, which allowed patenting of drugs and pharmaceutical products. However, after independence in 1947, the Indian Patents and Designs Act of 1911 was amended during 1970–1972 to preclude the patenting of pharmaceutical and agrochemical products while allowing the process patents for periods of 5 to 7 years. Most interestingly, the Patents Act allowed compulsory licensing of such processes/products to encourage generics production 3 years after the date of issuance but with payment of an agreed upon royalty. Such arrangements changed in 1995 when India became a founding member of the World Trade Organization (WTO) and a signatory of the WTO's Trade-Related Aspects of Intellectual Property Rights (TRIPS). However, the TRIPS agreement mandated India to provide patent protection for drugs and pharmaceutical products of WTO member countries by January 2005, prompting the Indian Parliament to pass the Second Patents Amendments Act in 2002, allowing a 20 year patent term like the rest of the WTO member countries, which became effective as of January 2005. The act, however, considerably diluted certain protections, particularly in the area of traditional knowledge and benefit sharing, while promoting compulsory licensing on grounds of public health emergency for manufacturing and export of potential drugs to countries with insufficient or no manufacturing capacity. Another feature of the 2005 Patent Act amendment, known as section 3(d), is the exclusion of patent protection for slightly improved forms or the mere discovery of any new form or new uses of known pharmaceutical drugs which do not result in the significant enhancement of the known efficacy of that drug. This aspect of section 3(d) prompted court cases in India such as the case involving *F. Hoffman-La Roche Ltd* v. *Cipla Ltd* in the High Court of Delhi (April, 2009) when the Indian company Cipla marketed a generic version of Roche's anticancer drug Tarcerva. In denying the request from Roche for an injunction, the Delhi High Court cited section 3(d) for a lack of significant efficacy of the new drug over the old version, without defining what is 'significant efficacy' under the Indian law. Section 3(d) was also used to deny in 2005 Novartis' application for a patent for the drug Glivec (also known as Gleevec), widely used in treating chronic myeloid leukaemia (CML) and gastrointestinal stromal tumours (GIST). Glivec was the polymorphic salt form of patented imatinib mesylate free base and was claimed to have better bioavailability, thus prompting the company to apply for an Indian patent. Both the Madras High Court and the Indian Supreme Court denied patent eligibility of the Novartis patent, citing obviousness of the claims of a previously-issued US patent known as the Zimmerman patent.[6]

In addition to section 3(d), other parameters such as compulsory licensing have also played a major role in patent controversies often reaching to the Intellectual Property Appellate Board (IPAB) and the High Court/Supreme Court of India. For example, in October/November 2012, the IPAB revoked a 2006 patent granted to Roche for a hepatitis C-treating drug Pegasys on the ground of obviousness and lack of significant efficacy. Another patent controversy involved Merck's diabetes drugs Januvia and Janumet, the orally-taken patented gliptin family drugs that were also produced by Glenmark Pharmaceuticals and marketed at a 30% cheaper rate as Zitamed and Zita. Similarly, the Indian Patent Office in 2012 revoked the exclusive patent rights of an anticancer drug Nexavar held by Bayer to allow a local generic maker to market the drug at a significantly reduced price. IPAB also revoked the Glaxo Smith Kline (GSK) patent on the breast cancer drug Tykerb, citing obviousness and a lack of full description. Other examples include revocation of the patents on Pfizer's anticancer drug Sutent issued in 2007 on the ground of a lack of inventive steps and Allergen's eye drugs Ganfort and Combigen for glaucoma. These decisions created major conflicts between the multinational corporations that held the patents and the generic makers in India that took advantage of the Indian patent laws that are designed not to promote innovation but to help bring low-cost drugs and pharmaceuticals to the Indian market.

3.9 Patenting Human Stem Cells: Beyond the Legal Issues

Ever since the development of pluripotent stem cells such as embryonic stem cells (ESC) with the ability of self-renewal and differentiation to multiple lineages (Loring and Campbell, 2006) including human ESC (hESC), the question of how to protect such scientific innovations, that are crucial for tissue regeneration and engineering in seriously ill patients, has been a major headache for patent attorneys, the technology developers and the court system. Besides complexities of such ideas and efforts, the ethical issues surrounding the manipulation of human eggs (oocytes) and embryos have been a major deterrent in the rapid advancement of the field. Significant progress has, nevertheless, been made, even though Congressional actions in the USA have limited the governmental funding of generating the large number of cell lines needed for the treatment of people who in turn need to repair their damaged organs such as the heart, liver or kidney. To avoid and to reduce the concerns pertaining to the use of human embryos, a new technique, induced pluripotent stem cells (iPSC), was introduced in 2006 that involved initially the introduction of 24 genes in mouse fibroblasts to induce pluripotency (Takahashi and Imanaka, 2006), but was later brought down to four genes *Oct3/4*, *Sox2*, *Klf4* and *c-Myc*. Subsequent improvements brought down the number to one (*Oct4*) using neural stem cells and the efficiency of pluripotent cell generation to reach almost 100%. Since the original method involved an oncogenic gene *c-Myc*, further improvements allowed the use of chemicals for reprogramming mouse fibroblasts to iPSC and the use of nuclear transfer using fibroblasts from type I diabetes patients, indicating its potential for therapeutic use in the treatment of degenerative diseases such as type I diabetes or Parkinson's, with minimal immune rejection but with the prospect of appropriate drug screening for drug efficacy determination for the donor. For organ transplantation that requires billions of terminally differentiated cells, the unlimited self-renewal of ESCs and iPSCs may provide the cell populations enriched for a particular organ such as the pancreas.

Yet another recent way to produce pluripotent stem cells from human oocytes without going through the fertilization process is a technique known as parthenogenesis. In this process, human oocytes are treated with certain chemicals or electrical current that stimulate the process of differentiation giving rise to a parthenote and leading to the formation of blastocysts but no further differentiation. Pluripotent stem cells that can give rise to many different kinds of human tissues under the right cues can then be extracted from the blastocysts. All three processes, ESC, iPSC and parthenogenesis, however, require the participation of a large number of human oocytes, whether fertilized or not, which are the central ingredients of potential life as embryos, thus creating a sense of uneasiness in the minds of many people as interfering in natural life processes for economic gain. Life is sacred to many individuals and religious groups, even though what exactly constitutes life remains murky. This problem is further exacerbated when such life processes tend to become a market-place commodity requiring patent protection. Since 1980, patenting genetically manipulated life forms, requiring human intervention, have been legal in the USA as a result of the US Supreme Court decision in the case *Diamond* v. *Chakrabarty*[7] and many patents have been issued on genetically manipulated microorganisms, plants and animals (Chakrabarty, 2002, 2003a). However, in the EU, under the doctrine of *ordre public* or morality, only lower forms of life such as microorganisms are patentable. Thus stem cells, including human ESCs, have been patented in the USA but not in the EU, highlighting the differences in the two systems. However, important questions are now being raised on the patent eligibility issues of pluripotent stem cells, with special reference to a recent court case in the UK involving patent eligibility of pluripotent stem cells derived through parthenogenesis that throws light on the definition of what constitutes a human embryo, and, therefore, a form of potential human life. It is important to note that a recent US patent,[8] entitled 'Gamete donor selection based on genetic calculations' issued in September 2013, provides some insights on how parents in the future may be able to select genetic traits of their children from prospective donors (eye colour, athletic ability, etc.), raising questions of both moral and legal importance on patenting of human genetic traits.

3.10 Patent Eligibility Issues Involving Stem Cells Derived From Human Embryonic and Induced Pluripotent Stem Cells and Parthenogenesis

As major investments using embryonic stem cell therapies are being made, for example in the phase I clinical trials for the treatment of macular degeneration, which often leads to blindness, an important question is the patent eligibility of such human embryonic stem cells (hESC)s, as held by the Wisconsin Alumni Research Foundation (WARF). A central argument follows the recent US Supreme Court decision in June 2013 that isolated and purified human genes such as *BRCA1/BRCA2*, patented by the University of Utah Research Foundation and Myriad Genetics, are not patent eligible since they are products of nature without human intervention, as discussed earlier. Whatever the merit of such a decision (Chakrabarty, 2013), the argument that patented isolated and purified stem cells are essentially identical to the stem cells occurring in human blastocysts resonates with many people who are either against patenting any part of human body or interfering in the natural processes of human development. A way to circumvent such arguments is not to rely on the patent eligibility of stem cells themselves but on the methods of their use as therapeutic agents in regenerative medicine. Certainly the 2013 US Supreme Court decision in the University of Utah/Myriad Genetics patent case will cast a long shadow in the coming days on the advancements of stem cell research, whether such cells are derived through *in vitro* fertilization, induction through the use of cytokines or other agents such as Oct, Klf, Myc, or through the parthenogenetic mode of human oocyte differentiation and stem cell extraction.

Stem cell patents have a long and interesting history. Even though human somatic cell nuclear transfer using adult cells has been reported, the original Thomson patent assigned to WARF was on hESC using human embryos. Although initially funded by government grants, the Dickey amendment of the US Congress prohibited federal funding on any research involving human embryos, which prompted WARF to seek such funding from industrial partners such as Geron Corporation, that subsequently received FDA approval in 2009 to conduct clinical trials with hESCs. Since the issue of its three patents[9] on the derivation and potential use of hESCs in regenerative medicine, WARF has been challenged repeatedly for its assertion of its ownership and licensing rights of its patents that have been upheld by the courts, albeit with certain limitations of its claims based on the obviousness of its initial patent claims. WARF's initial patent application involving hESC generation and potential application, while successful in the EU, was later rescinded, but re-instated with a narrow set of claims in 2008 by the Enlarged Board of Appeals of the European Patent Office (EPO) under Rule 28 of the European Patent Convention involving *ordre public* or morality clause. A consumer watchdog group has been trying since 2006 to overturn the WARF stem cell patent 7,029,913 but has been declared to lack standing by the Court of Appeals for the Federal Circuit in 2014. A major advantage of hESC is that patients can provide their own nucleus for somatic cell nuclear transfer, thus minimizing rejection of the transplanted organ or regenerated tissue. Such methods have also led to the development of a disease-specific diploid-state hESC line from one adult donor with type I diabetes and a healthy donor control by the introduction of the nuclei of adult skin cells to unfertilized donor oocytes by somatic cell nuclear transfer for treating diabetic patients with their own insulin-producing cells to minimize immune rejection. The challenge of using adult cells to convert them to an embryonic state is to change the nature of the DNA from terminally differentiated adult cells such as dermal fibroblasts through genetic rearrangements to an embryonic state, using caffeine as an inhibitor of premature cell division. A large number of eggs from women are used in the process after their DNA is replaced by the donor nuclei and triggering cell division. The challenge is still the use of large number of eggs, as is common, for example, in the final birth of Dolly the sheep in 1997, and in making banks of matching human leukocyte antigen (HLA) cell lines to generate complete HLA haplotype match for a large segment of the population.

In contrast to the US patent laws, the EU patent laws with the *ordre public* and morality clause does not allow patenting of higher forms of life, as mentioned earlier, including the hESCs. The celebrated case is that of German researcher Oliver Brustle, whose patent application on the generation of hESC granted in 1999 was challenged by Greenpeace in 2005 and ultimately rejected on moral grounds in 2011 throughout the EU under the *ordre public* clause in the EU's Directive on the Legal Protection of Biotechnological Inventions. While the situation is different in the USA with regard to patentability that allowed the issuance of the WARF patents on hESCs, there is significant opposition to the issuance of such patents even in the USA. The discontent is not just the perceived limitation of research in this important area by academic sectors and clinicians, but also the ethical issues pertaining to the use of human oocytes and their fertilization, leading to the formation of blastocysts from which stem cells are derived. Many people and organizations are uneasy, and often offended, in the so-called temporary monopoly that the patent system provides to ownership rights to human life, including a potential large scale worldwide trade in human eggs, particularly in poor countries. This has led to strong opposition to the WARF patents, and patents over hESCs in general. Consequently, many researchers have tried to circumvent the use of human oocytes in preference to adult cells that might be more acceptable to people, leading to the success in recent years in the production of induced pluripotent, but not totipotent, stem cells (iPSCs). Such cells can be generated from cultured adult cells from a biopsy sample in presence of certain inducers as mentioned earlier, that allow the cells to differentiate into a variety of other cells such as heart, nerve or brain. Thus a patient's own cells from a biopsy sample can be induced to become pluripotent and can be used in repairing a damaged tissue without the fear of significant immune rejection. A problem of iPSC is often a lack of the stability and pluripotency of the cells, although the ethical issues surrounding the destruction of embryos are avoided.

3.11 Patenting Parthenogenesis-Derived Stem Cells: Defining an Embryo

Aside from iPSC, yet another way to generate pluripotent stem cells without requiring the fertilization of an oocyte through introduction of a donor nucleus is through the process of parthenogenesis. In this process, the activation of the oocyte to reach the blastocyst stage, but no further, is accomplished through chemical treatments or electrical stimulation of the oocytes without requiring any external donor nucleus. Such an activated oocyte, known as a parthenote and the process as parthenogenesis, contains only the maternal nuclear chromosome but no paternal DNA and is known not to undergo full fertilization to give rise to an embryo. Thus parthenotes are not totipotent. However, a parthenote is capable of developing into a blastocyst type of structure with an outer trophectoderm and an inner cell mass with the ability of a pluripotent cell, giving rise to many differentiated tissue cell types. In an interesting case in the UK, a company International Stem Cell Corporation (ISCC) filed to the UK IPO (UK Intellectual Property Office) two patent applications, GB 0621068.6 entitled 'Parthenogenetic activation of oocytes for the production of human embryonic stem cells' and a second one GB 0621069.4 entitled 'Synthetic cornea from retinal stem cells' for the registration of national patents. Both the applications involved method claims on the isolation of pluripotent stem cells from parthenogenetically-activated oocytes and their differentiation into specific corneal cells. It was pointed out that cells from a parthenote are only pluripotent, never totipotent, since their genomic imprinting is defective because of the absence of paternal DNA required for full development of an embryo. The UK IPO in August 2012 refused to register the applications based on the Brustle ruling by the Court of Justice of the European Union (CJEU) that the patents involved production of blastocyst-derived stem cells, which, under the Article 6(2)(c) of the Biotech Directive 98/44/EC and Article 5(1), are products of the human body at various early stages of its formation and are therefore not a patentable invention. On appeal from ISCC, the High Court of Justice (England & Wales), Chancery Division (Patents Court, UK), in the case *International Stem Cell Corporation* v. *Comptroller General of Patents, Designs and Trade Marks*, referred the case to CJEU on the interpretation of Article 6(2)(c) of the directive on the correct definition of human embryo. The need was to fully differentiate between fertilized ovum, ovum fertilized through donor nuclear transfer and parthenotes to decide if parthenogenetically-activated oocytes, that cannot undergo any further differentiation beyond the blastocyst stage to form a human embryo, can be designated as cells capable of commencing the process of development as a human being. The ISCC claimed that the Brustle decision (EU:C:2011:669) did not apply in the ISCC patent registration filing since parthenogenetically-derived oocytes are incapable of giving rise to a fertilized ovum and therefore should be patent eligible under Directive 98/44. Additionally, ISCC made changes in its patent application to exclude any further genetic manipulation that might allow the parthenote to develop into a fertilized ovum. Unable to reach a verdict on its own, the UK High Court of Justice, Chancery Division, dealing with patent-related issues, referred the case to the CJEU for a preliminary ruling on whether such a method of creating a parthenote for the extraction of pluripotent stem cells violates the provisions of Article 6(2)(c) of Directive 98/44 and remains unpatentable subject matter. In December 2014, the CJEU rendered its verdict pointing out that in the Brustle decision, the court was led to believe that a non-fertilized human ovum created through parthenogenesis was somewhat similar to a fertilized ovum and was classified as a potential embryo. However, more recent scientific evidence points to the inability of a parthenote to undergo any further development beyond the blastocyst stage and ISCC's amended application for registration not to use any further genetic manipulation ensures that the parthenote will not become an embryo with totipotent character. The CJEU therefore recommended to the UK High Court that unless further evidence to the contrary is available, the parthenote does not *per se* constitute a human embryo under Article 6(2)(c).

3.12 The Impact of the Court of Justice of the European Union Decision

The CJEU decision (Davey *et al.*, 2015) differentiating between a parthenote and an embryo begs two questions: if a human parthenote is not a potential human life but a human embryo is, what exactly is the characteristic of life that is present in the embryo formed from blastocyst during fertilization but not in the parthenote-derived blastocyst from which stem cells are harvested, the question of the contribution of paternal DNA aside? A second vexing question is the definition of human. Since genetically engineered or any other form of human cannot be patented either in the EU or in the USA, but genetically manipulated animals (oncomouse, for example) can be patented in the USA, what exactly is the difference between an animal, say a monkey, and a human being? This is not just a rhetorical question, since a patent application for a part human–part animal chimera was filed to the USPTO in 1998 but was rejected because of the 13th amendment of the US Constitution prohibiting slavery and ownership of human beings (Chakrabarty, 2003c). Nevertheless, the pertinent question, both moral and legal, is how much human characteristics, including certain number of human genes, must be present in an animal to give it the legal status of a human? The problem is more acute in the EU where the *ordre public* and morality clause prevents the patenting of not only humans but human cells or organs including human stem cells. The 1980 *Diamond* v. *Chakrabarty* US Supreme Court decision[10] led not only to the patenting of genetically engineered life forms such as bacteria, plants and animals but also stem cells such as human bone marrow stem cells,[11] followed by the human embryonic stem cells by Thomson *et al.* (1998) as mentioned earlier. However, as also mentioned earlier, no patent on human stem cells has been allowed in the EU so far, although the CJEU decision appears to open a crack in the door of the EU morality clause. It should be noted, however, that this decision by the CJEU may actually open up a can of worms for future decision makers. The CJEU decision is based on the fact that the blastocysts derived from a human oocyte that underwent cell division under electrical stimuli or chemical treatment without a paternal nucleus are incapable of further differentiation to give rise to a human embryo because the paternal nucleus is important for full genetic shuffling. In therapeutic cloning, with enucleated human oocytes (but with their mitochondrial DNA), and transfer of donor nucleus, the fertilization is supposed to lead to development of a human embryo irrespective of the paternal or maternal nature of the transferred nucleus. In somatic cell nuclear transfer, will the transfer of a nucleus from a female human donor to an enucleated human egg fail to produce an embryo because of the lack of paternal DNA and full genetic imprinting? Although there are many technical reasons why primate cloning, as opposed to mouse, sheep or cattle, has been unsuccessful so far (Simerly *et al.*, 2003) and will be resisted by public opposition and governmental regulations including the question if any deformed or destroyed embryo during the experimental process may be legally designed as human mutilation or murder, it will be interesting to evaluate various options pertaining to the methods of embryonic or parthenogenesis-related development of blastocysts and their further differentiation, if any, to produce viable embryos, and any future patent eligibility issues (Davey *et al.*, 2015).

Notes

1 US Supreme Court, *Parker* v. *Flook*, 437 U.S. 584 (1978), No. 77-642; argued 25 April 1978, decided 22 June 1978.

2 US Supreme Court, *Diamond* v. *Diehr*, 450 U.S. 175 (1981), No. 79-1112; argued 14 October 1980, decided 3 March 1981.

3 US Supreme Court, *Gottschalk* v. *Benson*, 409 U.S. 63 (1972), No. 71-485; argued 16 October 1972, decided 20 November 1972.

4 US Supreme Court, *Diamond* v. *Diehr*, 450 U.S. 175, 182 and 185 (1981).

5 US Supreme Court, *Diamond* v. *Diehr*, 450 U.S. 175 (1981), No. 79-1112; argued 14 October 1980, decided 3 March 1981.

6 Jurg Zimmermann, 28 May 1996. *Pyrimidine derivatives and processes for the preparation thereof.* US5521184 A

7 US Supreme Court, *Diamond* v. *Chakrabarty*, 447 U.S. 303 (1980), No. 79-136; argued 17 March 1980, decided 16 June 1980.

8 Anne Wojcicki, Linda Avey, Joanna Louise Mountain, John Michael Macpherson, Joyce Yeh-hong Tung, 24 September 2013. *Gamete donor selection based on genetic calculations.* US8543339 B2.

9 James A. Thomson *Primate embryonic stem cells.* 1 December 1998, US5843780 A;13 March 2001, US6200806 B1; 18 April 2006, US7029913 B2.

10 US Supreme Court, *Diamond* v. *Chakrabarty*, 447 U.S. 303 (1980), No. 79-136; argued 17 March 1980, decided 16 June 1980.

11 Tsukamoto, A., Baum, C., Aihara, Y. and Weissman, I., 29 October, 1991. *Human hematopoietic stem cell.* US5061620 A.

References

Berger, M.F., Hodis, E., Heffernan, T.P., Deribe, Y.L., Lawrence, M.S., Protopopov, A., Ivanova, E., Watson, I.R., Nickerson, E., Ghosh, P., Zhang, H., Zeid, R. *et al.* (2012) Melanoma genome sequencing reveals frequent PREX2 mutations. *Nature* 485, 502–506.

Chakrabarty, A.M. (2002) Patenting of life forms: from a concept to reality. In: Magnus, D., Caplan, A. and McGee, G. (eds) *Who Owns Life?* Prometheus Books, Amherst, NewYork, pp. 17–24.

Chakrabarty, A.M. (2003a) Patenting life forms: yesterday, today and tomorrow. In: Scott Kieff, F. (ed.) *Perspectives on the Properties of the Human Genome Project*. Elsevier Academic Press, New York, pp. 3–11.

Chakrabarty, A.M. (2003b) Environmental biotechnology in the postgenomics era. *Biotechnology Advances* 22, 3–8.

Chakrabarty, A.M. (2003c) Crossing species boundaries and making human-nonhuman hybrids: moral and legal ramifications. *American Journal of Bioethics* 3, 20–21.

Chakrabarty, A.M. (2012) Bacterial proteins: A new class of cancer therapeutics. *Journal of Commercial Biotechnology* 18, 4–10.

Chakrabarty, A.M. (2013) Patenting human genes and mutations: A personal perspective. *Journal of Commercial Biotechnology* 19, 777–780.

Couzin-Frankel, J. (2010) Chasing a disease to the vanishing point. *Science* 328, 298–300.

Davey, S., Davey, N., Gu, Q., Xu, N., Vatsa, R., Devalaraja, S., Harris, P., Gannavaram, S., Dave, R. and Chakrabarty, A. (2015) Interfacing of science, medicine and law: The stem cell patent controversy in the United States and European Union. *Frontiers in Cell and Developmental Biology*, doi: 10.3389/fcell.2015.00071.

Fialho A.M. and Chakrabarty A.M. (2012) Patent controversies and court cases: Cancer diagnosis, therapy and prevention. *Cancer Biology & Therapy* 13, 1229–1234.

Hasina, R., Mollberg, N., Kawada, I., Mutreja, K., Kanade, G., Yala, S., Surati, M., Liu, R., Li, X., Zhou, Y., Ferguson, B.D., Nallasura, V. *et al.* (2013) Critical role for the receptor tyrosine EphB4 in esophageal cancers. *Cancer Research* 73,184–194.

Loring, J.F. and Campbell, C. (2006) Intellectual property and human embryonic stem cell research. *Science* 311, 1716–1717.

Marumoto, T., Tashiro, A., Friedmann-Morvinski, D., Scadeng, M., Soda, Y., Gage F.H. and Verma, I.M. (2009) Development of a novel mouse glioma model using lentiviral vectors. *Nature Medicine* 15,110–116.

Simerly, C., Dominko, T., Navara, C., Payne, C., Capuano, S., Gosman, G., Chong, K.-Y., Takahashi, D., Chace, C., Compton, D., Hewitson, L. and Schatten, G. (2003) Molecular correlates of primate nuclear transfer failures. *Science* 300, 297.

Takahashi, K. and Imanaka, S. (2006) Induction of pluripotent stem cells from mouse embryonic and adult fibroblast cultures by defined factors. *Cell* 126, 663–676.

Thomson, J.A., Itskovitz-Eldor, J., Shapiro, S.S., Waknitz, M.A., Swiergiel, J.J., Marshall, V.S. and Jones, J.M. (1998) Embryonic stem cell lines derived from human blastocysts. *Science* 282,1145–1147.

Warso, M.A., Richards, J.M., Mehta, D., Christov, K., Schaeffer, C., Bressler, L.R., Yamada, T., Majumdar, D., Kennedy, S.A., Beattie, C.W. and Das Gupta, T.K. (2013) A first-in-class, first-in-human phase I trial of p28, a non-HDM2-mediated peptide inhibitor of p53 ubiquitination in patients with advanced solid tumours. *The British Journal of Cancer* 108,1061–1070.

4 Checks and Balances in Biotechnology-Related Patents: In Agreement to the Indian Patents (Amendment) Act, 2005

Om Prakash[1]* and Poonam C. Singh[2]

[1]*Maharana Pratap Govt P.G. College, Hardoi, Uttar Pradesh;* [2]*CSIR-National Botanical Research Institute, Lucknow, Uttar Pradesh, India*

4.1 Introduction

A new idea or information may be kept confidential indefinitely; however, it is equally unmanageable to keep the same idea a secret while reaping its benefits. The intellectual property (IP) protections provide confidentiality, management and provide commercial opportunities to a new idea with a cost for a defined period. India has a vibrant culture, unique lifestyle, lots of traditional knowledge and a dynamic economy, which provides unique knowledge and innovations leading to IP (Birtchnell, 2013). This national asset needs to be protected and commercialized after a qualified evaluation for the benefit of both inventor and people. In India the competent authority is 'The Indian Patent Office' (IPO), which carries out the evaluation following the provisions of 'The Patents (Amendment) Act, 2005' (IPAA-2005).[1] Though the fundamentals of patent laws are the same all over the world, procedural differences exist, which make them stringent or liberal. However, patent evaluation is more of a techno-legal evaluation than an exact scientific evaluation.

In a historic decision, the Controller General (CG) of India 'The bullet Kurian', issued a compulsory licence in favour of Natco Pharmaceuticals to manufacture and sell a generic version of Nexavar (a kidney/liver cancer drug by the generic name of Sorafenib Tosylate from Bayer), with a payment of 6% royalty to Bayer besides other commitments from Natco Pharmaceuticals under Section 84 of IPAA-2005 (Nair *et al.*, 2014). This settlement is an example of a strong enforcement mechanism needed for the protection of these rights and the provisions that IPAA-2005 provides to balance between the interests of inventors/owners on one hand and the interests of the society on the other. Therefore the objective of the laws defined in the IPAA-2005 is:

1. To encourage scientific research, new technology and industrial progress.
2. Accelerate the technological and industrial development of the country.
3. To promote an inventor to disclose his inventions instead of keeping them as a trade secret.
4. To offer a reward for the expenses of developing inventions to the stage at which they can be commercially exploitable.
5. To provide inducement for capital investment in new lines of production.

These objectives of the legislation of a country inspire individuals to gain recognition for their creativity and material reward for their reproducible and marketable inventions which henceforth encourages innovation that endorses a better living.

The Indian Patents Act (Khader, 2007) discusses different aspects of the act such as the laws applicable on patents, key provisions of India's

*Corresponding author, e-mail: ops92002@gmail.com

new Patents Act, the IPAA-2005, comparative analysis and critiques of the IPAA-2005. Features such as 'evergreening' for patents, re-grant opposition, post-grant opposition and the intricacies of compulsory licensing schemes are also discussed in detail. It defines the capacity and capability of the IPO to implement the newly strengthened patents regime. The role of the Indian judicial system is to interpret the act in case of litigations which may further be implemented by the IPO.

Biotechnology deals with living subject matters and manipulation of their genomic materials. Such genetic alterations have been constantly debated because of their influence and deep impact on the environment, human, animal and plant life. It also involves serious ethical issues. Therefore, while continuing with an invention, requisite care should be taken in escorting the primary or intended use or economic profitability of the subject matter such that it is not conflicting with public order, ethics or harmful to human, animal or plant life or the environment. Usually the biotechnology applications comprise the claims relating to the following subject matters (Controller General, 2013):

- Polynucleotides or gene sequences (product and/or process);
- Polypeptides or protein sequences (product and/or process);
- Vectors (e.g. plasmids) (product and/or process);
- Gene constructs or cassettes and gene libraries;
- Host cells, microorganisms and stem cells (product and/or process), transgenic cells;
- Plant and animal tissue culture (product and/or process);
- Pharmaceutical or vaccine compositions comprising microorganisms, proteins, polynucleotides (product and/or process);
- Antibodies or antigen-binding fragments thereof (monoclonal or polyclonal);
- Diagnostic kits and tests; and
- Diagnostic tests (products/process) such as a test for the detection of a mutation in a gene/protein, which might be associated with a particular condition such as protein expression or a disease.

4.2 Checks and Balances Offered by Sections of The Indian Patents (Amendment) Act, 2005

The patents right is granted according to the general principles as laid down in IPAA-2005. Section 83, Chapter XVI of IPAA-2005 checks the granting of patents right such that it is not granted merely to acknowledge the right to accession but is expected to be practically implemented within Indian territory. The inherent features of this section are to:

- Encourage transfer and dissemination of technology;
- Ensure that patented inventions are available at reasonably affordable price;
- Ensure mutual benefit to producer and user of the technology;
- Not impede the protection of public health and nutrition; and
- Not in any way prohibit central government in doing so for societal welfare.

The 'Checks' imposed are resilient due to the 'Flexibilities' given in Chapter II of IPAA-2005. Provisions of Sections 3 and 4 of this chapter must be explored before putting an application for grant of a patent before IPO. Sections 3(a), 3(b), 3(c), 3(f), 3(g), 3(h), 3(i), 3(j), 3(k), 3(l), 3(m), 3(n), 3(o) and 3(p) and Section 4 (inventions related to defence) offer a complete check which may be considered as non-bailable without any balancing act for the grant of a claim(s) (only claim(s) are granted in an application) for all fields of subject matter, that is if claimed subject matter falls under said section/sections, patent cannot be granted. However, exceptions have been included in the interest of the society/inventor given in different sections of Section 3. Sections 3(d) and 3(e) impose certain checks which may be non-patentable even if the disclosures have passed the tests of the fixed criteria, which are 'Novelty', 'Non-obviousness' and 'Industrial Applicability'. However, provisions in Sections 3(d) and 3(e) offer balances for 'Potential Inventions'. Sections 10(4) and 10(5) also offer some checks which are not as stringent as in Sections 3 and 4 since they can be overcome by careful drafting/addition/division of the application. Some non-patentable inventions are covered in detail in Section 3 and Section 4 of Chapter II. Details of these sections, their checks and balances with examples are discussed below.

4.2.1 Section 2(1)(j)

This section describes the invention and provides a complete check without any flexibility, that is, if any subject matter is unable to overcome the provisions of this section, the patent will not be granted.

It is important for inventors to consider and overcome the checks imposed in Sections 3 and 4 before applying 2(1)(j). According to Section 2(1)(j) of the act, an 'invention' is described as a new product or process developed after incorporating an inventive step and is capable of industrial application. If the claims filed are described as:

- Use of compounds in the treatment of disease X;
- Use of compound A in the process of preparing B;
- Use of compound A in the composition of Y; or
- A product of a known substance for the treatment of a new disease (which is nothing but use/application claim) then such types of claims are not considered as inventions since the claimed subject matter neither pertains to product nor to process.

In addition to 2(1)(j), an objection as per Section 3(i) would also be invoked. It may be noted that sometimes such claimed inventions relate to the second use of already known compounds that are already present in the public domain, in such cases along with 2(1)(j), Section 3(d) will also be invoked. This can be further understood from an example where known derivatives of fumaric acid were claimed for a second medical indication. It was observed that IPAA-2005 checked the claimed subject matter on two counts (i) the claims relate neither to a new product nor to a process and the compounds of the invention were admittedly known (Section 2(1)(j)); and (ii) they relate to a new use of a known substance (Section 3(d)). After amending the above claim to the product claim Section 3(i) was removed but sections lacking in novelty (2(1)(j)) and 3(d) remained. However, the application was rejected on the basis of novelty criteria by The Intellectual Property Appellate Board (IPAB).[2]

4.2.2 Section 3(b)

This section checks inventions contrary to morality. To evade checks imposed by this section the proposed use or economic profitability of an invention should be carefully considered so that the subject matter is not conflicting to public order, anti-ethical or judgmental to human, animal or plant life or health or to the environment. A few restrictive examples may further clarify the issues:

- A process for cloning human beings or animals;
- A process for modifying the germ line of human beings;
- A process for modifying the genetic identity of animals, which is likely to cause them suffering without any substantial medical or other benefit to man or animal, and also animals resulting from such process;
- A process for preparing seeds or other genetic materials comprising elements that might cause adverse environmental impact, such as terminator gene technology; and
- Uses of human embryos for commercial exploitation.

One of the most classical examples falling in this section is that of 'terminator technology' in hybrid seeds. The patent had already been granted in the USA and filed in 78 countries including India (November 1995). Due to oppositions from farmers and non-government organizations throughout the world and India and considering that it will terminate the farmers' independence and threaten food security and impose ecological issues, the Indian Government banned the entry of any seed material carrying the 'terminator gene' and decided to reject the patent to Delta & Pine Land Company (Phillip, 1999) and in due course of time the subject matter was rendered 'non-patentable technology' and included in Section 3(b) on ethical grounds.

4.2.3 Section 3(c)

The subsection checks the claims related to discovery of any living thing or non-living substance occurring in nature. As per Subsection 3(c) of the act, mere discovery of a scientific principle or formulation of an abstract theory or discovery of any living being or non-living matter occurring in nature is not a patentable subject matter. Products comprising of microorganisms, nucleic acid sequences, proteins, enzymes, compounds, etc., directly isolated from nature, are not patentable subject matter (check). However, processes of isolation of these products can be considered as a patentable subject matter (balance). To understand this section, consider a claim on an isolated peptide that inhibits parasitaemia in malaria-infected red blood cells and intracellular replication of a malaria parasite in malaria-infected human red blood cells and is a structural equivalent of a cupredoxin or cytochrome. It was observed that the disclosure made in the claim did not clearly indicate what modifications/alterations/deletions were made in the wild-type peptides. No intervention of human intellect was observed in the claimed subject matter and therefore it was checked under Section 3(c) of the Act. The term 'isolated' used in claims refers to discovery not invention under the meaning of this act. Thus, the subject matter of the claim is considered to be an 'isolated non-living substance occurring in the nature' and the functional features for the said isolated peptide are considered inherent to cupredoxin or cytochrome proteins, and therefore is not patentable as per Section3(c) of the Act.

4.2.4 Section 3(d)

The section deals with claims/patents related to 'mere discovery of new form of known substance which does not result in enhancement of known efficacy of that substance or mere discovery of any new property or new use of a known substance or of the mere use of a known process, machine or apparatus is not a patentable invention unless such known process results in a new product or employs at least one new reactant'. This introductory sentence of the section itself shows the following checks,'the mere discovery of':

- A new form of a known substance;
- Any new property of a known substance; or
- A new use for a known substance or a process, machine or apparatus.

The balance is also provided in the same introductory sentence:

- Enhancement of the known efficacy of that substance;
- A known process results in a new product; or
- A known process employs at least one new reactant.

These balances provide an opportunity to convert an otherwise non-patentable subject matter into a patentable one. In this way the IPAA-2005 promotes 'Potential Inventions' and discourages just workshop modifications. For the purposes of this clause, salts, esters, ethers, polymorphs, metabolites, pure form, particle size, isomers, mixtures of isomers, complexes, combinations and other derivatives of known substance shall be considered to be the same substance (complete check), unless they differ significantly in properties with regard to efficacy (balance). For example, in an application, Abraxis Bioscience, LLC, USA claimed 'a composition comprising paclitaxel coupled to an anti-SPARC antibody or fragment thereof binding specifically to osteonectin and a pharmaceutically acceptable carrier' under the application topic 'Treatment methods utilizing albumin binding proteins as targets'. After many pre-grant oppositions and arguments (opponent: Natco Pharma Limited, Hyderabad), it was assessed that the applicant had not submitted any data showing comparative efficacy between the paclitaxel and paclitaxel coupled to an anti-SPARC antibody or any fragment of the components which binds specifically to osteonectin. Therefore, the Controller admitted with the opponent that the invention paclitaxel coupled to an anti-SPARC antibody or fragment thereof which binds specifically to osteonectin is not patentable under Section 3(d) of the IPAA-2005 due to the absence of enhanced efficacy. Had there been mention of 'treatment method' it would have also invoked Section 3(i). The Controller did not continue with the application for the grant of patent under Section 15 (powers of controller) of the IPAA-2005, citing claims lack the inventive step under Section 2(1)(ja) of the IPAA-2005 and the claimed composition is not patentable under Section 3(d) of the Patents Act 1970.[3]

4.2.5 Section 3(e)

The section deals with 'mere admixture resulting only in aggregation of the properties or a method of making such mere admixture'. According to this section a claim is not patentable if the efficacy is just additive of the independent properties of the constituents (provides the check). It should have some functional interaction producing a new or improved result and/or must show the synergistic effect (2+2>4), only then it is a subject matter for patent (provides the balance). Unacceptability of mere admixtures is a well-accepted principle of patent law. For example, consider an innovative combination of enzymes, fats and growth-promoting molecules to control plant-parasitic nematodes along with dormant spores of naturally occurring fungi, *Paecilomyces lilacinus* and *Arthrobotrys* sp. The subject matter of claim was checked under the Section 3(e) of the act. Upon analysis, it was found that the claim is directed to a composition of two known fungal species which are known for biocontrol activity against nematodes. Besides, the specification is unclear on advantages or the synergistic effect of a combinative effect of these two fungal species over the sum of their individual effects. Therefore, the subject matter of the claim is checked under Section 3(e) of the act. Had the applicant shown a synergistic effect of the mixture over the sum of the individuals, the claim would have been considered.

4.2.6 Section 3(h)

Section 3(h) of the act deals with 'a method of agriculture or horticulture', which is not considered as patentable subject matter. While considering patentability under Section 3(h), it should be understood that conventional methods performed on actual open fields should be taken as method of agriculture/horticulture. For example, a method of producing a new form of a known plant, irrespective of the modification involved in conditions under which natural phenomena would pursue their inevitable course, is not patentable. For example, subject matter such as a method of growing leguminous plants as inter-cropping for improving fertility of soil by augmenting nitrogen content of the soil is an agriculture method and, hence, falls within the scope of Section 3(h) of the act. Similarly, a method of producing improved soil from the soil with nematodes by treating the soil with a preparation containing specified phosphorathioates (claimed by Virginia Carolina Chemical Corporation application 1958 RFC 38) was not patentable under Section 3(h). A method of producing mushroom plant (64/Cal/79) and a method for cultivation of an alga (445/Del/93) have also been rendered not patentable under Section 3(h) (Barbosa and Grau-Kuntz, 2010).

4.2.7 Section 3(i)

The section deals with 'method of treatment and diagnosis'. According to Section 3(i) of the act, any process used in the treatment of human beings using medicinal, surgical, curative, prophylactic, diagnostic, therapeutic or other means is not patentable under the meaning of the act. This check also extends to the treatment of animals to: (i) make them free of disease; or (ii) increase their economic value or that of their products. Exclusion of methods involving diagnosis, therapy and surgery from patentability in different countries has been discussed in detail by Asif (2013). It is observed that IPO has drafted its Section 3(i) rules keeping in mind the basic human rights, the medical oath taken by medical practitioners to work for the welfare of mankind and contradictory patents law to benefit an inventor. However, the wider aspects in which IPO interprets the exclusions in this section than the patent office/courts of other countries like the USA and UK, are that it excludes a larger number of inventions from patentability. As per the Manual of Patent Office Practice & Procedure, Section 3(i) states that the following provisions are not patentable in the interest of patients.

1. Medicinal methods: for example, a process of oral administration of medicines, or administering medicines through injections, or applying them topically or through a dermal patch.
2. Surgical methods: for example, cataract removal through stitch-free incision.
3. Curative methods: for example, a method to clean plaque from teeth.
4. Prophylactic methods: for example, vaccination methods.
5. Diagnostic methods: for example, disease diagnosis, the process of investigating disease history and symptoms and by applying tests to identify the nature of a medical illness including the general physical state or fitness of an individual, are considered diagnostic and hence not patentable.
6. Therapeutic methods: the term 'therapy' includes prevention as well as treatment or cure of disease. Therefore, the process relating to therapy may be considered as a method of treatment and as such is not patentable.
7. Any method of treatment of animals to render them free of disease or to increase their economic value or that of their products: for example, treatment methods to enhance wool yield of sheep or a method to artificially induce the body mass of birds and animals reared for meat.
8. Further examples of subject matters excluded under this provision are: any operation on the body, which requires the skill and knowledge of a surgeon and includes treatments related to cosmetics, pregnancy termination, induction of labour, control of oestrus or menstrual regulation, male sterilization methods, artificial insemination, embryo transplants, removal of organs, skin or bone marrow from a living donor, treatments for experimental and research purposes and any therapy or diagnosis practised on the human or animal body.

However, the checks provided by these exclusions are balanced by the following provisions of the act.

1. Application of substances to the body for purely cosmetic purposes is not therapy.
2. Patent may be obtained for surgical, therapeutic or diagnostic instrument or apparatus.
3. Furthermore, manufacture of artificial limbs and taking measurements thereof on the human body are patentable.

Several examples falling under Section 3(i) have been discussed by Kanakanala (2007). Consider a claim of 'group of drug response markers P21, REV3L, FGF5, PTK7, POLH, P27 and SSTR2 as a method of detection of a gene signature with at least two of the said drug response markers for monitoring drug response in a patient suffering from cancer treated with a combination of Gemcitabine and P1446A'. In this claim the subject matter is not patentable since it is directed to a method of diagnosis of human beings or animals, which are legally barred from patentability under Section 3(i) of the act. Now, consider Patent Application No. 693/KOL/2007 in which a controversy was raised, 'if a device meant for detection of antibodies to HIV and p24 antigen of HIV in human serum or plasma' was excluded under Section 3(i). The opponents disputed that the applicant had not revealed the diagnostic facet of the apparatus/accessory/mechanism. On analysis it was observed by the examiner that the said invention was a device and not a diagnostic or therapeutic process/method, therefore the opposition raised was not viable and Section 3(i) was not found to be applicable. Sometimes the claims are so drafted that a combination/composition of drugs in certain dosage forms is claimed, but the claimed subject matter relates to application or administration of individual drugs in simultaneous, sequential or concomitant manner. In such cases, although the claims are directed to a combination/composition of drugs, the claimed invention resides in the method of administration of individual drugs in the said manner and thus, it falls within the scope of Section 3(i) of the act.

4.2.8 Section 3(j)

According to Section 3(j) of the act, plants and animals in whole or any part thereof other than microorganisms but including seeds, varieties and species and essentially biological processes for production or propagation of plants and animals are not patentable inventions. It may be noted that microorganisms are excluded from the non-patentability list in Section 3(j), however while considering patentability of microorganisms it is essential to follow provisions of Section 3(c), which states that naturally occurring microorganisms constitute discovery of living things and therefore not patentable unless they are modified or their isolation process is claimed under the meaning of the act. Section 3(j) of the act also objects to the claims related to essentially biological processes of growing plants, germination of seeds and of development stages of plants and animals. For example, if a patent claims a method of producing pure hybrid seeds, plants and crops, with the following steps: (i) producing a male fertile parent; (ii) producing a male sterile female parent; (iii) breeding the said male and female parents; and (iv) harvesting seeds from the female parent which contain pure hybrid seeds, then the claims are rendered to be non-patentable owing to crossbreeding being an essentially biological process for producing pure hybrid seeds under Section 3(j) of the act.

4.2.9 Section 3(k)

The section relates to 'computer programs *per se* and algorithms, mathematical or business methods'. According to Section 3(k) of the act, a mathematical or business method or computer programs *per se* or algorithms are not patentable inventions. A draft manual with guidelines on the types of claim allowed in respect of software-related inventions was released in 2008 and accepted in March 2011 by IPO. Software is usually protected by copyright, which has to be purchased by the user (Chingalea and Raob, 2015). Bioinformatics has recently emerged from the combination of information technology and biotechnology. Due to the complex nature of bioinformatics it is difficult to decide what form of the IP protection such as patent, copyrights and trade secrets should be applied for protecting bioinformatics databases or software. A bioinformatics database is usually not a strong candidate for patent protection and a majority of countries protect it through other forms of IP such as copyright and trade secret. As a mere composition of information, a bioinformatics database is not eligible for patent protection but patent protection may be extended to it 'if it is not a mere catalogue, but is more along the lines of data processing system that has the ability to convert the raw data into a tangible result' (Singh, 2014). In the biotechnological field, computer-related claims include software for data analysis and generating models for hypothetical drugs, proteins, peptides, enzymes and others. For example, a computer-assisted method is claimed to generate a compound to inhibit the glutamine formation active site activity of a glutamine synthetase polypeptide. The said test compound is capable of inhibiting the interaction between an (i) adenylated or (ii) de-adenylated catalytic triad site of the glutamine formation active site and a γ-glutamyl phosphate intermediate. The method involves: (i) providing a three-dimensional structure of a glutamine formation active site of a glutamine synthetase polypeptide; and (ii) designing a test compound capable of inhibiting the interaction between the glutamine formation active site and a γ-glutamyl phosphate intermediate, based on the three-dimensional structure. It was assessed that the claimed method relates to a method of designing the inhibitory compound based on three-dimensional structures assisted by a mathematical method or computer program *per se* and does not lead to a real product. Therefore the subject matter of the claim falls within the scope of statutorily non-patentable inventions under Section 3(k) of the act.

4.2.10 Section 3(p)

The section deals with 'traditional knowledge related inventions'. According to Section 3(p) of the act, an invention, which, in effect, is traditional knowledge (TK) or which is an aggregation or duplication of known properties of a traditionally known component or components, is not a patentable subject matter. All the practices traditionally being carried from generation to generation may or may not be present in document form and are therefore unavailable for reference, especially for an applicant/examiner. The Council of Scientific and Industrial Research (CSIR) has been allotted the task to codify Indian TK for public reference and it is wonderfully being compiled. For the examination of TK-related subject matters, separate guidelines have been issued by the Office of the Controller General of Patents, Designs & Trade Marks (CGPDTM) to prevent misappropriation of India's traditional medicinal knowledge classified in approximately 25,000 subgroups for Ayurveda, Unani, Siddha and Yoga at international patent offices by the Traditional Knowledge Digital Library (TKDL). The preventive measures taken by the CSIR-TKDL unit have successfully prevented approximately 200 cases to date against biopiracy. India is now being recognized as a global leader in the area of TK protection.

One of the most recent successes of TKDL is discussed herewith. The claim relates to a method of hair treatment comprising the steps to provide a hair-building solid agent for altering and/or maintaining the electrostatic charge of the hair and/or skin such that it has a substantial negative polarity, for use on, or with, one or more of hair, skin and hair-building solids. Other related claims included a medicinal composition of the hair-building solid agent used for treating hair loss that contained natural ingredients such as turmeric, pine bark and green tea. During the processing of the application the European Search Opinion raised objections based on novelty (Section 2(1)(ja)), which was killed based on the closest prior art, which stated that 'All compounds mentioned as further ingredients in the composition are well-known in the fields of hair treatment as described as such in the present application'. Later the applicant filed amended claims and arguments against the European Search Opinion. A pre-grant opposition by a third party (TKDL) was filed, which provided prior art evidence showing that the treatment had been known and used for a very long time through Indian systems of medicines like Ayurveda and Unani. Objections were raised to the amended claims based on method of medical treatment, and clarity (Section 3(i)), stating 'The comments developed in this [the third party observations] filing concern mainly the subject-matter of original claims 10, 11 and 12, which relate to the introduction of pharmacologically active ingredients'. Taking into account the third-party observation the examination report of the Patent Application EP2361602 (US 13/026,062)[4] filed in 2011 was given on 29 June 2015 and the company withdrew the application (CSIR, 2015).

4.2.11 Section 10(4)

This section deals with the most important techno-legal document called 'Complete Specification' (CS). This document fully and particularly describes the invention and reveals the best method of performing the invention. Disclosure of technical subject matter and the claims need to be very clear as they are not the same but are mutually supportive and in agreement with each other. The technical disclosure of the specification constitutes the essential component of the patent system, which may be considered equivalent to the patent itself. The claims on the other hand notify the public the area of breach, that is, it provides the legal boundary of the invention. Application where the claims are broad and indeterminate and of a speculative character are treated as not supportive by the description. While filing the document for CS it should be ensured that the following are carried out.

1. The invention, its operation or use and the method by which it is to be performed should be fully and particularly described.
2. The best performing method of the invention known to the applicant should be disclosed for which the applicant claims protection.
3. The CS should end with a claim(s) defining the scope of the invention for which the protection is claimed.
4. While dealing with biological materials it is required to submit the biological material with an international depository authority and make its reference in the CS, if applicable.
5. CS should be accompanied by an abstract.

It may be noted that insufficient specification cannot become sufficient on the basis of knowledge generated in the state of the art after the filing date. While deciding the sufficiency of disclosure an applicant must ensure that the disclosure made by him/her is sufficient for a skilled person to reproduce the invention following the directions given in the specification.

Another important feature is 'Claims having laundry list', which are exemplified by specifications disclosing a list of a wide range of unrelated diseases as potential future therapeutic or diagnostic targets of a claimed gene or the protein that it encodes. For example, such claims are generally made when the activity of a said protein/peptide/enzyme has not been fully characterized, and therefore any potential uses are speculative. Similarly, analogues or variants of polynucleotides or polypeptide sequences, in the form of additions, substitutions or deletions, could extend to an almost infinite number of predictable variants. Therefore, if there is no evidence in the specification as filed that the gene or polypeptide or its analogue is of therapeutic or diagnostic use in each different disease claimed, then the specification is insufficient even if it may be identified in the future by following the applicant's process. When a DNA sequence is claimed to hybridize with a specific probe having certain activity, the claim will not be supported if the hybridization conditions are not specifically disclosed to repeat the experiment successfully. Similarly, sufficient data and methodology are required to support claims to antibodies that may have therapeutic or diagnostic potential for the target protein in a specific disease.

4.2.12 Section 10(5)

This section deals with the presentation of technical subject matter where one patent is granted for a single invention; patenting of multiple inventions in one patent is discouraged. A single application may contain more than one independent claim, however, the claims should be related such that they form a single inventive concept. Applications which include multiple dependent claims but are not related are termed to have 'Lack of Unity of Invention' and applications are liable to be rejected. The section provides the balance by allowing amendments to the 'parent application' to claim a single invention. The additional inventions are recommended to be filed as 'divisional application(s)'. It may be noted that an applicant is voluntarily allowed to file a divisional (a new complete) application as to avoid rejection. The provision is in favour of the applicant since the applicant is able to maintain the priority rights of divisional application(s) from the date of the parent invention while filing at any time before the grant of the patent. However, each divisional application is treated as a new application which should be independent of the parent application, have its own set of CS and requires separate fees for its filing and separate request for examination. This provision in the IPO and other countries helps an applicant in postponing the expenses involved in filing the patent application (Van Zeebroeck *et al.*, 2006; Teaminvntree, 2015). For example, in a gene technology-related invention, usually applicants claim more than one polynucleotide and/or polypeptide sequence. There are two ways of accenting lack of unity in one's application: (i) before consideration of prior art: it is applicable if the claims do not share the same technical features and thus fall in separate groups; and (ii) after the search of the prior art: if the claims share technical features but fail to make a contribution over the prior art (Balasubramanian, 2014).

4.3 Conclusion

The first patent was granted to George Alfred DePenning from Calcutta by The Government of India on 3 March 1856 for his invention 'An Efficient Punkah Pulling Machine'. Since then the Intellectual Property System of India has undergone several amendments, which has resulted in the present IPAA 2005 relating to the Patents, Designs, Trade Marks and Geographical Indications. This makes the system stand on a firm techno-legal ground and plays a crucial role in the industrialization of the country and acts as an important player in the transfer and dissemination of technology. India's present IP laws are in line with international treaty's constraints and commitments as well as India's policy of objectives proportionate with the socio-economic priorities of the country. Besides reinforcing the acts of the patent laws, parallel changes have been made to modernize the organization, continuously strengthening the infrastructure and human resources, introducing more transparency as well as improving the legal procedures and the spread of public awareness.

With the emergence of the technology and biotechnology boom the practice of IPR has moved into newer and unexplored territories. Accordingly, the provisions of Section 3(d) were discussed, debated and decided in various forums in favour of the Indian legislature. However, many of the Indian Government's Patent Acts rules such as that of TKDL have been constantly under controversy and their implications debatable (Mueller, 2007). India's Patents Act is highly restrictive and is in sharp contrast with the acts of the USA and many other countries, be this in relation to TKDL or broadly permitting the patenting of any new use for a known product (so long as that new use meets the requirements of novelty, non-obviousness and utility). China's herbal medicine market is booming due to broader availability of patent protection. In India, natural herb-based remedies, however innovative, scientific, method or product, may be protected since ayurvedic therapies and natural herbal remedies have been practised in India for thousands of years, courtesy of TKDL. Through its stringent rules, the Indian government has tried to balance between the interests of inventors and innovators on one hand and the interests of society on the other.

Members of the Paris Convention and Patent Cooperation Treaty accept the general guidelines referred to as minimum criteria along with the 'Flexibilities' for granting patent right to the applicant. Each member country can follow its own definitions of the 'Flexibilities' in respect to the fixed criteria (Novelty, Non-obviousness, Industrial Applicability or Utility). Regarding 'Flexibilities', the Indian legislature has introduced 'Chapter II' in the Patents Act, wherein the Sections 3 and 4 have been directing the executing officials of the Patents Act not to grant patent right even if the disclosures have passed the tests of the fixed criteria. According to Section 83 (Chapter XVI) of the Indian Patents Act the patent right is not granted to the patentee merely to enjoy monopoly to import the patented article but is granted to work the invention in Indian territory; to encourage transfer and disseminate technology, which shall be conducive to social and economic welfare, and to balance rights and obligations: it does not impede the protection of public health and nutrition; and does not in any way prohibit central government to undertake measures for promoting public health.

With the Indian government becoming active with respect to patents, in future India should be playing important roles to bring the patent laws of different countries into compliance with each other. Such steps will maximize the benefit to the inventors all over the globe, allowing patent laws to protect the interests of the inventor, public and the country. As for the inventors, follow the indomitable and innovative spirit of former Indian President Dr A.P.J. Abdul Kalam, whose message of life was 'Dream, Dream, Dream, Dreams transform into thoughts, and thoughts result in action.' Change these actions into patentable technologies to fulfil his dream of 'igniting young minds' and be a part in the economic growth of India.

Notes

1 Ministry of Law and Justice (Legislative Department), New Delhi: The Patents (Amendment) Act, 2005 (No. 15 of 2005). An Act further to amend the Patents Act, 1970 (4th April, 2005).

2 *Ag. Fumapharm* v. *the Controller of Patents and Designs Patent Office and Others*. Intellectual Property Appellate Board (IPAB), case no. OA 6 of 2009/PT/KOL & Miscellaneous Petition No 34/2011 in OA/6/2009/PT/KOL. Patent no IN/PCT/2002/00543 'Fumaric acid derivatives effective against mitochondrial diseases'.

3 Abraxis Bioscience. *Treatment methods utilizing albumin-binding proteins as targets*. Application No. 4572/CHENP/2006. Date of decision – 1 June 2015.

4 Elliot Isaacs, 11 August. *Hair building solid agent.* U.S. Patent Application US 13/026,062, patent US20110195039 A1, filed 11 February 2010.

References

Asif, E. (2013) Exclusion of diagnostic, therapeutic and surgical methods from patentability. *Journal of Intellectual Property Rights* 18, 242–250.

Balasubramanian, S. (2014) India: unity of invention with a focus on pharmaceutical and biotechnological inventions. Available at: http://www.mondaq.com/india/x/311150/Patent/Unity+Of+Invention+ (accessed 24 September 2015).

Barbosa, D.B. and Grau-Kuntz, K. (2010) Exclusions from Patentable Subject Matter and Exceptions and Limitations to the Rights–Biotechnology. SCP/15/3 Annex III, pp. 1–100. Available at: http://www.wipo.int/edocs/mdocs/scp/en/scp_15/scp_15_3-annex3.pdf (accessed 24 September 2015).

Birtchnell, T. (2013) *Indovation: Innovation and a Global Knowledge Economy in India*, 1st edn. Palgrave Macmillan, Basingstoke, UK.

Chingalea, R. and Raob, S.D. (2015) Software patent in India: A comparative judicial and empirical overview. *Journal of Intellectual Property Rights* 20, 210–222.

Controller General (2013) Guidelines for examination of biotechnology applications for patent. Issued by Office of the Controller General of Patents, Designs and Trade Marks, 2013. Available at: http://www.ipindia.nic.in/whats_new/biotech_Guidelines_25March2013.pdf (accessed 24 September 2015).

CSIR (2015) TKDL Major Milestones – Recent (Year 2014–2015). Available at: http://www.csir.res.in/%5Ccsir%5Ctkdl%5CTKDLMilestones_2014.asp (accessed 24 September 2015).

Kanakanala, K.C. (2007) Diagnostic method patent model – Patent incentives and socio-ethical concerns. *Journal of Intellectual Property Rights* 12, 104–110.

Khader, F.A. (2007) *The Law of Patents: With a Special Focus on Pharmaceuticals in India*. Lexis Nexis Butterworths, Nagpur, India. ISBN: 978-81-8038-150-8.

Mueller, J.M. (2007) The tiger awakens: the tumultuous transformation of India's patent system and the rise of Indian pharmaceutical innovation. *University of Pittsburgh Law Review* 68 (3). Available at SSRN: http://ssrn.com/abstract=923538 (accessed 24 September 2015).

Nair, G.G., Fernandes, A. and Nair, K. (2014) Landmark pharma patent jurisprudence in India. *Journal of Intellectual Property Rights* 19, 79–88.

Phillip, P. (1999) *Censored, 1999: The News That Didn't Make the News–The Year's Top 25 Censored Stories*. Seven Stories Press, New York, pp. 37–40.

Singh, K.K. (2014) Intellectual property protection in bioinformatics and open bio development. *Asian Biotechnology and Development Review* 16, 25–45.

Teaminvntree (2015) Law governing inclusion of multiple inventions in a single patent application. Available at: http://www.invntree.com/blogs/law-governing-inclusion-of-multiple-inventions-in-a-single-patent-application (accessed 24 September 2015).

Van Zeebroeck, N., de la Potterie, B.V.P. and Guellec, D. (2006) Claiming more: the increased voluminosity of patent applications and its determinants. *Research Policy* 38, 1006–1020.

5 Intellectual Property in the Biotechnology Sector: The Importance of 'Star Scientists' in the Entrepreneurship and Universities Environment

TOMAS GABRIEL BAS*

Institute of Innovation based on Science, University of Talca, Talca, Chile

5.1 Introduction

Molecular biology, with the integration of the chemical industry, formed what became known as modern biotechnology. Moreover, the development of biopharmaceutical products since 1920 has resulted in the emergence of small and medium-sized firms, in addition to the classic multinational companies that are very intensive in R&D. Historically, biotechnology firms have been largely involved in the pharmaceutical, environment, agrochemical and food industries. These two sectors (biotechnology and pharmaceutical) develop oligopolistic structures formed in an atmosphere of very high competition. Biotechnology is not an industry but a set of specific activities and technologies such as biomaterials, combinatorial chemistry, DNA markers, genetic engineering, monoclonal antibodies and recombinant DNA. These technologies create new products, new processes for existing products as well as new organisms for environmental cleaning or human consumption. Most major biotechnology firms are spin-offs from research universities or from other firms previously spun-off from academic institutions. In the most developed countries, the institutions provide the firms with adequate infrastructure and resources (Niosi and Bas, 2003). The generation of epistemological knowledge is one of the most important components, through the complexity caused by the use of large molecules, compounds of which are based on highly complex biological three-dimensional structures and a much higher molecular mass such as insulin, measuring 5.8 kDa, or growth hormone with 22.1 kDa, or the most complex and large molecules the monoclonal antibody with 150 kDa. Whereas chemical pharmaceutical drugs, comprising a chemical structures basis, have a much smaller size and simpler molecular complex, where the chemical synthesis process is already known. Such is the case, for example aspirin, measuring 0.18 kDa. Other complexities concern the asymmetric between large multinationals and small firms and finally, high transaction costs to medium and long term, that make their development extremely complicated (Bas and Niosi, 2007). In these industries about 10 years are needed to obtain a new molecule, with an investment of between US$600 and US$1300 million. Today, there are some 8100 specialized biotechnology firms in the world, and some 60% of them are located in the USA, Spain, South Korea, Canada, Germany, India and the UK.

Universities are responsible for the generation of knowledge par excellence. However, the commercialization of intellectual property (IP) and technology transfer has been strengthened recently with an environment of sustainable business culture, becoming an adjunct of economic growth of

*Corresponding author e-mail: tbas@utalca.cl

nations. Besides their prominent role in teaching and conducting research for publication, academics can also be key factors in innovation through the technology transfer of their inventions through to commercialization (Abd Rahim *et al.*, 2015). Since the 1970s, university research has played an important role in both the promotion and dissemination of scientific knowledge and growth of the biotechnology industry, which then has turned to entrepreneurship. At the same time, proactive behaviour by organizational members has been linked to positive outcomes at both the individual and organizational levels. Traditionally, the stock of knowledge has been available within the scientific community, through journals, databases, conferences. However, due to the growing need for financing academia, university scientists have been increasingly involved in issues related to the creation of spin-offs and therefore need to protect them by patenting. The renewed search for knowledge in general is highlighted by the skills and abilities of scientists to solve or anticipate problems that may arise over time. In short, this is reinforced in some way by the familiarity that scientists have with a particular field of knowledge, which, in the case of biotechnology, is molecular biology (Lo Storto, 2013; Agarwala and Shah, 2014). The Bayh-Dole Act of 1980 (PL 96-517) creates the basis on which a public university, as well as a non-profit institution, could choose to exercise the ownership of an invention, without prejudice to the utility that the state could exert over such invention because the state funded the research through taxes. The property rights that universities have on these inventions result in some commitments, such as attempting to commercialize the invention, preferably through trading companies (Ferguson and Kaundinya, 2014).

The world of entrepreneurship, similar to universities and scientists, to be successful depends on the skills of the managers who run them and the creativity of the 'brains' responsible for conducting R&D. In this sense, we can define entrepreneurship like an activity involving the discovery, evaluation and exploitation of opportunities, in order to introduce new goods and services, new organizational structures, new markets, processes and materials in ways that possibly did not exist before. Entrepreneurship is based on the following assumptions: it requires the existence of opportunities; differences exist between people; the relationship to risk; and it is a process that braids reports with innovation activities and organizing (Pesqueux, 2011). Business functionality can be conceptualized as the discovery of opportunities and the subsequent creation of new economic activity, often through the creation of a new organization (Reynolds, 2005).

This chapter is organized as follows: first, different theories are analysed, such as entrepreneurship, growth and economic development. Intellectual property is then analysed, as a tool for protecting creativity of firms and universities, followed by the task of 'star scientists'. Finally, results of patents and scientific publications in biotechnology are analysed.

5.2 Theory and Entrepreneurship

The entrepreneurship term began to take shape as we interpret it today from the evolutionary studies on innovation made by Joseph Schumpeter (1934), who proposed the concept of 'creative destruction' and 'entrepreneurship'. For him, the notion of entrepreneurship is not a closed concept, because it covers numerous variables involving both internal and external components from the company in relation to productivity, growth, investment, risk, governance and culture. It is an ecosystem that is in constant evolution and interaction between the parties, and therefore far from a stipulated stability. Schumpeter defines the entrepreneur as the coordinator of production and agent of change through his famous slogan 'creative destruction'. In this sense, to Schumpeter the entrepreneur is primarily generating innovation. However, many economists do not consider the importance and influence of the entrepreneur in the early stages of economic development, but rather in the later stages of development, where economic growth is driven by knowledge and competence (Ács and Naudé, 2013; Soriano and Huarng, 2013).

On the other hand, 'neoclassical theory' from the perspective of Alfred Marshall (in Schumpeter, 1954) tries to explain the equilibrium conditions in the markets under the assumptions of the existence of perfect knowledge and free information. Perfect competition, uniformity in the existence of goods and the free entry and exit, make this theory somewhat anachronistic versus the evolutionary theory, where it is analysed from an opposite point of view from Schumpeter's theory. The principles of Marshall (Schumpeter, 1954) attempt to show that markets operate under certain assumptions of perfect competition, in which there is no chance of excess profits and therefore there is no exploitation of labour in the production process. Marshall suggests that large-scale production is essential to economic progress and economic innovation (Schumpeter, 1954), while postulating that the equilibrium is obtained through the many different players in the market; competition, therefore, occurs in a non-monopolistic market and perfect balance. His theories consider the establishment of the balance of supply and demand in the market for goods and services, whereas Schumpeter considered 'creative-destruction', no balance, changing markets, skills and resources as hubs of innovation.

Many authors have contributed to the analysis of this topic, among whom are Robert Solow, who developed the concept of technological progress in macroeconomic growth, while Paul Romer integrated knowledge with human capital as determinants of growth and development. It is no surprise that innovation, knowledge and human capital, all related concepts with entrepreneurship, are difficult to integrate in a single economic model, whether neoclassical or evolutionist. Kirzner (1997) represents the neo-Austrian approach to the entrepreneur, who focused on answering the question: what is the process that leads an economy towards equilibrium? Kirzner says that initially the economy is out of balance and competition between employers who are in a state of 'alert' can lead to a balance. Unlike neoclassical economists, Kirzner believes that markets are not always optimal, no fully informed agents for change occur, employers need incentives, and these incentives come from the difference between agents in terms of information and all knowledge. Baumol (1993) refers to two types of employer: (i) entrepreneur-business organizer; and (ii) entrepreneur-innovator. The first category fits into the classicism described by Say (1803) and Knight (1921), while the second is covered by the evolution described by Schumpeter (1934).

The literature indicates that firms and universities do not have the same components or the same skills or routines. Firms and universities generate a variety of different dynamic capabilities, as if each were equipped with limited rationality or influenced by the sociocultural nature of learning and collective knowledge (Srivastava and Frankwick, 2011). Firms and universities do not have identical resources and do not react in the same way, whether from internal or external stimuli, which means that certain firms grow much faster than others in the same sector (Rothaermel and Deeds, 2004). According to Sternberg (2000), the theory of economic development is based on the types of relationships that lead economies to develop or not. These relationships are based on the creation of skills and knowledge to enhance innovation capabilities. Therefore, the applications of new knowledge and new technologies influence innovation and at the same time are essential in entrepreneurship. On the other hand, growth theory emphasizes the role that shows the accumulation of knowledge as a determinant of economic developments (Bas and Niosi, 2007). Many authors (Hall and Jones, 1999; Easterly and Levine, 2003) indicate that differences between countries are generally associated with the accumulation of productive factors (capital and labour), whereas the competitiveness of these countries is more linked to the innovative progress. Therefore, we can see that the application of new knowledge is vital to innovation, which simultaneously affects economic development (Nonaka *et al.*, 1996). Sveiby (2001) uses the knowledge concept as a synonym for individual competition. This leads us to believe that competition is the combination of knowhow added to the ability to think in a different relationship between individual actors within a social system. Numerous studies indicate that different economic levels between countries are associated with the accumulation of productive factors (abilities to develop skills), such as those derived from innovation and capabilities of the star scientists of the firms' or universities' activities (Niosi and Bas, 2001; Bas and Kunc, 2009).

Growth theories tend to recognize that knowledge generation and thus the innovation that accompanies this play a crucial role in determining the trajectories of many of the national and regional economies (Solow, 1956; Temple, 1999). Entrepreneurship, which is generated through the dynamics of innovation with the development of new processes, products and organizational structures, tends to concentrate in regions where knowledge is produced within an ecosystem where there is a continuous interaction between different crucial actors, such as public and private firms, government agencies, financing agencies and universities (Finegold *et al.*, 2004).

On the other hand, some researchers (Waguespack and Binir, 2005) have focused their work on the sources of competitiveness developed through IP (patents), which stimulates the conditions for development of a sustainable innovation (Bas *et al.*, 2008) and scientific publications. In many developing countries, innovations practised by numerous companies, especially multinationals, are mainly associated with the adaptation of products, technologies, business strategies and existing organizations and that have been previously created in developed countries. Accordingly, the creation of fresh knowledge, new products, technology and innovations is not associated with the development of original knowledge. The fundamental incorporation of knowledge, technology and competitiveness is essential to improve the capacity to innovate and increase its impact in the enterprises, in the countries and in the universities (Katz, 2000). To innovate is to use the knowledge and knowhow to generate significant changes to improve products, processes, services, relationships, practices and processes of firms, which necessarily creates additional value. Thus, we can understand that the main determinant of economic growth is innovation, through the generation of knowledge, which takes us to the development of successful enterprises. This type of growth is reflected in the socio-economic returns, as a result of externalities or indirect effects generated by them. IP exists in order to protect the innovation of the enterprises and the copyright of scientific publications.

5.3 Intellectual Property

The WIPO Copyright Treaty (WCT) concerns the protection of works and the rights of authors in their works in the digital environment. In addition to the rights under the Berne Convention, authors benefit from some property rights. Regarding the object of protection, the treaty covers two categories of objects of copyright: (i) computer programs, whatever the mode or form of expression; and (ii) compilations of data or other material (databases). One of the features most used by the inventors is the patent, although there are others. To obtain a patent, an invention must be new, original and industrially applicable. A patent grants a bundle of exclusive rights for a limited time, usually 20 years, which allows the inventor to prevent others from making, using, selling, offering for sale or importing the patented invention without permission. Other protection mechanisms include copyright, which protects literary works, and in this case the rights belong to the scientific work of academics that use them. In this sense, Fehder *et al.* (2014) evaluated the role of formal IPR in the configuration of the downstream demand for knowledge that is given initially through scientific publication, in fields where research is generated and is used across different institutional bridges (comparatively between academia with industry). For scientific discoveries with potential commercial application, researchers may seek to establish a formal IP protection to enable researchers to access knowledge described in a particular journal.

Lei *et al.* (2009) estimate that a system of IP can promote inventive skills in science and help promote the transformation of research achievements in marketed products. But the restrictions associated with access can reduce the use of inventions by other scientists. It is well known that biological scientists report increasing difficulties associated with access to research tools, but only if the tools are built into the physical property controlled by others and not easily duplicated.

There are four main types of legal IP protection.

1. Patents – new or improved products and processes that are susceptible to being applied industrially.
2. Hallmarks of the brand – for goods and services, allowing distinctions to be made between different operators.
3. Designs for product appearance – of all or part of a product resulting from the features of, in particular, the lines, contours, colours, shape, texture or materials of the product itself or its ornamentation.
4. Copyright for material – literary and artistic material, music, films, sound recordings and broadcasts, including software and multimedia. Furthermore, moral rights are the author's right to the integrity of creation in its original form, the attribution, and for the prevention of unfavourable associations with the creation of the author.

IP helps to promote investment in new and improved products and processes, and allows innovators to make a return on their investment. It also plays an important role in the dissemination and use of the knowledge and innovations contained for example in patent documents and copyright of one publication. This enables further technological developments, allows creative industries to grow and flourish and fosters the creation and development of the market for ideas. The increasing importance of IP for economic performance at the micro and macro levels, its centrality to the knowledge economy and its importance for global value chains are changing how market actors use IP and the way policy makers perceive the role of IP systems and IP rights.

In an effort better to balance private and social benefits, and thus contribute to both economic growth and the welfare of societies, IP rights systems worldwide have in recent years been at the centre of business and policy discussions, and have been undergoing a number of changes. However, the ability of companies to take strategic decisions allowing them to perform and compete, and the possibility to design policies aimed at shaping IP systems and aligning them with societal goals, require robust and reliable evidence. In this respect IP data represent a unique source of information that can be used to inform a wide range of policy- and business-relevant issues, including: science and technology developments, R&D and innovation, entrepreneurship and enterprise dynamics, branding and reputation, and economic development and growth. IP data also reveal important facts about how IP systems themselves are performing, and thus help better to shape them.

IPR is vital to protect what innovative companies produce. Innovation processes depend largely on the generation of knowledge and therefore the right knowledge management in companies. IPR is used not only to protect innovation but also to defend a competing product and also for seeking new funds to better position the innovative company (Candelin-Palmqvist *et al.*, 2012). The increased use of IPR in scientific research always leads to academic and political debates. This highlights the advantages of IPRs, suggesting that this may facilitate creativity, improve financing and the exchange of knowledge and skills that otherwise would remain secret and therefore outside the trade forums. In fact, in the context of university research funded primarily with public funds, it has been suggested that IPR offers important incentives for the production and commercialization of inventions for commercial purpose (Murray and Stern, 2007).

Two key factors to be considered in the generation of knowledge are the intellectual capital and the star scientists. The literature has long debated the concept of intellectual capital, but there is an agreement to consider it a process that is built around pillars, such as human capital. Meantime, human capital includes tacit knowledge, individual skills, experiences, attitudes and the capacity for innovation and learning. In this context, hiring people with good knowledge and skills, their training, their frequent adaptation of new technologies and forms of organization, have become a necessity for businesses, not only for their operation and development but also for differentiation from competitors and possibilities for R&D, to protect what it has built and innovated (Bchini, 2015).

Oettl (2012) estimates that the current conceptualization of 'star scientists' is that it focuses exclusively on productivity. Scientists may be useful in a number of ways. They can influence the formation and quality of new ideas through discussion, feedback, encouragement and criticism. The area of research that examines the utility of individuals in organizations is well known. The definition of a 'star scientist' is based solely on individual productivity; in other words, we define stars so they produce themselves. For the author, this one-dimensional star-rating of scientists is vague, since innovation is usually conceived as a collaborative process through recombination of existing knowledge and ideas, rather than the discovery of something fundamentally new. The author defines a star scientist as an individual with high productivity and high utility, where the environment can influence the formation and quality of new ideas through discussion, feedback, encouragement and criticism.

5.4 Results

To analyse the biotechnology sector of this paper, different databases were studied with the intention of making a comprehensive diagnosis of this sector and its importance in generating entrepreneurship, patents and publications. For the construction and analysis of the results, they were considered from 2004 to 2014 for a more consolidated and representative of diagnostic results.

Figure 5.1 represents the publications by application areas in the biotechnology sector and illustrates the intervention of various fields of action, including biochemistry, genetics and molecular biology with 31% of total publications, followed by 17% for immunology and microbiology, chemical engineering has 16%, medicine 11%, engineering 9%, agricultural and biological sciences with 9% and 7% for environmental science.

The decision of a scientist, in relation to developing an entrepreneurship as well as bringing to market the fruits of their innovative activity, depends on numerous variables such as their own research and his own labour ecosystem. However, many scholars, whether by training, interests or different skills and knowledge, most likely are not content only with publishing the results of their creations, they also attempt to commercialize the production of their findings, depending on the supply and demand that they may be able to detect (Jain *et al.*, 2009; Rasmussen *et al.*, 2014). However, multiple factors

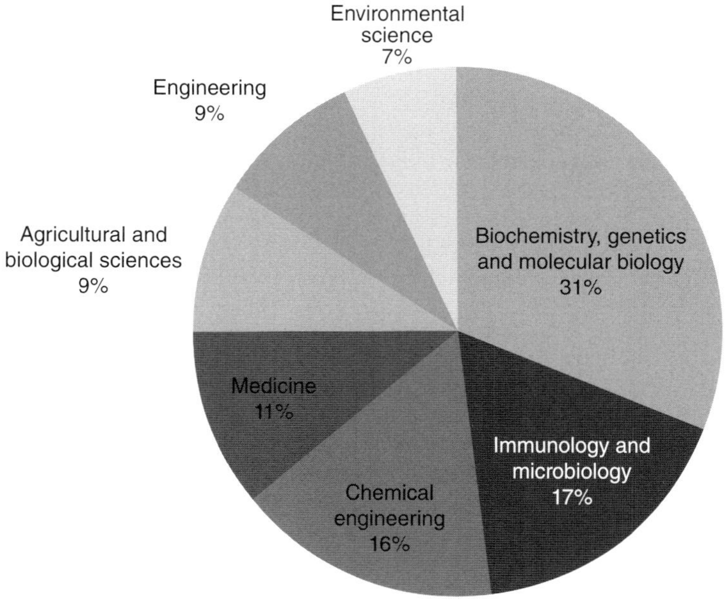

Fig. 5.1. Biotechnology publication areas (Author elaboration from Scopus data, 2014, http://www.scopus.com).

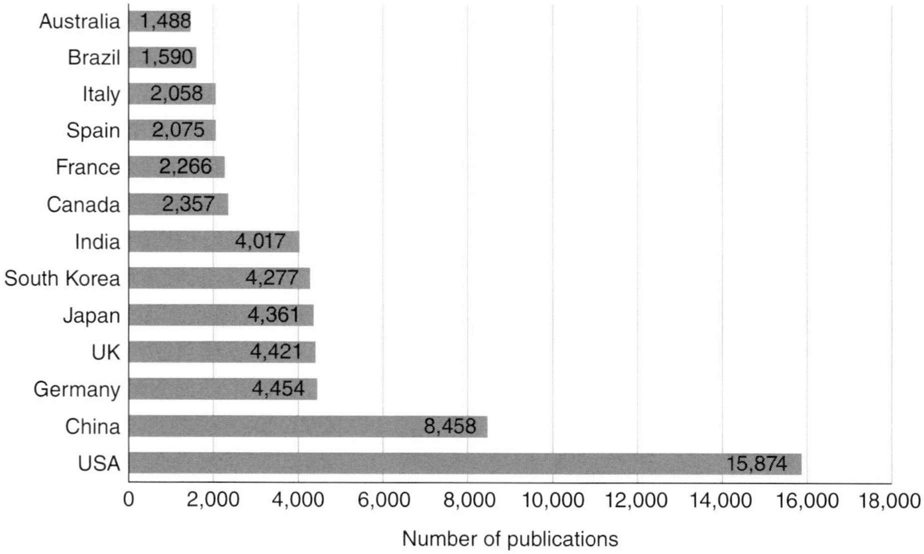

Fig. 5.2. Biotechnology publications by country (Author elaboration from Scopus data, 2014, http://www.scopus.com).

have been identified as determinants in making decisions that can influence the transition that can lead to a scientist having an intention to launch into the business world. This action is not only dependent on the interests and scientific capabilities of the academic, but also institutional aspects come into play, the scope of research funding, the culture of the university/department and the influence that academic peers can exercise within the workplace.

Figure 5.2 indicates the number of publications by country; the USA leads with 15,874 publications, followed by China with 8458, Germany with 4454, the UK with 4421, Japan with 4361, South Korea with 4277, India with 4017 and

Canada with 2357 publications. This shows that there is the highest concentration of publications in certain countries such as the USA and China with a quarter of the total publications, Germany, the UK, Japan, South Korea and India form another 25%, and very far the smallest quarter consists of Canada, France, Spain, Australia, Brazil and Italy.

Figure 5.3 shows that South Korea had only one 'star scientist', and he monopolized the total scientific publications in this country with 13% of the total, followed by Japan with also only one star scientist with 12% of the total publications. The USA adds two top scientists, attaining 18% of total author publications. In the world of biotechnology, as in other areas of research, there is a core of researchers called 'star scientists', who stand out above the norm, and dominate most of the findings in a certain area as shown in Figure 5.3.

Figure 5.4 shows the highest assignees of biotechnology patents dominated by research centres

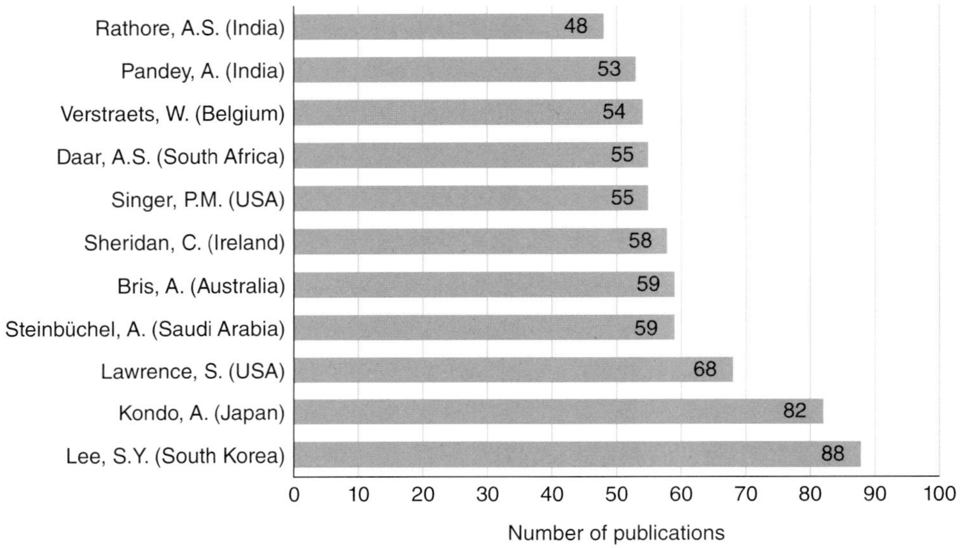

Fig. 5.3. Top authors in biotechnology publications (Author elaboration from Scopus data, 2014, http://www.scopus.com).

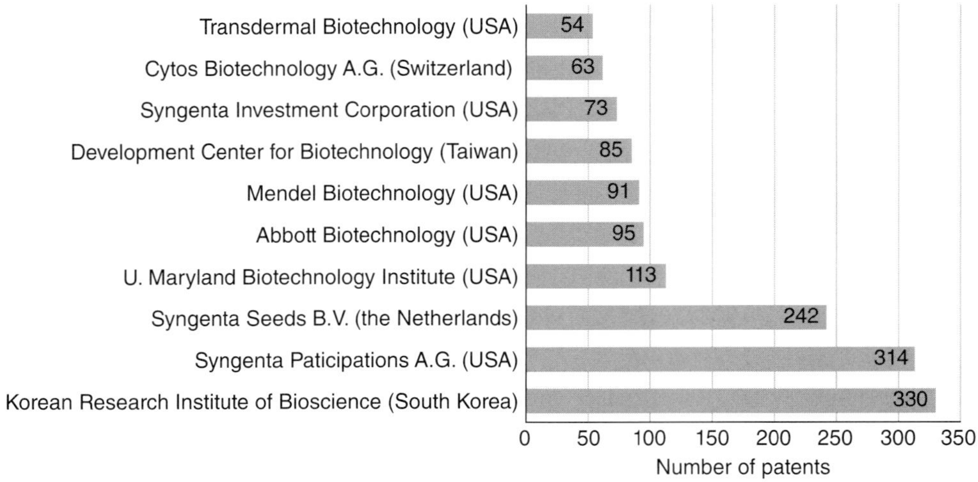

Fig. 5.4. Top biotechnology patents (Author elaboration from USPTO, 2014, http://www.uspto.org).

and enterprises. The leader is South Korea with 330 patents (Korean Research Institute of Biosciences), followed by the USA and the Netherlands (Syngenta) with 314 and with 242, respectively. The USA emerges albeit with more force, particularly Syngenta in many varieties of conformation.

5.5 Conclusion

There is no correlation between the number of patents by country and the number of scientific publications, nor a correlation between the number of patents by country and the nationality of the authors of publications. The deduction would be that there had been a correlation between the number of patents per country versus the number of publications by country of origin of the author. Nevertheless, it is not possible to identify consistently this affirmation, which could explain the degree of importance given to entrepreneurship and to a patent in certain countries in opposition to the greatest importance given to scientific publications by academics from other countries. Either way, it is important that entrepreneurship, academia and IP in the biotechnology sector is essential to development and innovation. It is important for entrepreneurs and for academics, and moves the wheel of innovation through IP.

The current literature on the scientific stars shows some reservations regarding the influence on the progress of science produced by them. It is important to note that, in general, these researchers emphasize above average and they are able to make a difference from the rest of the regular researchers both by their skills, their abilities and their 'brains'. However, it is fair to say that direct and indirect colleagues and assistants, as well as a good availability of funding, institutional support policies, etc., play a crucial role in the success of these 'scientific stars'. We must not forget another important point is the fact that these scientists are also creative people and are able to give a boost in all that relates to their field of research and higher education lays the foundation. However, it is important to note that without the contribution of a 'star scientist' it would be much more complex to operate efficiently than if it were to be done under the supervision of these scientific stars, while conversely, the scientific star would able to function relatively well without employees that would have the power to surround themselves with new skills. So even if literature seeks to make a reclassification of competencies and the final results of productivity (entrepreneurship, scientific publication) of the star scientist, these still mark the horizon of productivity in science and innovation; of course, there is no denying the contribution of their direct and indirect environment, institutions and infrastructure, which leads to excellent scientific publications or an innovation that will lead to a successful patent or even an entrepreneurship.

References

Abd Rahim, N., Mohamed, Z. and Amrin, A. (2015) Commercialization of emerging technology: the role of academic entrepreneur. *Procedia–Social and Behavioral Sciences* 169, 53–60.

Ács, Z. and Naudé, W. (2013) Entrepreneurship, stages of development, and industrialization. In: Szirmai, A., Naudé, W.A. and Alcorta, L. (eds) *Pathways to Industrialization in the 21st Century*. Oxford University Press, Oxford, UK.

Agarwala, R. and Shah, S. (2014) Knowledge sources of entrepreneurship: firm formation by academic, user and employee innovators. *Research Policy* 43, 1109–1133.

Bas, T.G. and Kunc, M. (2009) National systems of innovations and natural resources clusters: evidence from copper mining industry patents. *European Planning Studies* 17, 1861–1879.

Bas, T.G. and Niosi, J. (2007) The issue of asymmetrical growth in Specialized Biotechnology Firms in the USA and the UK. *International Journal of Biotechnology* 9, 87–100.

Bas, T.G., Amoroso, E. and Kunc, M. (2008) Innovation, entrepreneurship and clusters in Latin America natural resource: Implication and future challenges. *Journal of Technology Management & Innovation* 3, 52–65.

Baumol, W.J. (1993) Formal entrepreneurship theory in economics: existence and bounds. *Journal of Business Venturing* 3, 197–210.

Bchini, B. (2015) Intellectual capital and value creation in the Tunisian manufacturing companies. *Procedia Economics and Finance* 23, 783–791.

Candelin-Palmqvist, H., Sandberg, B. and Mylly, U.M. (2012) Intellectual property rights in innovation management research: A review. *Technovation* 32, 502–512.

Easterly, W. and Levine, R. (2003) Tropics, Germs, and Crops: How Endowments Influence Economic Development. *Journal of Monetary Economics* 50(1), 3–39.

Fehder, D., Murraya, F. and Stern, S. (2014) Intellectual property rights and the evolution of scientific journals as knowledge platforms. *International Journal of Industrial Organization* 36, 83–94.

Ferguson, S. and Kaundinya, U. (2014) Licensing the technology: biotechnology commercialization strategies using university and Federal labs. In: Shimasak, L. (ed.) *Biotechnology Entrepreneurship Starting Managing, and Leading Biotech Companies.* Elsevier, pp. 185–206.

Finegold, D., Wong, P.-K. and Cheah, T.-C. (2004) Adapting a foreign direct investment strategy to the knowledge economy: the case of Singapore's emerging biotechnology cluster. *European Planning Studies* 12, 7.

Hall, R. and Jones, C. (1999) Why do some countries produce so much more output per worker than others? *Quarterly Journal of Economics* 114, 83–116.

Jain, J., George, G. and Maltarich, M. (2009) Academics or entrepreneurs? Investigating role identity modification of university scientists involved in commercialization activity. *Research Policy* 38, 922–935.

Katz, J. (2000) The dynamics of technological learning during the import substitution period and recent structural changes in the industrial sector of Argentina, Brazil and México. In: Kim, L. and Nelson, R.R. (eds) *Technology Learning and Innovation. Experiences of newly industrializing economies*. Cambridge University Press, Cambridge, UK, pp. 307–334.

Kirzner, I. (1997) Entrepreneurship discovery and the competitive market process: an Austrian approach. *Journal of Economic Literature* 35, 60–85.

Knight, F.H. (1921) *Risk, Uncertainty and Profit.* Houghton Mifflin, New York.

Lei, Z., Juneja, R. and Wright, B. (2009) Patents versus patenting: implications of intellectual property protection for biological research. *Nature Biotechnology* 27, 36–40.

Lo Storto, C. (2013) Knowledge search strategies in the biotechnology domain: a patent benchmarking analysis. *Procedia – Social and Behavioral Sciences* 75, 353–358.

Murray, F. and Stern, S. (2007) Do formal intellectual property rights hinder the free flow of scientific knowledge? An empirical test of the anti-commons hypothesis. *Journal of Economic Behavior & Organization* 63, 648–687.

Niosi, J. and Bas, T.G. (2001) The competence of regions. Canada's clusters in Biotechnology. *Small Business Economics* 17, 31–42.

Niosi, J. and Bas, T.G. (2003) Biotechnology megacentres: Montreal and Toronto regional systems of innovation. *European Planning Studies* 11(7), 789–803.

Nonaka, I., Umemoto, K. and Senoo, D. (1996) From information processing to knowledge creation: a paradigm shift in business management. *Technology in Society* 18, 203–218.

Oettl, A. (2012) Reconceptualizing stars: scientist helpfulness and peer performance. *Management Science* 58(6), 1122–1140.

Pesqueux, Y. (2011) *Entrepreneur, entrepreneuriat (et entreprise): de quoi s'agit-il?* Available at: https://hal.archives-ouvertes.fr/hal-00567820 (accessed 20 December 2015).

Rasmussen, E., Mosey, S. and Wright, M. (2014) The influence of university departments on the evolution of entrepreneurial competencies in spin-off ventures. *Research Policy* 43, 92–106.

Reynolds, P. (2005) Understanding business creation: Serendipity and scope in two decades of business creation studies. *Small Business Economics* 24, 359–364.

Rothaermel, F. and Deeds, D. (2004) Exploration and exploitation alliances in biotechnology: A system of new product development. *Strategic Management Journal* 25, 201–221.

Say, J. (1803) Traité d'économie politique: ou, simple exposition de la manière dont se forment, se distribuent et se consomment les richesses. [Treatise on Political Economy: On the Production, Distribution and Consumption of Wealth.] Kelley, New York (1964) (1st edn: 1827).

Schumpeter, J. (1934) *The Theory of Economic Development*. Harvard University Press, Cambridge, Massachusetts.

Schumpeter, J. (1954) *History of Economic Analysis* (Elizabeth Boody Schumpeter (ed.). Oxford University Press, New York.

Solow, R. (1956) A contribution to the theory of economic growth. *The Quarterly Journal of Economics* 70, 65–94.

Soriano, R. and Huarng, K.H. (2013) Innovation and entrepreneurship in knowledge industries. *Journal of Business Research* 66, 1964–1969.

Srivastava, P. and Frankwick, G. (2011) Environment, management attitude and organizational learning in alliances. *Management Decision* 49, 156–166.

Sternberg, R. (ed.) (2000) *Handbook of Intelligence*. Cambridge University Press, New York.

Sveiby, K. (2001) A knowledge-based theory of the firms to guide strategy formulation. *Journal of Intellectual Capital* 2, 23–36.

Temple, J. (1999) A positive effect of human capital on growth. *Economics Letters* 65, 131–134.

Waguespack, D. and Birnir, J. (2005) Foreignness and the diffusion of ideas. *Journal of Engineering and Technology Management* 22, 31–50.

6 Intellectual Property in Agricultural Biotechnology: From Patent Thickets to Generics

MONICA ALANDETE-SAEZ, DAVID J. JEFFERSON AND ALAN B. BENNETT*

The Public Intellectual Property Resource for Agriculture, University of California Davis, Davis, California

6.1 Introduction

Scientific advances in many fields have been treated historically as public goods, and this was particularly true in agriculture. Universities and other public-sector institutions were the leaders in developing improved crop varieties that were transferred to farms through cooperative extension services (Conway and Toenniessen, 1999). However, in North America and other world regions, this model has shifted rapidly in the last few decades. Changes have been due largely to the increased utilization of formal intellectual property (IP) protections for agricultural technologies and plant varieties by public sector institutions, as well as the development of a research-intensive private sector that now makes major contributions towards enhancing agricultural productivity (Kowalski *et al.*, 2002).

Arguments justifying the expanded use of formal IP rights for agricultural biotechnology-based products are frequently based on the fact that significant investments of time and financial resources are required to develop new transgenic crops, and the costs of obtaining regulatory approvals to market such crops are commensurately high. Additionally, the bureaucratic procedures surrounding regulatory approvals for genetically modified (GM) crops are daunting, and the scope of approvals is territorially limited country-by-country (Graff *et al.*, 2009). In the face of these costs, it is argued, the time-limited exclusivity provided by the IP right affords the investor an opportunity to recoup costs associated with research and development. Indeed, it is unlikely that the agricultural biotechnology industry would have developed in the absence of a strong framework for IP protection.

The contemporary system of agricultural biotechnology innovation demands a balance of exclusive and non-exclusive access to proprietary technologies to effectively support new crop development, and to provide both commercial growers and subsistence farmers with the best genetic technology possible to cultivate productive, nutritious and marketable crops. Major issues confronting researchers today include the ownership of enabling IP and the rights to practise or use certain technologies. The issues of technology access and IP rights (IPR) are intertwined with the costs associated with obtaining regulatory approvals for GM crops. We explore these matters throughout the following chapter.

6.2 What is Intellectual Property?

Intellectual property encompasses a legally-created bundle of rights that applies to ideas or 'products of the mind', and gives the owner a set of rights comparable to those associated with tangible property. The modern conceptualization of IP and its

*Corresponding author, e-mail: abbennett@ucdavis.edu

concomitant tensions are encapsulated in Thomas Jefferson's (Jefferson, 1987) insight that:

> [i]f nature has made any one thing less susceptible than all others of exclusive property, it is the action of the thinking power called an idea, which an individual may exclusively possess as long as he keeps it to himself... Inventions then cannot, in nature, be a subject of property. [However] society may give an exclusive right to the profits arising from them, as an encouragement to men to pursue ideas which may produce utility.

Jefferson's idea that society should provide a legal mechanism through which inventors could enjoy the exclusive right to profits derived from their ideas was subsequently integrated into the US Constitution, Article I, Section 8, which states that 'The Congress shall have power ... to promote the progress of science and useful arts by securing for limited times to authors and inventors the exclusive right to their respective writing and discoveries.' This passage forms the basis for IPR in the USA and has become the cornerstone of the innovation process in that country as well as in many other territories throughout the world.

Several forms of IP exist today, including plant and utility patents, plant breeders' rights (plant variety protections), copyright, trademarks and trade secrets. In agricultural biotechnology, the dominant forms of IP are patents and plant breeders' rights, and these are the primary focus of this chapter. Patents are available for 'industrial property', i.e. novel, inventive and useful commercializable creations, and grant a set of rights including the ability to exclude others from using the invention. Patents confer time-limited rights, granted by a national government for a specified period, which in most jurisdictions is 20 years. Additionally, the rights granted through patents are geographically limited to the specific countries in which protection is obtained. These limitations illustrate the significant differences between the rights granted through patents and those associated with regimes providing rights for real property ownership.

The monopoly resulting from the patent holder's right to exclude others provides a powerful economic tool. As such, in order for the right to be granted, the invention must meet rigorous standards of novelty, non-obviousness (in the USA; in other jurisdictions this requirement is known as 'inventive step') and utility. That is, the invention must be original and not previously known in the 'prior art', it must not be an obvious extension of previously known information and the invention must have some useful purpose. The standard of novelty has important implications for researchers. Given that the primary means of scientific communication is through broad publication, researchers who seek to patent their inventions must carefully time their disclosures in order to preserve novelty.

Very few countries grant patents for plants. The USA is the only jurisdiction providing for a specific plant patent, which grants protection for asexually-reproducing species, while sexually-reproducing species are protectable via US utility patents. A small number of territories include plant varieties within the ambit of patentable subject matter for utility patents, including Australia and Japan. However, most countries provide an IP right for new varieties via plant breeders' rights (PBR) regimes. Similar to patents, PBR (in the USA, 'plant variety protections') grant a limited period of exclusive rights over new plant varieties that meet certain criteria (typically novelty, distinctiveness, uniformity and stability). Like patents, PBRs are territorial in nature, and empower the right holder to exclude others from activities such as making, using, selling, offering for sale, importing or exporting the protected variety. Most commonly, PBR protection lasts for between 15 and 25 years, depending on the legal framework and species of the variety protected.

The overarching international framework for PBR is provided by the International Union for the Protection of New Varieties of Plants (UPOV). This intergovernmental organization passed its first act in 1961, which was subsequently revised several times. Of the two current versions of UPOV, dating from 1978 and 1991, the 1991 Convention contains relatively stronger IP protections for new plant varieties. Over the past two decades, there has been a global trend towards the 'ratcheting-up' of the rights associated with PBR, which is illustrated by the fact that in the 1990s about 20 territories were members of UPOV, while this number exploded to 72 members by 2015 (Jefferson *et al.*, 2014). At least some of the increase in UPOV membership is attributable to the fact that the installation of an effective system for PBR is frequently a precondition for developing countries to enter into bilateral investment treaties with wealthy nations. This has been the case, for instance, in countries such as Peru and Chile (Jefferson, 2014).

6.3 Intellectual Property in Agricultural Biotechnology

The historical impact of public sector research in agricultural development has been significant. In the USA, this history grew largely out of the Morrill Land-Grant Acts of 1862 and 1890, which, among other provisions, established a system of agricultural colleges. These colleges, many of which have subsequently become universities with broad-ranging academic programmes, have spearheaded the development of improved crop varieties in the USA. These new varieties were then transferred to farms and to the agricultural industry through government-sponsored cooperative extension services.

Internationally, the system of crop research centres sponsored by the Consultative Group on International Agricultural Research (CGIAR) has, since 1971, led to the development of new crop cultivars and agronomic practices that are delivered to farmers – many of whom are smallholders in the developing world – as public goods for the purpose of supporting global food production. Over time, the CGIAR model has evolved, and currently the rate of change is accelerating. At the core of this evolution is the increasingly important role of IP protections over agricultural inventions, as well as the development of a research-intensive private sector in agricultural biotechnology.

Thus, both North American and global agricultural systems are currently experiencing a paradigm shift. While research efforts were led historically by public sector institutions, with the resulting products delivered to agriculturalists for free as public goods, the contemporary R&D system is increasingly dominated by private companies who protect their research results through IP mechanisms and regard these products as private assets. This paradigm shift has been accomplished primarily through a much more intensive use of IP systems to protect agricultural innovations in the past 30–40 years.

By the late 1990s, the proliferation of IP rights in biotechnology had led to the development of a pair of hypotheses related to the under- or over-utilization of resources. According to one hypothesis, a 'tragedy of the commons' results when individuals overuse shared resources, such as common pastures, because they have no incentive to conserve or extend the life of the resource (Hardin, 1968). By analogy, Heller and Eisenberg (1998) described a 'tragedy of the anti-commons', which, as the result of the proliferation and fragmentation of IP ownership across multiple owners, prevents any single institution or company from assembling all of the necessary rights to produce a product, resulting in the under-use (or non-use) of resources. Interestingly, whereas patents and other forms of IP are commonly understood as mechanisms to encourage investment in research and development, the appearance of anti-commons has the opposite effect – it essentially blocks future innovation.

A prominent example of the complexity resulting from fragmented IP ownership and the potential for anti-commons to arise is exemplified in the case of 'Golden Rice', which involved the development of β-carotene-enriched rice. To create Golden Rice, public-sector researchers used at least 40 patented or proprietary methods and materials belonging to a dozen or more different owners in the gene transfer process (Kryder *et al.*, 2000). The motivation underlying the development of this transgenic rice variety was to combat vitamin A deficiency in the developing world, with the hope to broadly market the product in many countries in South and East Asia. Yet the commercialization of Golden Rice was initially constrained by private sector owners of IP related to transgenic technologies. Interestingly, issues related to the development of transgenic humanitarian crops continue to evolve. Instead of IP-related constraints, subsequent generations of Golden Rice are primarily facing obstacles related to regulatory approvals and public perception.

As illustrated by the example of Golden Rice, a portfolio of patented technologies must be compiled through in-licensing or potentially via strategic corporate mergers and acquisitions in order to develop a new genetically engineered crop. Several large corporations have effectively employed these methods to develop and utilize platforms of proprietary technologies to develop new varieties of major commercial crops. However, work on crops of less economic value has progressed slowly, and speciality crops have rarely benefited from biotechnological improvement (Clark *et al.*, 2004). Nevertheless, several GM speciality crops – including apple, banana, chickpea and citrus – are currently undergoing field testing in multiple countries, including Canada, Switzerland, Germany, Uganda, India and the USA (James, 2014).

Based on the requirement that a large number of proprietary technologies be assembled in order to produce a new genetically engineered crop and the fragmentation of IP ownership over these technologies, it appears that the pre-conditions for the development of anti-commons exist in the agricultural biotechnology sector. In addition, the recently observed stagnation of the development of new agricultural biotechnology products may be at least in part due to the emergence of an anti-commons (Graff *et al.*, 2009). However, it could also be argued that the high costs and bureaucratic obstacles associated with obtaining regulatory approval for GM crops are also factors related to the appearance of an agricultural biotechnology anti-commons.

Navigating the complex IP landscape of a research project in agricultural biotechnology requires both analytical tools and specialized legal and technical training (Fenton *et al.*, 2007). Such analysis demands current knowledge of both the law and the state of the relevant science, as well as access to databases containing information related to both patents and the scientific literature. The result of these analyses typically takes the form of what is known as a 'freedom to operate' (FTO) opinion. An FTO study contains a legal assessment about whether a research project or the development of a new product can proceed with a low, or tolerable, likelihood that associated activities will not infringe existing patents or other types of IP rights (Alandete-Saez *et al.*, 2015).

It is important to note that the FTO determination is not absolute but rather reflects an evaluation of risk, since there is typically some uncertainty surrounding the interpretation of patent claims, as well as ambiguity as to whether new IP may issue or be discovered at a later date. The FTO opinion may lead to a range of options: identifying in-licensing targets, considering the substitution of technologies, deciding to ignore the potential infringement, investing in work-around technologies, or perhaps deciding to abandon the project altogether.

Private firms are likely to engage in FTO analysis because any infringement risk may directly affect their ability to develop new products and their ultimate profitability. Additionally, public and not-for-profit private institutions are becoming increasingly aware of the necessity for better freedom to operate information in early stages of R&D projects. This is particularly true for research projects undertaken by universities or not-for-profit research centres for the specific purpose of developing new crops for use in developing countries. In these cases, it is critical that IP considerations be taken into account early in the research process to avoid waste of resources, which are often scarcely available for humanitarian projects.

6.4 Roles of the Public and Private Sectors

Historically, most agricultural inputs such as fertilizers, machinery and pesticides were provided by the private sector, while the public sector developed and delivered new crop varieties directly to farmers who often saved and reused seeds from growing season to growing season (Bennett *et al.*, 2013). Because seeds from conventionally bred non-hybrid varieties can easily be saved and replanted with little impact on productivity of the resulting crop, for-profit companies in the past had little incentive to invest in improved crop varieties. However, with the introduction of hybrid breeding techniques and more recently genetic engineering, it became possible to control the use of new crop varieties, both through biological and legal means. As a result, private seed companies have dramatically increased their R&D activities surrounding the development of new varieties, and today most farmers in developed countries purchase improved and proprietary planting material from the private sector.

Essentially, over the past three decades a bifurcated division of labour in the production of agricultural biotechnologies has emerged between the public and private sectors. While the latter has focused on applied research and commercialization, especially of field crops with large market potential, the former has concentrated on generating basic knowledge or advanced germplasm. Due to factors such as the high commercial value and widespread cultivation of certain crops, internal consolidation of the global seed industry and ever-stronger IP regimes granting exclusive rights over new crop varieties, multinational seed corporations have accumulated tremendous wealth and influence. This is particularly true in the area of GM crops, largely because the regulatory costs associated with bringing a product to market are prohibitive for most public sector and non-profit institutions.

The few GM crops that have been approved for global markets are those with high commercial value, because it is unlikely that applicants seeking regulatory approval will endeavour to undertake this process unless a significant return on investment can be expected. As a result, transgenic crops with low commercial but high social value – such as those designed to combat malnutrition or abiotic stresses related to climate change in developing countries – have been largely neglected by researchers. Some have suggested that the public sector institutions should play a stronger role in the development of GM varieties for less-profitable markets or humanitarian uses (de Janvry and Dethier, 1985; Sunding and Zilberman, 2001).

In so doing, it would be advisable for these institutions to utilize the same mechanisms for IP protection that their private sector counterparts employ, but as a means to translate their public service missions into practice rather than to constrain broad access. The past several years have witnessed several prominent, illustrative cases of public–private partnerships that employ strategies involving protection of agricultural IP combined with mechanisms to facilitate availability of protected technologies. For instance, the Water Efficient Maize for Africa (WEMA) project involves a collaboration between the African Agricultural Technology Foundation (AATF; a private NGO), the International Maize and Wheat Improvement Centre (CIMMYT; a public research institution) and Monsanto (a publicly-traded multinational corporation) for the purpose of developing drought-tolerant and insect-protected maize through a variety of techniques – including genetic engineering – with the goal to make these varieties available royalty-free to smallholder farmers in sub-Saharan Africa through African seed companies (AATF, 2010).

Public sector and non-profit research institutions could employ similar strategies for technology transfer that include non-exclusive and royalty-free licences for research and humanitarian uses, while offering royalty-based licences for commercial use. Such an approach can enable these research institutions to meet simultaneous goals of broad access to a public good and recuperation of investments of time and money associated with R&D. We strongly recommend that all research institutions – especially those in the public and non-profit sectors – craft their technology transfer policies to guarantee that developers of agricultural biotechnologies designed to serve the poor or for speciality crops have access to public-sector enabling technologies, even if they are patent protected.

6.5 Recent Development in Global Intellectual Property Laws

As the economic, political and social issues surrounding the relationship between IP and agricultural biotechnology have evolved over the past several decades, so have the international legal frameworks themselves. Changes have come in the form of the redefinition of patentable subject matter in various jurisdictions, more rigorous regimes for plant breeders' rights and the recent enactment of biosafety laws in several countries.

The patenting of plant and animal genes has been particularly controversial during the last decade. In former rulings the US Patent and Trademark Office (USPTO) had concluded that an isolated and purified DNA molecule having the same sequence as a naturally occurring gene was eligible for patent protection because this gene would not exist in its isolated form in nature. However, on 13 June 2013, a significant reform to US patent law occurred with the issuing of an opinion by the Supreme Court in relation to biological innovations. In *Association for Molecular Pathology* v. *Myriad Genetics* ('*Myriad*') (2013),[1] the Court determined that isolated DNA sequences were not patentable subject matter, and that such isolated polynucleotide sequences were barred by the 'product of nature' exclusion to patentability.

A major implication of *Myriad* was the immediate invalidation in the USA of thousands of 'gene patents' that claimed isolated nucleotide sequences of genomic DNA. Although the court in *Myriad* held that complementary cDNA sequences could be patent-eligible, the impact of the decision on agricultural biotechnology research is substantial, given the extent to which such work relies on isolating particular genomic regions for the purposes of subsequent genetic transformation in host organisms.

In Australia the issue of patentability of genetic material is also steeped in litigation. In September 2014, the Federal Court of Australia, which hears appeals, upheld a claim for isolated DNA – essentially the same patent that was at issue in the USA *Myriad* decision. The case, *D'Arcy* v. *Myriad Genetics Inc. & Genetic Technologies Limited*,[2] was appealed to the High Court and oral arguments began in June 2015. In contrast to US law, under which laws of nature, natural phenomena and abstract ideas act as *de facto* statutory exceptions to patentability, in Australia an invention is *prima facie* patentable if it is a 'manner of manufacture' (Sherman, 2015). Thus, in the Australian case, the key issue will be whether the action of isolating a gene is a manner of manufacture, effectively equivalent to creating a non-naturally occurring molecule. If the High Court adopts this reasoning, it will effectively ratify a significant expansion of patentability of plant genetic material in Australia.

Meanwhile, in Europe the scope of patentable subject matter has recently been effectively broadened, in that plants or seeds obtained through conventional breeding methods have been judicially recognized as protectable with utility patents. In early 2015, the highest court of the European Patent Office (EPO) considered two cases arising out of patent applications for tomato and broccoli varieties and declared that plant products, including fruits, seeds and parts of plants, are patentable, even if they are obtained through essentially biological breeding methods involving crossing and selection.[3] It is still possible that the European Union Commission could take measures to overturn the EPO's interpretation of EU law; however at the time of writing the decision stands. Although the EPO cases did not directly address the protectability of GM crops – which remain patentable in Europe – these decisions could have implications for further consolidation of the control of agricultural markets by a few large seed companies, since many of the patent applications for conventionally bred varieties will likely be submitted by these firms.

6.6 Regulation and Status of Commercialized Genetically Modified Crops

In addition to the recent legal evolutions in IP protections discussed above, the past 20 years have witnessed substantial shifts in the regulation of GM crops. Much like regimes granting IP rights, transgenic plants are administered on a jurisdiction-by-jurisdiction basis, at either regional or national levels. Currently, GM traits (known as 'events') have been approved in 40 territories worldwide, though in some jurisdictions GM crops are permitted to be imported but not grown (ISAAA, 2015). Furthermore, several territories are currently in the process of approving research field trials of transgenic events. Seven of these countries are located in sub-Saharan Africa, an indicator that this region will likely represent a sizable new market for GM crops in the near future (James, 2014).

Interestingly, diverging trends appear to be emerging surrounding the regulatory approvals, cultivation and commercialization of GM crops in developed versus developing countries. For instance, in 2014 developing countries planted more biotech crops than industrial countries for the third consecutive year (James, 2014). Of the 28 states that planted GM crops that year, 20 were developing nations. Also in 2014, Bangladesh initiated the commercialization of its first approved biotech event, which it accomplished in less than 1 year. Vietnam and Indonesia began commercialization of biotech crops in 2015. Finally, Kenya, Uganda, Mozambique and Tanzania are all expected to release their first approved GM varieties in 2017 (James, 2014).

Nevertheless, civil society debate surrounding biotech crops continues in many world regions. For instance, in 2015 the European Parliament approved a new directive that permits European Union member states to restrict or ban the cultivation of GM crops at a national level, even if EU regulators have declared the particular events in question to be safe for cultivation.[4] However, this law does not apply to the free circulation or import of GM seeds and plant-propagating material. Another recent proposal by the European Commission would allow EU member states to entirely ban the use of GMOs in food and feed (EC Press Release, 2015). Unlike the former initiative, this latter proposal has not yet been approved by the European Parliament and has been met with substantial criticism and resistance (Teffer, 2015). Currently, Monsanto's MON810 maize is the only GM crop grown anywhere in Europe, and its cultivation has already been banned in eight EU countries (James, 2014).

While cultivation of GM crops continues at a massive scale in countries with industrialized agricultural sectors such as the USA, Canada and Australia, regulatory requirements for biosafety assessment, often cumbersome in their administration, may impede the commercialization of new transgenic events. For instance, in the USA some commentators have bemoaned the fact that regulatory criteria are not evidence based, and have not been reformed to accommodate experience from decades of research on and commercialization of transgenic crops, nor expanding knowledge of plant genome structure and dynamics (Bradford *et al.*, 2005). This has led to a system – at least in the USA – that simultaneously over-regulates crops and technologies that have proven track records of safety, and fails to provide oversight of crops that should be considered as genetically engineered yet prematurely attain 'deregulated' (i.e. deemed safe) status (Camacho *et al.*, 2014).

6.7 Emergence of 'Agbiogenerics' and the Intellectual Property–Regulatory Complex

Intellectual property and regulatory frameworks governing the development and commercial release of GM crops are both complicated and beset by a host of scientific, legal, ethical and social issues. Yet an area of even greater future controversy will likely be the 'IP-regulatory complex' that is emerging as patents over key agricultural technologies expire. The last US patent granting rights over the first blockbuster agricultural biotechnology product – the original glyphosate herbicide-tolerance trait marketed as Monsanto's 'Roundup Ready' in soybeans – expired in April 2015. Scientists, breeders and farmers are no longer vulnerable to any infringement lawsuits that might be brought by Monsanto for the unauthorized use, manufacture or commercialization of this trait. However, the key question remains of how regulatory approvals will apply to generic versions of this product, or to similar products that incorporate the previously patented trait (Jefferson *et al.*, 2015).

The uncertainty over how 'agbiogenerics' will be regulated is further complicated by the fact that vast disparities exist between the regulatory regimes for GM crops in different jurisdictions. In the USA, once a gene inserted into a crop has been deregulated, that crop may be used indefinitely. In contrast, most other countries require regulatory approvals for transgenic events to be renewed periodically, the timing of which also varies by jurisdiction. For instance, in China applications must be resubmitted every 3 years, in Korea every 5 and in Japan and Europe every 10 years (Grushkin, 2013).

Finally, market considerations convolute the situation even further. Entities that formerly enjoyed IP rights over GM products have an incentive to remain in control of regulatory filings, even once the relevant patents have expired. Regulatory approvals can frequently provide a source of market exclusivity, since the more costly and complex the process is, the greater the barrier to entry it poses for competing products. Thus, former right holders who still own the proprietary data required to attain regulatory approvals threaten to extend their exclusive control of regulated articles, even after their patent rights have expired.

The original Roundup Ready, glyphosate tolerance event for soybeans (RR1) is an illustrative case, not only because it has recently become generic, but also since it demonstrates how an off-patent trait can be incorporated into crops whose markets would have previously been perceived as too small to merit investment. Recently, the University of Arkansas in the USA released a new soybean variety that incorporates the RR1 trait, in addition to other characteristics that are adaptive to the particular growing conditions in Arkansas. The university – a public research institution – intends to make their variety available to local farmers at a lower cost than they would be required to pay Monsanto for access to the second generation of Roundup Ready soybeans, and to allow farmers to save seed from season to season (Miller, 2014).

The release of the University of Arkansas' variety was enabled by Monsanto's commitment to maintain all regulatory approvals worldwide for the RR1 trait until 2021. Monsanto cites the importance of giving farmers and the industry time to develop plans and mechanisms to assure continuing import approval support in export markets as the primary motivation behind its commitment, a surprisingly altruistic move by a company known to proactively assert its rights over its proprietary technologies. In the absence of Monsanto's response, the university would have been required to shoulder these costs, which it likely could not bear. Furthermore, even if costs were not a prohibitive factor for some research institutions, they could still be effectively prevented from the ability to commercialize products that incorporate a given generic trait in jurisdictions that require periodic renewals, if the former right holder decides not to share its proprietary regulatory data.

At present, the multinational seed industry is developing a set of agreements known as the AgAccord, representing an attempt to navigate the complex issues that will likely arise surrounding the maintenance of regulatory approvals for transgenic traits following patent expiration. This agreement is not yet fully in force due to the current lack of sufficient signatories, who could be the owners of proprietary regulatory data, as well as research institutions and farmers who plant generic biotech crops. Some commentators have condemned the AgAccord as an industry-centric solution, which in fact it currently is, given the present absence of signatories from the public sector. Others have called for governmental intervention, via legislation that would provide a regulatory 'safe harbour' to protect pre-commercial activities reasonably related to the development of follow-on generics against patent infringement (Schonenberg, 2014).

There could be certain advantages to a legislative solution to the issue of maintenance of regulatory approvals. Industry self-regulation may be inherently biased towards private sector interests and subject to scant, if any, public oversight. Furthermore, even assuming the AgAccord represents a viable alternative, legal uncertainty would remain, for instance surrounding how regulators would respond when an already-approved yet off-patent event is stacked with another transgenic trait. Yet any governmental fix would also likely be afflicted with various shortcomings, including bureaucracy and political debate, lack of expertise and the territorially limited nature of national legislation.

Yet the issue of maintenance of regulatory approvals for GM crops is one of extra national importance, with enormous implications for markets worldwide. There is a daunting possibility for major trade disruptions if approvals are not maintained in territories like Brazil or Argentina, which are two of the world's largest producers of soybeans after the USA, given that the vast majority of the soybean crop in these countries is comprised of GM varieties (Jefferson *et al.*, 2015). Meanwhile, China alone imports 65% of global soybean imports, over 90% of which is biotech (James, 2014). It is difficult to fathom the extent to which the global soybean-based food, feed and fuel supply would be impacted if, overnight, varieties incorporating the glyphosate tolerance trait were no longer permitted to be grown, imported or exported. In the future, it might be desirable for countries with extensive GM crop cultivation to unite to create an international legal framework for the management of proprietary regulatory data when patents claiming regulated transgenic events expire. Such a treaty mechanism could offer a central, public sector alternative to industry-driven attempts at self-regulation.

6.8 Conclusion

The conception of IP as a tool to foster innovation has been influential for over two centuries, but has become a much more prominent feature of research in the life sciences and agricultural biotechnology, in particular, over the last 25 years. This trend is unlikely to be reversed in the near term, and the relevance of the use of IP mechanisms to protect intangible assets related to developments in the life sciences continues to increase. However, robust and sustained innovation in agricultural biotechnology, as in many technology sectors, requires a balance of both exclusive and non-exclusive access to proprietary technologies. This balance should ensure that the fundamental research tools are broadly available to support research and development in many areas of application, and at the same time provide the appropriate level of exclusivity for specific trait or trait/crop combinations that would allow the developer of new varieties to recoup their investments.

The public sector has a role to play alongside agricultural biotechnology companies, particularly in providing research tools and broad enabling technologies and in addressing biotechnology applications for speciality crops whose market size may not justify commercial investment. The most powerful approaches, however, may come from public–private partnerships that mobilize proprietary technologies to address agricultural biotechnology product developments that have a high social but low commercial value, including strategies to feed some of the world's poorest populations.

Finally, in addition to issues surrounding IP rights *per se*, over the past two decades further legal issues have arisen surrounding the commercialization of biotechnological innovations. Due to the pending expiration of patents covering many technologies related to the first generation of GM crops, questions surrounding how best to maintain regulatory approvals of GM traits have taken prominence.

Given these myriad changes to IP and other legal regimes – as well as evolutions in research and development activities surrounding agricultural biotechnologies – it remains critical that all actors involved in this technological space remain apprised of new trends. However, notwithstanding any of the aforementioned shifts, many of the overarching goals of biotechnological innovation should remain the same. Thus, balanced and equitable access to proprietary technologies should be promoted, and all sectors should endeavour to advance the state of agricultural innovation for the global good.

Notes

1 *Association for Molecular Pathology et al.* v. *Myriad Genetics Inc. et al.* 133 S. Ct. 2107, 569 US (2013).

2 *D'Arcy* v. *Myriad Genetics, Inc* [2014] FCAFC 115.

3 Enlarged Board of Appeal of the European Patent Office, Case No. G 0002/12 (25 March 2015); Enlarged Board of Appeal of the European Patent Office, Case No. G 0002/13 (25 March 2015).

4 Directive (EU) 2015/412 of the European Parliament and of the Council (11 March 2015).

References

African Agricultural Technology Foundation (2010) AATF marks second anniversary since formation of WEMA project. Press Release (19 March 2010). Available at: http://www.aatf-africa.org/userfiles/PressRelease-WEMAAnniversary.pdf (accessed 23 September 2015).

Alandete-Saez, M., Chi-Ham, C.L., Graff, G.D., Boettiger, S. and Bennett, A.B. (2015) Intellectual property in agricultural biotechnology: strategies for open access. In: Stewart, C.N. Jr (ed.) *Plant Biotechnology and Genetics: Principles, Techniques and Applications*. Wiley, Hoboken, New Jersey.

Bennett, A.B., Chi-Ham, C.L., Barrows, G., Sexton, S. and Zilberman, D. (2013) Agricultural biotechnology: economics, environment, ethics and the future. *Annual Reviews on Environment and Research* 38, 249–279.

Bradford, K.J., Van Deynze, A., Gutterson, N., Parrott, W. and Strauss, S.H. (2005) Regulating transgenic crops sensibly: lessons from plant breeding, biotechnology and genomics. *Nature Biotechnology* 23, 439–444.

Camacho, A., Van Deynze, A., Chi-Ham, C.L. and Bennett, A.B. (2014) Genetically engineered crops that fly under the US regulatory radar. *Nature Biotechnology* 32, 1087–1091.

Clark, D., Klee, H. and Dandekar, A. (2004) Despite benefits, commercialization of transgenic horticultural crops lags. *California Agriculture* 58, 89–93.

Conway, G. and Toenniessen, G. (1999) Feeding the world in the twenty-first century. *Nature* 402, C55–C58.

de Janvry, A. and Dethier, J.J. (1985) Technological innovation in agriculture: the political economy of its rate and bias. *Consultative Group on International Agricultural Research (CGIAR)* Study Paper No. 1.

European Commission Press Release (2015) More freedom for Member States to decide on the GMOs use for food & feed (Brussels, 22 April 2015). Available at: http://europa.eu/rapid/press-release_IP-15-4777_en.htm (accessed 23 September 2015).

Fenton, G.M., Chi-Ham, C.L. and Boettiger, S. (2007) Intellectual property freedom to operate: the law firm's approach and role. In: Krattiger, A., *et al*. (eds) *Intellectual Property Management in Health and Agricultural Innovation: A Handbook of Best Practices*. MIHR, Oxford, UK and PIPRA, Davis, California, pp. 1363–1384.

Graff, G.D., Zilberman, D. and Bennett, A.B. (2009) The contraction of agbiotech product quality innovation. *Nature Biotechnology* 27, 702–704.

Grushkin, D. (2013) Threat to global GM soybean access as patent nears expiry. *Nature Biotechnology* 31, 10–11.

Hardin, G. (1968) Tragedy of the commons. *Science* 162, 1243–1248.

Heller, M.A. and Eisenberg, R.S. (1998) Can patents deter innovation? The anti-commons in biomedical research. *Science* 280, 698–701.

International Service for the Acquisition of Agri-Biotech Applications (ISAAA) (2015) GM Approval Database. Available at: http://www.isaaa.org/gmapprovaldatabase/default.asp (accessed 23 September 2015).

James, C. (2014) Global status of commercialized biotech/GM crops in 2014. *International Service for the Acquisition of Agri-Biotech Applications*, ISAAA Brief No. 49.

Jefferson, D.J. (2014) Development, farmers' rights, and the Ley Monsanto: the struggle over the ratification of UPOV 91 in Chile. *IDEA – The Intellectual Property Law Review* 55, 31–78.

Jefferson, D.J., Camacho, A.B. and Chi-Ham, C.L. (2014) Towards a balanced regime of intellectual property rights for agricultural innovations. *Journal IP Rights* 19, 395–403.

Jefferson, D.J., Graff, G.D., Chi-Ham, C.L. and Bennett, A.B. (2015) The emergence of agbiogenerics. *Nature Biotechnology* 33, 819–823.

Jefferson, T. (1987) Thomas Jefferson to Isaac McPherson. In: Kurland, P.B. and Lerner, R. (eds) *The Founders' Constitution*. University of Chicago Press, Chicago, Illinois. Available at: http://press-pubs.uchicago.edu/founders/documents/a1_8_8s12.html (accessed 23 September 2015).

Kowalski, S.P., Ebora, R.V., Kryder, D. and Potter, R.H. (2002) Transgenic crops, biotechnology and ownership rights: what scientists need to know. *Plant Journal* 31, 407–421.

Kryder, R.D., Kowalski, S.P. and Krattiger, A.F. (2000) The intellectual and technical property components of pro-vitamin A rice (GoldenRice™): a preliminary freedom-to-operate review. Brief No. 20. ISSAA, Ithaca, New York.

Miller, F. (2014) Arkansas releases first Roundup Ready soybean. University of Arkansas Press Release (3 December 2014). Available at: http://arkansasagnews.uark.edu/8273.htm (accessed 23 September 2015).

Schonenberg, B.M. (2014) Twenty years in the making: transitioning patented seed traits into the generic market. *Marquette Law Review* 97, 1039–1083.

Sherman, B. (2015) Before the high court – D'Arcy v Myriad Genetics Inc: patenting genes in Australia. *Sydney Law Review* 37, 135–146.

Sunding, D. and Zilberman, D. (2001) The agricultural innovation process: research and technology adoption in a changing agricultural sector. *Handbook of Agricultural Economics* 1, 207–261.

Teffer, P. (2015) Agriculture MEPs call for rejection of GMO plan. *EU*observer. Available at: https://euobserver.com/environment/130100 (accessed 23 September 2015).

7 Bioprospecting for Improving Soil Health and Crop Productivity: Indian Patent Landscape

Vivek Srivastava, Puneet Singh Chauhan, Sankalp Misra, Swati Sharma, Aradhana Mishra and Chandra Shekhar Nautiyal*

CSIR-National Botanical Research Institute, Lucknow, India

7.1 Introduction

One of the major challenges for the 21st century will be an environmentally sound and sustainable crop production. Global plant production systems must be optimized to produce stable high yields from limited land under changing and variable climates. Demands for food, animal feed, and feedstocks for bioenergy and biorefining applications are increasing with population growth, urbanization and affluence. Current production methods in agriculture, such as improper use of chemical pesticides and fertilizers, create a long list of environmental and health problems (Gunnell *et al.*, 2007; Leach and Mumford, 2008). Furthermore, agriculture based on chemicals has made an adverse impact on the beneficial soil microbial communities, significantly lowering the microbial biodiversity (Mäder *et al.*, 2002). Moreover, challenges due to emerging, re-emerging and endemic plant pathogens continue to test our ability to safeguard plant growth and health worldwide (Miller *et al.*, 2009). Low input, sustainable, alternatives to chemical-derived fertilizers and pesticides are required to lessen input costs and maintain or increase yields, with potential biological solutions having an important role to play.

Therefore, in the need to adopt ecofriendly agricultural practices for sustainable agriculture, microorganisms are considered to be an important tool in overcoming problems associated with the excessive use of chemical fertilizers and pesticides (Lucy *et al.*, 2004; Bardi and Malusà, 2012; Bashan *et al.*, 2014; Owen *et al.*, 2015). Microbial cultures and products are now widely studied and used in agriculture and in the future enhancement and maintenance of soil fertility through microorganisms is expected to become a common practice (Singh *et al.*, 2011; Bhattacharyya and Jha, 2012; Vassilev *et al.*, 2013). The extensive research programmes on plant-beneficial microorganisms have resulted in the development of various types of bioinoculants, which can satisfy the growth requirements of crops (Bardi and Malusà, 2012; Malusá and Vassilev, 2014; Owen *et al.*, 2015).

There is a strong growing market for microbial inoculants worldwide with an annual growth rate of approximately 10%. The beneficial interdisciplinary research, e.g. on mass production, formulation, interactions and signalling with the environment will be the base for the future success of the biological control industry, as well as for innovative business management, product marketing and education. Overall, the use of microorganisms and the exploitation of beneficial plant–microbe interactions offer promising and environmentally friendly strategies for conventional and organic agriculture worldwide.

*Corresponding author e-mail: nautiyalnbri@lycos.com

7.2 Biofertilizers

The term biofertilizer or 'microbial inoculants' can be generally defined as a preparation containing live or latent cells of efficient strains of nitrogen fixing, phosphate solubilizing or cellulolytic microorganisms used for application of seed, soil or composting areas with the objective of increasing the numbers of such microorganisms and accelerating certain microbial process to augment the availability of nutrients in a form which can be assimilated by plants (NIIR, 2012). They are important components of integrated nutrient management. They form a cost effective, ecofriendly and renewable source of plant nutrients to supplement chemical fertilizers in sustainable agricultural systems.

Similarly, according to the US Environmental Protection Agency, biopesticides are derived from natural materials, such as animals, plants, bacteria and certain minerals, and are used widely for controlling insects and disease-causing pathogens. Biopesticides are categorized under microbial pesticides, plant-incorporated protectants and biochemical pesticides, which are produced through naturally occurring substances that control pests by non-toxic mechanisms. Microbial pesticides include bacteria, virus, fungus and other less common microorganisms. Biopesticides are usually inherently less toxic, generally affect only the target pest, are effective in very small quantities and easily biodegradable, thereby resulting in lower exposures and largely avoiding pollution problems. When used as a component of Integrated Pest Management (IPM) programmes, they can greatly control major pest menaces while crop yields remain high.

Biofertilizer products contain living cells of different types of microorganisms, which, when applied to seed, plant surface or soil, colonize the rhizosphere or the interior of the plant and promote growth by converting nutritionally important elements (nitrogen, phosphorus) from an unavailable to available form through biological processes such as nitrogen fixation and solubilization of rock phosphate (Rokhzadi *et al.*, 2008). Beneficial microorganisms in biofertilizers accelerate and improve plant growth and protect plants from pests and diseases (El-Yazeid *et al.*, 2007). In an era of sustainable crop production, plant–microbe interactions in the rhizosphere play a pivotal role in transformation, mobilization, solubilization, etc. of nutrients from a limited nutrient pool, and subsequently uptake of essential nutrients by plants to realize their full genetic potential.

7.3 Important Microorganisms Used in Biofertilizer

Biofertilizers enrich the soil environment with all kinds of micro- and macronutrients via nitrogen fixation, phosphate and potassium solubilization or mineralization, release of plant growth-regulating substances, production of antibiotics and biodegradation of organic matter in the soil (Sinha *et al.*, 2010). Application of biofertilizers as seed or soil inoculants enables them to multiply and participate in nutrient cycling and benefit crop productivity (Singh *et al.*, 2011). The rhizosphere is a thin layer of soil immediately surrounding plant roots that is an extremely important and active area for root activity and metabolism, where microbe populations are stimulated by root activities (Hiltner, 1904). This original concept has now been extended to include the soil surrounding a root in which physical, chemical and biological properties have been changed by root growth and activity (McCully, 2005). A large number of microorganisms such as bacteria, fungi, protozoa and algae coexist in the rhizosphere. Bacteria are the most abundant among them. Plants select those bacteria contributing most to their fitness by releasing organic compounds through exudates (Lynch, 1990), creating a very selective environment where diversity is low (Marilley and Aragno, 1999; García *et al.*, 2001). Since bacteria are the most abundant microorganisms in the rhizosphere, it is highly probable that they influence the plant's physiology to a greater extent, especially considering their competitiveness in root colonization (Antoun and Kloepper, 2001; Barriuso *et al.*, 2008). Rhizobacteria inhabit plant roots and exert a positive effect ranging from direct influence mechanisms to an indirect effect. Therefore the bacteria inhabiting the rhizosphere and beneficial to plants are termed plant growth-promoting rhizobacteria (PGPR; Kloepper and Schroth, 1978; Kloepper *et al.*, 1989; Cleyet-Marcel *et al.*, 2001). According to their relationship with the plants, PGPR can be divided into two groups: symbiotic bacteria and free-living rhizobacteria (Khan, 2005). In the last few years, the number of studies pertaining to isolation, identification and characterization of PGPR has increased considerably due to their notable role for efficient ecosystem functioning. Various species of bacteria such as *Pseudomonas*, *Azospirillum*, *Azotobacter*, *Klebsiella*, *Enterobacter*, *Alcaligenes*, *Arthrobacter*, *Burkholderia*, *Bacillus*, *Rhizobium*, *Azospirillum*, *Pantoea* and *Serratia* have been reported to enhance plant growth.

7.4 Applications of PGPR

7.4.1 Biological nitrogen fixation

A number of bacterial species belonging to the genera *Azospirillum*, *Alcaligenes*, *Arthrobacter*, *Acinetobacter*, *Bacillus*, *Burkholderia*, *Enterobacter*, *Erwinia*, *Flavobacterium*, *Pseudomonas*, *Rhizobium* and *Serratia* are associated with the plant rhizosphere and are able to exert a beneficial effect on plant growth (Egamberdiyeva, 2005; Tilak *et al.*, 2005). Plants play an important role in selecting and enriching the types of bacteria by the constituents of their root exudates. Thus, the development of the bacterial community in the rhizosphere depends on the nature and concentrations of organic constituents of exudates and the corresponding ability of the bacteria to utilize these as sources of energy (Curl and Truelove, 1986). There is a range of bacterial communities present in the soil rhizosphere, rhizoplane and internal plant tissues (Hallmann *et al.*, 1997). However, rhizospheric bacterial communities have proficient systems for the uptake and catabolism of organic compounds present in root exudates (Barraquio *et al.*, 2000). Since associative interactions of plants and microorganisms must have come into existence as a result of co-evolution, the use of rhizobacteria as bioinoculants must be pre-adapted, so that it fits into a long-term sustainable agricultural system. Therefore PGPR offers an attractive way to replace chemical fertilizers, pesticides and supplements commonly used as inoculants for improving the growth and yield of agricultural crops (Ashrafuzzaman *et al.*, 2009). The use of beneficial microorganisms as biofertilizer and biopesticides can reduce chemical fertilizer applications and consequently lower production cost. Application of PGPR in order to enhance productivity may be a feasible alternative to organic fertilizers, which also helps in reducing pollution and preserving the environment (Ashrafuzzaman *et al.*, 2009). PGPR or a consortia of PGPR and arbuscular mycorrhizal fungi (AMF) can improve the nutrient use efficiency of fertilizers and allow reduced application rates of chemical fertilizers (Adesemoye *et al.*, 2009). Out of 180 × 10^6 t/year globally, 80% of biological nitrogen fixation is contributed by symbiotic associations while the remainder comes from free-living or associative systems (Graham, 1999). The enrichment of soil by the ability to reduce and obtain such appreciable amounts of nitrogen from the atmospheric reservoir is confined to bacteria and archaea (Young, 1992). These include symbiotic nitrogen-fixing (N_2-fixing) forms, viz. *Rhizobium*, the obligate symbionts in leguminous plants and *Frankia* in non-leguminous trees, and non-symbiotic (free-living, associative or endophytic) N_2-fixing forms such as *Cyanobacteria*, *Azospirillum*, *Azotobacter*, *Acetobacter diazotrophicus* etc.

Symbiotic nitrogen fixers

Biological nitrogen fixation by bacteria has been extensively studied under symbiotic and non-symbiotic nitrogen fixation. Members from the genera *Rhizobium*, *Bradyrhizobium*, *Azolla* and *Frankia* actively account for symbiotic nitrogen fixation while *Azotobacter*, *Azospirillum*, *Bacillus* and *Pseudomonas* contribute non-symbiotically. *Frankia* forms root nodules on more than 280 species of woody plants from eight different families (Schwintzer and Tjepkema, 1990) and forms effective symbiosis with the species of *Alnus* and *Casuarina* (Wheeler and Miller, 1990).

RHIZOBIUM. When colonization of the roots of a non-leguminous plant by rhizobia takes place in a non-specific manner, the strains from this genus may behave as PGPR. In India, conducted field trials have shown that depending on the legume, soil and agroclimatic conditions, nearly 50% of nitrogenous fertilizer could be saved through rhizobial inoculations with considerable increase in yield (Rewari and Tilak, 1988; Tilak, 1993). It has been also revealed that under field conditions inoculation of *Rhizobium* sp. not only causes a greater increase in growth and yield, but also the number of nodules per root system is significantly higher in plants inoculated with *Rhizobium* sp. compared to plants without *Rhizobium* sp. (Akhtar and Siddiqui, 2009). In addition to their beneficial N_2-fixing activity with legumes, rhizobia can improve plant phosphate nutrition by mobilizing inorganic and organic phosphate (Alikhani *et al.*, 2006). Consortia of *Rhizobium* with phosphate solubilizing bacteria (PSB) revealed a synergistic effect on symbiotic parameters and grain yield of mungbean. Phosphate-solubilizing bacteria improve the competitive ability and symbiotic effectiveness of inoculated *Rhizobium* sp. in lentil under field conditions (Kumar and Chandra, 2008). The P-solubilizing strains and the N_2-fixing bacterial strains have great potential in being formulated and used as biofertilizers (Cakmakc *et al.*, 2007).

BRADYRHIZOBIUM. *Bradyrhizobium* species are gram-negative bacilli (rod shaped) with a single subpolar or polar flagellum. They are a common soil-dwelling microorganism that can form symbiotic relationships with leguminous plant species where they fix nitrogen in exchange for carbohydrates from the plant. Like other rhizobia, they have the ability to fix atmospheric nitrogen into forms readily available for other organisms to use. They are slow growing in comparison to *Rhizobium* species, which are considered fast-growing rhizobia. Studies performed with co-inoculation of *Bradyrhizobium* and certain PGPR indicated a positive effect on symbiotic nitrogen fixation by enhancing both root nodule number or mass, dry weight of nodules, yield components, grain yield, soil nutrient availability and increasing the nitrogenase activity (Polenko *et al.*, 1987; Son *et al.*, 2006) and enhances the nodulation and nitrogen fixation in *Glycine max* at a low root-zone temperature (Zhang *et al.*, 1996; Dashti *et al.*, 1998).

AZOLLA AND FRANKIA. A number of bacterial endophytes have the ability to form symbioses with plants and to fix bio-available nitrogen within unspecialized tissues of the host in the absence of nodulation as seen in the legume–rhizobia interaction. For example, cyanobacteria can form associations with a range of plants from different clades including *Gunnera*, cycads, lichens and *Azolla* (Santi *et al.*, 2013) and form heterocysts, specialized structures creating a micro-aerophilic environment suitable for nitrogen fixation with the nitrogenase enzyme (Berman-Frank *et al.*, 2003). *Azolla* biofertilizer is used for rice cultivation in different countries such as Vietnam, China, Thailand and the Philippines. Field trials indicated that rice yields increased by 0.5–2 t/ha due to *Azolla* application (Gupta, 2004). *Frankia* forms root nodules on more than 280 species of woody plants from eight different families (Schwintzer and Tjepkema, 1990) and forms effective symbiosis with the species of *Alnus* and *Casuarina* (Wheeler and Miller, 1990).

Non-symbiotic nitrogen fixers

Non-symbiotic nitrogen fixation has a great agronomic significance. One main limitation is the availability of carbon and energy source for the energy-intensive nitrogen fixation process. However, this limitation can be overcome by moving closer to or inside the plants, i.e. diazotrophs present in the rhizosphere, rhizoplane or those growing endophytically. Some important non-symbiotic nitrogen-fixing bacteria include *Azoarcus* sp., *Gluconacetobacter diazotrophicus*, *Herbaspirillium* sp., *Azotobacter* sp. (Vessey, 2003; Barriuso and Solano, 2008), *Achromobacter*, *Acetobacter*, *Arthrobacter*, *Azospirillum*, *Azomonas*, *Bacillus*, *Beijerinckia*, *Clostridium*, *Corynebacterium*, *Derxia*, *Enterobacter*, *Klebsiella*, *Pseudomonas*, *Rhodospirillum*, *Rhodopseudomonas* and *Xanthobacter* (Saxena and Tilak, 1998).

AZOTOBACTER. The family Azotobacteriaceae comprises of two genera, *Azomonas* (non-cyst forming) with three species (*A. agilis*, *A. insignis* and *A. macrocytogenes*) and *Azotobacter* (cyst forming) comprising six species, *A. chroococcum*, *A. vinelandii*, *A. beijerinckii*, *A. nigricans*, *A. armeniacus* and *A. paspali* (Tchan, 1984; Tchan and New, 1984). *Azotobacter* is generally regarded as a free-living aerobic nitrogen-fixer. *Azotobacter paspali*, which was first described by Dobereiner and Pedrosa, has been isolated from the rhizosphere of *Paspalum notatum*, a tetraploid subtropical grass, and is highly host specific. Various crops in India have been inoculated with diazotrophs, particularly *Azotobacter* and *Azospirillum* (Saxena and Tilak, 1999). Application of *Azotobacter* and *Azospirillum* improves the yield of both annual and perennial grasses (Biswas *et al.*, 1994). *Azotobacter* strains could affect seed germination and seedling growth in a plant (Shaukat *et al.*, 2006) and have been reported to increase seed germination of *Cicer arietinum*, *Phaseolus mungo*, *Vigna catjung* and *Zea mays* (Saikia and Brezbaruah, 1995). It also showed that wheat yield increased up to 30% with *Azotobacter* inoculation (Kloepper and Beauchamp, 1992; Gholami *et al.*, 2009).

AZOSPIRILLUM. This group of free-living rhizobacteria encompasses ten species, the classification of each one of its species is according to its particular biochemical and molecular characteristics: *A. lipoferum* and *A. brasilense* (Tarrand *et al.*, 1978); *A. amazonense* (Magalhães *et al.*, 1983); *A. halopraeferens* (Reinhold *et al.*, 1987); *A. irakense* (Khammas *et al.*, 1989); *A. largimobile* (Dekhil *et al.*, 1997); *A. doebereinerae* (Eckert *et al.*, 2001); *A. Oryzae* (Xie and Yokota, 2005); *A. melinis* (Peng *et al.*, 2006) and recently *A. canadensis* (Mehnaz *et al.*, 2007). *Azospirillum* was first isolated from cereals and most of its initial inoculation has been done on the main cereal crops and non-cereal species successfully. *Azospirillum* strains have no preferences for crop plants or weeds or for annual or perennial plants and can be successfully applied to plants that have no previous history of *Azospirillum* in their roots. Therefore *Azospirillum* is a general root colonizer and not a plant-specific bacterium (Bashan and Holguin, 1997; Bashan *et al.*, 2004). Members of the genus *Azospirillum* fix nitrogen under micro-aerophilic conditions and are frequently associated with the root and rhizosphere of a large number of agriculturally important crops and cereals. When azospirilla are established in the rhizosphere they usually, but not always, promote the growth of plants (Okon, 1985; Tilak and Subba Rao, 1987; Bashan *et al.*, 2004). While they exhibit N_2-fixing capability (~1–10 kg N/ha), the enhancement of yield is mainly pertained to improved root development due to the production of growth-promoting substances and consequently increased rates of water and mineral uptake (Dewan and Subba Rao, 1979; Okon and Kapulnik, 1986; Fallik *et al.*, 1994). Isolation and characterization of bacterial isolates from endorhizosphere of sugarcane (*Saccharum* sp.) and rye grass (*Lolium perenne*) revealed that *Azospirillum* isolates from sugarcane and rye grass exhibited maximum nitrogenase activity among *Bacillus*, *E. coli* and *Pseudomonas* (Gangwar and Kaur, 2009).

ACETOBACTER. *Acetobacter* is an important inoculant for sugarcane and successfully colonizes sugarcane varieties in India where the chemical N fertilization is completely avoided for at least 2 successive years and replaced by organic manures (Ashbolt and Inkerman, 1990). The Acetobacteriaceae family includes genera *Acetobacter*, *Gluconobacter*, *Gluconoacetobacter* and *Acidomonas*.

BACILLUS. *Bacillus* is the most abundant genus in the rhizosphere and its PGPR activity has been known for many years (Probanza *et al.*, 2002). The metabolites that are released by these strains strongly affect the environment by increasing nutrient availability of the plants (Charest *et al.*, 2005; Barriuso and Solano, 2008). Natural occurrence in the immediate vicinity of plant roots enables *B. subtilis* to maintain stable contact with higher plants and promote their growth. Inoculation of *Bacillus licheniformis* on tomato and pepper shows considerable colonization and can be used as a biofertilizer without altering normal management in greenhouses (García *et al.*, 2004). Evaluation of a consortium of *Bacillus* spp. on the first developmental stages of two micropropagated bananas revealed that this bacterial consortium can be described as a prospective way to increase plant health and survival rates in commercial nurseries (Jaizme-Vega *et al.*, 2004). *Bacillus* is also proven to increase the yield, growth and nutrition of raspberry plants under organic growing conditions (Orhan *et al.*, 2006). *Bacillus megaterium* is very reliable in improving different root parameters (rooting performance, root length and dry matter content of root) in mint (Kaymak *et al.*, 2008). The inoculation of phosphate-solubilizing bacteria (PSB) *Bacillus megaterium* var. *phosphaticum* and potassium solubilizing bacteria (KSB) *Bacillus mucilaginosus* in nutrient-limited soil showed consistent increase in mineral availability, uptake and plant growth of pepper and cucumber, suggesting its potential use as fertilizer (Han *et al.*, 2006; Supanjani *et al.*, 2006). *Bacillus pumilus* can be used as a bio-inoculant for biofertilizer production to increase the crop yield of wheat varieties (Hafeez *et al.*, 2006). In addition, PSB such as species of *Bacillus* and *Paenibacillus* (formerly *Bacillus*) have been applied to soil to particularly enhance the phosphorus status of plants (Brown, 1974).

PSEUDOMONAS. *Pseudomonas* sp. is omnipresent in agricultural soils and has many traits that make them well suited as PGPR. The *Pseudomonas* strains that have been most effectively used belong to the fluorescent *Pseudomonas* sp. (FLPs). FLPs help in safeguarding soil health and are metabolically and functionally most diverse (Lugtenberg and Dekkers, 1999; Lata and Tilak, 2002).

The combined effect of *Pseudomonas fluorescens* inoculants and microbial fertilizer plays an efficient role in stimulating yield and growth traits of chickpea (Rokhzadi *et al.*, 2008). A significant increase in fresh and dry masses has been provided by the isolates from roots, shoots and rhizosphere soil of sugarcane (Mehnaz *et al.*, 2009). Field trials conducted with *Pseudomonas* strains led to a great increase in yield of legumes (Chen *et al.*, 2006). Specific strains such as *Pseudomonas fluorescens* and *Pseudomonas putida* have recently been used as seed inoculants on crop plants to promote growth and increase yields. These pseudomonads have been termed as PGPR since they rapidly colonize plant roots of potato, sugarbeet and radish, and cause considerable increase in yield up to 144% in field tests (Burr *et al.*, 1978; Kloepper *et al.*, 2004).

7.4.2 Phosphate solubilization

Phosphorus is the most important key element in the nutrition of plants next to nitrogen. It plays a vital role in almost all major metabolic processes in plants including photosynthesis, energy transfer, signal transduction, macromolecular biosynthesis and respiration (Khan *et al.*, 2010) and nitrogen fixation in legumes (Saber *et al.*, 2005).

A key advantage of beneficial microorganisms is to assimilate phosphorus for their own requirements, which in turn makes it available in its soluble form in sufficient quantities in the soil. *Pseudomonas*, *Bacillus*, *Micrococcus*, *Flavobacterium*, *Fusarium*, *Sclerotium*, *Aspergillus* and *Penicillium* have been reported to be active in the solubilization process (Pindi and Satyanarayana, 2012). A phosphate-solubilizing bacterial strain NII-0909 of *Micrococcus* sp. has polyvalent properties including phosphate solubilization and siderophore production (Dastager *et al.*, 2010). Similarly, two fungi, *Aspergillus fumigatus* and *A. niger*, were isolated from decaying cassava peels and were found to convert cassava wastes by the semi-solid fermentation technique to phosphate biofertilizers (Ogbo, 2010). *Burkholderia vietnamiensis*, a stress-tolerant bacterium, produces gluconic and 2-ketogluconic acids, which are involved in phosphate solubilization (Park *et al.*, 2010). *Enterobacter* and *Burkholderia* that were isolated from the rhizosphere of sunflower were found to produce siderophores and indolic compounds (ICs), which can solubilize phosphate (Ambrosini *et al.*, 2012).

A substantial number of microbial species exhibit P solubilization capacity; these include bacteria, fungi, actinomycetes and even algae. In addition to *Pseudomonas* and *Bacillus*, other bacteria reported as P-solubilizers include *Rhodococcus*, *Arthrobacter*, *Serratia*, *Chryseobacterium*, *Gordonia*, *Phyllobacterium*, *Delftia* sp. (Wani *et al.*, 2005; Chen *et al.*, 2006), *Azotobacter* (Kumar *et al.*, 2001), *Xanthomonas* (De Freitas *et al.*, 1997), *Enterobacter*, *Pantoea* and *Klebsiella* (Chung *et al.*, 2005), *Vibrio proteolyticus* and *Xanthobacter agilis* (Vazquez *et al.*, 2000). Furthermore, symbiotic nitrogenous rhizobia, which fix atmospheric nitrogen into ammonia and export the fixed nitrogen to the host plants, have also shown P solubilization activity (Zaidi *et al.*, 2009). For instance, *Rhizobium leguminosarum bv. trifolii* (Abril *et al.*, 2007) and *Rhizobium* sp. nodulating *Crotalaria* species improved plant P-nutrition by mobilizing inorganic and organic P (Sridevi *et al.*, 2007). Various P solubilizing bacteria have also been isolated from stressed environments, for example the halophilic bacteria *Kushneria sinocarni* isolated from the sediment of *Daqiao saltern* on the eastern coast of China, which may be useful in salt-affected agricultural soils (Zhu *et al.*, 2011).

7.4.3 Potassium solubilization

Potassium (K) is the third important plant nutrient. It plays a key role in the growth, metabolism and development of plants. Without adequate supply of potassium, plants will have poorly developed roots, grow slowly, produce small seeds and have lower yields (McAfee, 2008; White and Karley, 2010) and have increased susceptibility to diseases (Amtmann *et al.*, 2008; Armengaud *et al.*, 2010) and pests (Amtmann *et al.*, 2006; Troufflard *et al.*, 2010). Potassium-solubilizing bacteria (KSB) are able to release K from insoluble minerals (Parmar and Sindhu, 2013). A wide range of rhizospheric microorganisms are reported as K-solubilizers including *B. mucilaginosus* (Raj, 2004; Zhao *et al.*, 2008; Basak and Biswas, 2009; Zarjani *et al.*, 2013), *B. edaphicus* (Sheng, 2005), *B. circulanscan* (Lian *et al.*, 2002), *Burkholderia*, *A. ferrooxidans* (Sheng and He, 2006), *Arthrobacter* sp. (Zarjani *et al.*, 2013), *Enterobacter hormaechei* (KSB-8) (Prajapati *et al.*, 2013), *Paenibacillus mucilaginosus* (Liu *et al.*, 2012), *P. frequentans*, *Cladosporium* (Argelis *et al.*, 1993), *Aminobacter*, *Sphingomonas*, *Burkholderia* (Uroz *et al.*, 2007) and *Paenibacillus glucanolyticus* (Sangeeth *et al.*, 2012). These microbial strains have the ability to solubilize K from K-bearing minerals, but only a few bacteria, such as *B. mucilaginosus* and *B. edaphicus*, have high activity in mobilizing and solubilizing K from minerals (Lian *et al.*, 2002; Sheng, 2005; Zhao *et al.*, 2008; Rajawat *et al.*, 2012). Therefore bacteria have wide applications in mining, metallurgy, microbial fertilizer and feed (Lian *et al.*, 2002; Sheng, 2005; Zhao *et al.*, 2008). Potassium solubilizing microorganisms (KSM) such as the genera *Aspergillus*, *Bacillus* and *Clostridium* are found to be efficient in potassium solubilization in the soil and mobilize in different crops (Mohammadi and Sohrabi, 2012). KSB can dissolve K-minerals such as mica, illite and orthoclase in the soil through the production and excretion of organic acids or production of capsular polysaccharide (Sheng and He, 2006). Nevertheless, better nutrient efficacy has been reported in several consortia mixtures of *Rhizobium* sp. and PGPR (Sivaramaiah *et al.*, 2007), associations of nodule-inducing rhizobia, free-living N-fixing bacteria and AMF (Toro *et al.*, 1998; Barea *et al.*, 2002; Lisette *et al.*, 2003; Wu *et al.*, 2005; Adesemoye *et al.*, 2008; Wang *et al.*, 2011), formulations of different categories of bacteria such as *Rhizobium* and PSB (Alagawadi and Gaur, 1988), PGPR, PSB and *Rhizobium* (Prasad and Chandra, 2003), PSB and KSB (Han and Lee, 2005; Vassilev *et al.*, 2006), as well as different combinations of PGPR and AMF (Singh and Adholeya, 2003; Malusá *et al.*, 2007). Therefore to exploit these kinds of consortia for fertilization purposes will prove to be a great innovation in sustainable agriculture.

7.4.4 Biocontrol agents

PGPR are indigenous to soil and the plant rhizosphere. They play a major role in the biocontrol of plant pathogens and are able to restrain a broad spectrum of bacterial, fungal and nematode diseases. The application of PGPR has become a common practice in many regions of the world. Although considerable control of plant pathogens has been demonstrated by PGPR in laboratory and greenhouse studies, the results pertaining to field tests have been inconsistent. Some of these rhizobacteria may also form part of IPM programmes. Hence, application of PGPR in agriculture is possible as a biocontrol and biofertilizer (Siddiqui, 2006). The bacteria isolated from the rhizosphere of *Lolium perenne* are capable of acting as PGPR and as biocontrol agents (Shoebitz *et al.*, 2007). A major group of rhizobacteria with potential for biological control is the pseudomonads, as they possess many traits that make them well suited as biocontrol and growth-promoting agents (Kremer and Kennedy, 1996). Remarkable progress has been made in characterizing the process of root colonization by pseudomonads, the biotic and abiotic factors affecting colonization, bacterial traits and genes contributing to rhizosphere competence, and the mechanisms of pathogen suppression (Weller, 2007). The traits include the ability to grow rapidly under *in vitro* condition, rapid utilization of root exudates, colonization and multiplication in the rhizosphere, the production of a wide spectrum of bioactive metabolites (i.e. antibiotics, siderophores, volatiles, and growth-promoting substances), to compete aggressively with other microorganisms and adapt to environmental stresses. The major limitation of pseudomonads as biocontrol agents is their inability to produce resting spores (as do many *Bacillus* spp.), which complicates formulation of the bacteria for commercial use. Fluorescent pseudomonads, equipped with multiple mechanisms for biocontrol of phytopathogens and plant growth promotion, are being used widely as they produce a wide variety of antibiotics, chitinolytic enzymes, growth-promoting hormones, siderophores, HCN and catalase, and can solubilize phosphorus (Kraus and Loper, 1995; Banasco, 1998). *Pseudomonas fluorescens*, a plant growth-promoting rhizobacterium, is an efficient biocontrol agent in rice grown in saline soils of coastal ecosystems (Paul *et al.*, 2006). Cold-tolerant fluorescent *Pseudomonas* isolated from Garhwal Himalayas act as potent plant growth-promoting and biocontrol agents in pea (Negi *et al.*, 2005). *Pseudomonas fluorescens* also produces 2, 4-diacetyl phloroglucinol, which inhibits growth of phytopathogenic fungi (Nowak-Thompson *et al.*, 1994). *Pseudomonas stutzeri* produces extracellular chitinase and laminarinase, which lyses the mycelia of *Fusarium solani* (Mauch *et al.*, 1988). Strong antagonism is shown by one of the isolates of a fluorescent *Pseudomonas* sp. EM85 against *Rhizoctonia solani*, a causal agent of damping-off of cotton (Pal *et al.*, 2000). *Pseudomonas putida* has shown antifungal activity against *Macrophomina phaseolina*, hence proving a potential for the biocontrol of root-rot disease complex of chickpea. It has also been shown that anaerobic regulator (ANR)-mediated cyanogenesis contributes to the suppression of black root rot (Laville *et al.*, 1998).

Bacillus subtilis is also used as a biocontrol agent. This ubiquitous inhabitant of soil is widely recognized as a powerful biocontrol agent. In addition, due to its various properties such as broad host range, its ability to form endospores and produce different biologically active compounds with a broad spectrum of activity, *B. subtilis* as well as other *Bacilli* are potentially useful as biocontrol agents (Nagórska *et al.*, 2007). *Bacillus megaterium* from the tea rhizosphere is able to solubilize phosphate, produce IAA, siderophore and antifungal metabolites and thus it helps in the plant growth promotion and reduction of disease intensity (Chakraborty *et al.*, 2006). *Bacillus thuringiensis* and *Bacillus sphaericus* are capable of solubilizing inorganic phosphates and help in the control of lepidopteron pests (Seshadri *et al.*, 2007). The biocontrol activity of *Paenibacillus lentimorbus* NRRL B-30488, which was isolated from milk, against *Fusarium oxysporum* is reported by alteration and distortion of the hyphal cell wall through the action of its chitinase and β-1,3-glucanase enzymes (Das Gupta *et al.*, 2006). *Bacillus amyloliquefaciens* FZB42 produces three families of lipopeptides, surfactins, bacillomycin D and fengycins, which are well known secondary metabolites with antifungal activity (Koumoutsi *et al.*, 2004).

Arbuscular mycorrhizal fungi are ubiquitous in nature and constitute an integral component of terrestrial ecosystems, forming symbiotic associations with plant root systems of over 80% of all terrestrial plant species, including many agronomically important species. AM fungi play an important role in organic and/or sustainable farming systems that rely on biological processes rather than agrochemicals to control plant pathogens. They are particularly important in conferring bioprotection to plants against many soil-borne pathogens such as species of *Aphanomyces*, *Cylindrocladium*, *Fusarium*, *Macrophomina*, *Phytophthora*, *Pythium*, *Rhizoctonia*, *Sclerotinium*, *Verticillium* and *Thielaviopsis* and various nematodes (Harrier and Waston, 2004). AM fungi are also known to enhance plant uptake of other mineral nutrients (Bodker *et al.*, 1998).

Trichoderma spp. are widely used in biocontrol of plant pathogens and, at the same time, are recognized as plant growth promoters; an important role of auxin signalling for plant growth promotion has also been shown for *Trichoderma* spp. (Contreras-Cornejo *et al.*, 2009). There are several model organisms like *Trichoderma* that have been well-studied in their mode of action and regulation for plant-growth promotion (Harman *et al.*, 2004; Keswani *et al.*, 2014; Bisen *et al.*, 2015). *Trichoderma harzianum* strain T-22 is particularly important for agronomic purposes, as it is able to colonize the roots of most plant species across a wide range of soil types, with beneficial effects for plant growth and disease resistance (Sofo *et al.*, 2004; Tucci *et al.*, 2011). Furthermore, it has been successfully used for the biological control of many plant pathogens through chemiotropic mycoparasitic interactions with fungal or bacterial organisms as a target (Sofo *et al.*, 2004; Vitale *et al.*, 2012).

7.5 Recent Patents/Applications for Agriculturally Important Microorganisms

Patents provide rights to applicants to exclude others from using their invention. Patent data also form a critical source of information on the subject/technological area. Almost 80% of the information found in the patent literature is not found anywhere else. Guidelines for preparing patent landscape reports are given in WIPO (2015).

In the present study, the terms 'biofertilizer' or 'biopesticide' or 'bioinoculant' were used in the title, claim section of patent documents and the Orbit database (https://www.orbit.com). Since the study was focused on patents/applications in India, search term 'IN' (the code for India) was added to the publication country field. With these parameters, a total of 110 results were obtained when accessed on 17 September 2015 (Table 7.1).

These 110 patents applications are filed in India and these may be filed in other countries as well. Of these, 43 are filed in the Patent Co-operation Treaty (PCT), 34 in the USA, 30 in the EU and 25 in Japan. Selection of country for filing is largely dependent on the assessment of market, status of innovation and its adaptability in that field in the country. Budget is also a key consideration, as patenting activity involves expenditure (Fig. 7.1). In the PCT route an applicant can also later file a patent application to other countries that are member countries of the treaty. Filing trends of these 110 patents/application shows that developed countries have a higher interest in microbial inoculant-based inventions.

With reference to patent, the assignee field indicates who owns the invention and gives an idea about major players in the research area. In the

Table 7.1. List of patents/applications for terms 'biofertilizer' or 'biopesticide' or 'bioinoculant'.

Serial no.	Patent/publication no.	Patent/application title
1	IN2126/DELNP/2013	Lyophilized biopesticide effervescent granule and production method thereof
2	IN0277/DEL/2008	Novel biopesticide compositions and method for isolation and characterization of same
3	IN2765/DEL/2009	Novel mycorrhizae-based biofertilizer compositions and method for mass production and formulation of same
4	IN0923/DEL/2009	Fungal bio-agent and VAM encapsulated alginate beads and its application thereof as biocontrol agent and biofertilizer soil borne plant pathogens
5	IN1756/DEL/2011	A natural-source based biopesticide formulation for effective biocontrol and the procedure for synthesizing the same
6	N4055/DEL/2012	Novel Mycorrhizae-based biofertilizer consortium and ROC-mediated production of same
7	IN1328/KOL/2013	Method of improving elemental and nutritional content of plant seeds using *Bacillus* strain mcc0008 as a biofertilizer
8	IN0168/MUM/2008	Formulation of entomopathogenic fungus for use as a biopesticide
9	IN0325/CHE/2015	A nano-biofertilizer for improving seed germination and plant development
10	IN2369/DEL/2004	Novel weed and industrial waste based biofertilizer
11	IN0444/DEL/2002	A process for the production of blue green algal biofertilizer
12	IN0559/DEL/2009	Biopesticide formulation comprising of fungal bio-agents, botanical antagonists, peat powder and acacia gum for the control of plant pathogens and the method for the preparation
13	IN0227/DELNP/2004	New biopesticide using gene from *Erwinia pyrifoliae* wt 3, novel pathogen that affects Asian pear trees
14	IN0708/DEL/2009	Broad spectrum biopesticide formulation with fungal antagonist and entemopathogenic fungi
15	IN0251/CHE/2011	Biopesticide from weeds
16	IN189739	A method for the preparation of a biopesticide from the roots of *Decalpis hamiltonii*
17	IN0518/KOL/2011	Nitrate reducing microbial consortium as biofertilizer for plant growth promotion
18	IN0834/DEL/2009	Soil health rejuvenator and organic matter enhancer
19	IN0766/DEL/2009	Tapioca starch and soya dust base co-formulation for growth enhancement and carrier base for biofertilizer and biopesticides
20	IN1533/DEL/2004	A process for the preparation of a biopesticide
21	IN0857/DEL/1999	A biofertilizer and a process for the preparation thereof
22	IN1244/DEL/2002	The process of preparing a biopesticide formulation for use against coffee berry borer (cbb)
23	IN4054/DEL/2012	Novel root organ culture (roc) medicated ectomycorrhizae-based bioinoculums and method for production of same
24	IN1019/DEL/2007	A process for the production of organic formulation of biopesticide *Pseudomonas fluorescens*
25	IN1507/DEL/2009	A biopesticide formulation for controlling insect pests and fungal pathogens and process for preparation thereof
26	IN1534/DEL/2004	A biopesticide based on *Eucalyptus* plant extract
27	IN1241/DEL/2011	A novel base material for commercial production of fungal biopesticide and biofertilizer
28	IN0384/MUM/2002	Process for treatment of organic solid waste
29	IN1480/DEL/2004	A synergistic bioinoculant composition comprising bacterial strains of accession nos, nrrl b-30486, nrrl b-30487, and nrrl b-30488 and a method of producing said composition thereof
30	IN1263/DEL/2007	Formulation of a biofertilizer for colder regions
31	IN2991/DELNP/2006	Novel strain of *Bacillus* as a bioinoculant
32	IN0457/DEL/1998	A process for the storage of a VAM (vescicular arbuscular mycorrhiza) biofertilizer
33	IN0010/DEL/2012	A novel recombinant strain of *Trichoderma* useful for enhancing nutritional value and growth of plants
34	IN1451/CHE/2007	Bio capsule for influencing rhizosphere of plant system for better crop growth and agricultural productivity

Continued

Table 7.1. Continued.

Serial no.	Patent/publication no.	Patent/application title
35	IN2027/MUMNP/2010	Amino acid and nucleic acid sequences for controlling pathogens
36	IN1710/DEL/2009	A novel formulation of the plant growth promoting rhizobacteria with enhanced shelf-life and the method of its preparation
37	IN1532/DEL/2004	Process for the preparation of a biopesticide
38	IN1060/DEL/2013	Technology for native plant growth promoting bacterial (PGPB) consortium formulations, useful as biofertilizer cum biopesticide
39	IN1746/DEL/2006	Polymeric seed coats based on bioactive botanicals
40	IN0518/DEL/2011	Liquid bioinoculant of *Azotobacter chroococcum* and the process thereof
41	IN0437/CHE/2003	Plant growth promoter containing bio-extract
42	IN0437/CHENP/2003	Method for measuring the level of a medium in a reactor
43	IN0507/MUMNP/2007	Compositions for use against one or more pathogens
44	IN3456/DEL/2005	Biofertilizer cum biofungicide/biobactericide composition
45	IN0580/DEL/2009	Microbial formulation for widespread used in agricultural practices
46	IN0519/KOL/2011	A gyro mixer machine with external rear pulley and main-frame of steel construction
47	IN3072/DEL/2010	Insecticidal compounds from *Nothapodites foetida* and process for the extraction thereof
48	IN1839/DEL/2009	A process for integrated production of ethanol and seawood sap from *Kappaphycus alvarezii*
49	IN2004/DEL/2010	A novel cyst based liquid biofertilizer of azospirillum brasilense and method for producing thereof
50	IN1355/KOL/1996	A novel biopesticide and growth promoter concentration and its preparation
51	IN1179/KOL/2013	Microbial consortium for nitrate and phosphate sequestration for environmental sustenance
52	IN1755/DEL/2007	Process for commercial production of *Trichoderma* biopesticides using industrial tobacco waste
53	IN1971/KOLNP/2006	Proteins inducing multiple resistance of plants to phytopathogens and pests
54	IN0700/KOLNP/2015	Phosphate solubilizing rhizobacteria *Bacillus firmus* as biofertilizer to increase canola yield
55	IN1353/CHE/2004	A biopesticide based on nucleopolyhedro virus for the management of teak pest *Hyblaea puera*
56	IN3169/DEL/2010	A process of preparing a liquid biofertilizer formulation
57	IN1610/KOLNP/2013	Synergistic activity of peracetic acid and at least one sar inducer for the control of pathogens in and onto growing plants
58	IN3603/CHE/2010	Invert-emulsion formulation of a fungal antagonist for biological management of plant diseases
59	IN0031/DEL/2000	A process for the production of biogas using ligno-cellulosic fibrous waste
60	IN0032/DEL/2000	An improved process for the simultaneous production of biogas and fertillizer by high rate biomethanation of palm oil mill effluent
61	IN2533/DEL/2011	An innovative procedure for storing and packing of mycorrhizal based biofertilizers
62	IN1104/DEL/2007	Multi-utility herbal formulation for farming
63	IN2730/DEL/2009	A bioinoculant consortium for enhancing the active content of medicinal plant and its method thereof
64	IN1294/MAS/1996	Insecticidal matrix and process for preparation thereof
65	IN2641/CHE/2013	Mass multiplication technique for pink-pigmented facultative methylotrophs (ppfm)
66	IN0568/CHE/2012	Production process for improved yield of *Trichoderma* biomass
67	IN1547/DEL/2010	A novel formulation of stress tolerant liquid biofertilizer of phosphate-solubilizing bacteria
68	IN0469/CHE/2007	Coir pith based cyanobacterial biofertilizer using *Phormidium* sp. bdu5
69	IN0437/MAS/2003	Plant growth promoter containing bio-extract
70	IN0657/CHE/2006	A phytopesticidal formulation 'ponneem' for controlling insect pests
71	IN1126/DEL/2003	Efficient process for the preparation of neem based reduced azadirachtin(s) pesticides
72	IN188170	A process for preparing a biopesticide to resist termites
73	IN179400	A process for producing an entamophagas *Beauveria bassiana* biopesticide

Continued

Table 7.1. Continued.

Serial no.	Patent/publication no.	Patent/application title
74	IN180517	A process for the preparation of a new biopesticide formulation for the effective control of pests particularly teak defoliator and Epilachna beetle
75	IN0684/DEL/2010	Integrated process for the production of oil bearing *Chlorella variabilis* for lipid extraction utilizing by-products of Jatropha methyl ester (jme) production
76	IN256686	Phthalamide derivative, agricultural or horticultural pesticide, and use of the pesticide
77	IN0210/DEL/2007	Development of quality control kit for rhizobial liquid inoculants
78	IN2862/CHE/2013	Enzyme based formulation of chitinase from entomopathogenic fungi
79	IN3171/CHE/2012	Hydrophobic biopesticide from watermelon seed extract
80	IN3490/CHE/2010	Development of novel insecticidal wettable powder formulations of *Heterorhabditis indica* strain nbaii hi1 and *Heterorhabditis bacteriophora* strain nbaii hb5 for the biological control of white grubs and other insect pests, and the methods thereof for their preparation and use
81	IN0450/DEL/2009	A novel formulation of biocide for controlling mite of coconut
82	IN6075/DELNP/2008	High organic matter products and related systems for restoring organic matter and nutrients in soil
83	IN2962/MUM/2012	Mosquito control composition and method of using the same
84	IN0465/DEL/2012	Recombinant nitrogen fixing microorganism and uses thereof
85	IN0925/CHE/2010	Go green
86	IN3911/DEL/2012	A novel process and apparatus for eco-friendly treatment of ligninous and other hard-to-biodegrade biowaste
87	IN0802/CHE/2011	Aqueous liquid inoculant formulation of nitrogen fixing bacterial consortium
88	IN7792/DELNP/2011	Pesticidal compositions
89	IN4023/CHE/2012	A chitinase from *Brevibacillus laterosporus*, its production and use thereof
90	IN0998/DEL/2006	A process for extraction of bioactive psoralen compound
91	IN2150/DEL/2004	Light, heat and water resistant neem meliacin concentrates and products with controlled release
92	IN4088/DELNP/2014	Microbial inoculants and fertilizer compositions comprising the same
93	IN2475/DELNP/2012	*Fusarium* and fusarium mycotoxin biocontrol
94	IN7454/CHENP/2009	Stable organic-carrier-based microbial inoculants and method for producing the same
95	IN2187/MUM/2009	System and method for biological treatment of biodegradable waste including biodegradable municipal solid waste
96	IN1509/MUMNP/2002	*Streptomyces galbus* strain with insecticidal activity and method of using as an insecticide
97	IN0574/DELNP/2015	Safened herbicidal compositions including 4-amino-3-chloro-5-fluoro-6- (4-chloro-2-fluoro-3-methoxyphenyl)pyridine-2-carboxylic acid
98	IN4580/DELNP/2014	Plant growth-promoting microbes and uses thereof
99	IN1199/MUMNP/2012	Method for treating phytopathogenic microorganisms using surface-modified nanoparticulate copper salts
100	IN1867/DELNP/2012	N-(cyanophenyl) pyrazolecarboxamide aqueous formulation
101	IN9531/CHENP/2011	Use of anaerobic digestion to destroy biohazards and to enhance biogas production
102	IN0342/MUM/2010	Composition and method of preparation of bio fungicide based on *Trichoderma harzianum* for controlling soil borne diseases
103	IN0413/CHENP/2010	Insecticidal barrier with a durable lower part
104	IN0416/CHENP/2010	Insecticidal barrier partly with synergist
105	IN6947/DELNP/2008	Method for disrupting reproductive performance of arthropods
106	IN5700/DELNP/2008	Mixture of indoxacarb and dimehypo
107	IN1536/DELNP/2009	Composition based on *Bacillus* spp. and correlate genera and their use in pest control
108	IN2925/DELNP/2004	A synergistic fermented plant growth promoting, bio-control composition
109	IN1549/DEL/2003	Biopesticidal compositions
110	IN1322/KOLNP/2002	Regulatory element from a sugarcane proline rich protein and uses thereof

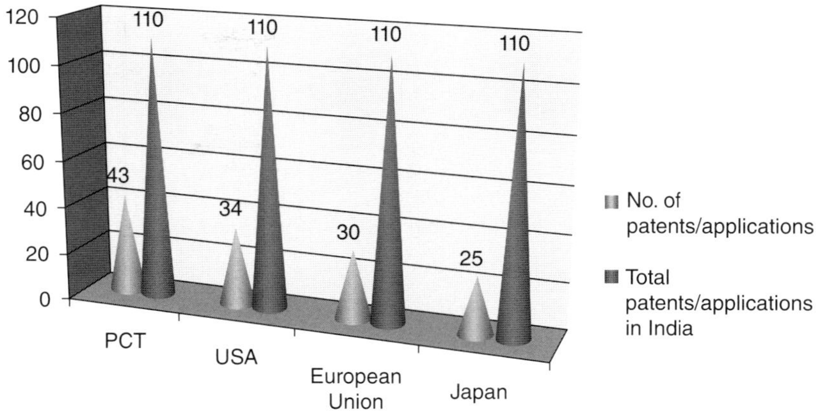

Fig. 7.1. Patents/application, for terms 'biofertilizer' or 'biopesticide' or 'bioinoculant'.

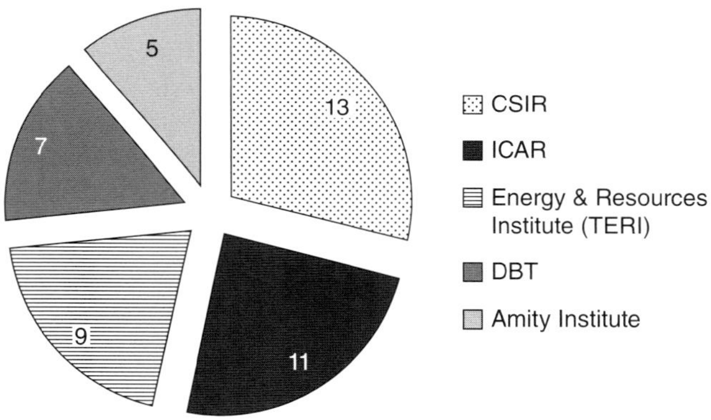

Fig. 7.2. Patents/application assignee for terms 'biofertilizer' or 'biopesticide' or 'bioinoculant'.

present study, the Council of Scientific and Industrial Research (CSIR) has 15 patents as assignee, the Indian Council of Agricultural Research (ICAR) has 11 patents; the Energy & Resources Institute (TERI) is placed third with nine patent applications and the Department of Biotechnology (DBT) and Amity University have seven and five applications, respectively. The remainder of the assignees have fewer applications, mostly one with each. From the analysis of names of assignees, it is evident that in India most of the patents and hence R&D in this area is being done by public-funded research institutions. CSIR, ICAR and DBT account for one-third of the total patent applications. TERI is also a non-profit organization. This shows that private R&D has not focused much in the area of bio-inoculants and most of the research is supported by government (Fig. 7.2). For better research and inventions, both the sectors, private and government, should invest more capital and human resources in the R&D of microbial bio-inoculants.

The filing trend in the application shows that for the period from 2010 onwards 29 applications have been filed. However, prior to this, from 2006 to 2010, the number of patent applications filed was much higher, i.e. 46. From 2001 to 2005 the number of patents filed was 21, in the period 1996 to 2000 there were only ten. This is also an indication that there is need for investment by different players to boost R&D, particularly in India, otherwise there is likely to be a declining trend in the number of patent applications filed (Fig. 7.3).

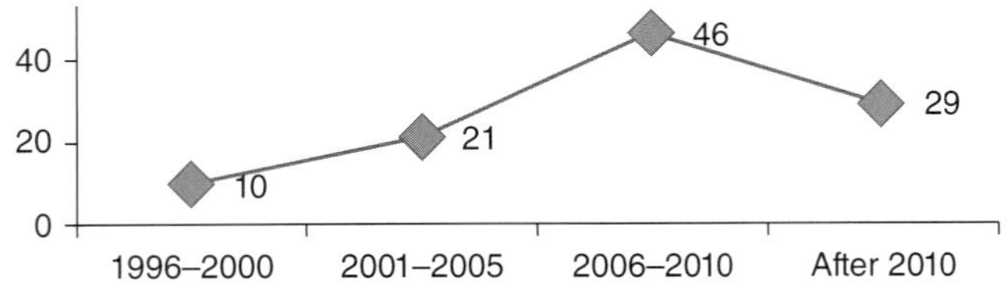

Fig. 7.3. Pattern for patents/application field for terms 'biofertilizer' or 'biopesticide' or 'bioinoculant'.

Out of 110 records the highest number of records are for patents/applications of microbes as biopesticides, followed by biofertilizers and bio-inoculants. Some of the hits have also resulted in records that are not directly related with any of the above mentioned categories, e.g. 'IN0413/CHENP/2010 Insecticidal Barrier with a durable lower part', which is an invention for a net with a mesh size preventing insects from entering.

Microbes have been used for making formulations with the objective of biofertilizer for growth and biopesticides for controlling pathogens. Commonly used microbes mentioned in patents are *Pseudomonas*, *Escherichia coli*, *Nitrobacter*, *Nitrosomonas*, *Bacillus* sp., *Trichoderma*, *Azotobacter*, *Serratia*, *Azospirillum*, *Strentophomonas*, blue-green algae, *Phormidium*, *Beauveria bassiana*, vesicular arbuscular mycorrihza (VAM), *Rhizobia*, virus-like Nucleopolyhedrovirus, *Rhizobacteria*, *Lactobacillus*, *Sphaerodes*, *Mycoparasitica*, *Streptomyces*, *Photorhabdus luminescens* and *Hebrospirillum*. In addition to microbes, higher plant parts or extracts of them have also been used for making biopesticides, among which are *Eucalyptus* leaves, neem, *Callistemon*, *Chenopodium*, *Argimone*, *Lantana*, *Phyllanthus*, *Embellica*, *Allium* and roots of *Decalpis*.

7.6 Patents/Application from CSIR-National Botanical Research Institute

One of the major R&D activities of the CSIR-National Botanical Research Institute, Lucknow (CSIR-NBRI), has been to develop (abiotic and biotic) stress-tolerant bio-inoculants for diverse soil and climatic conditions. These formulations are useful as a plant growth-enhancer for seed, soil and foliar applications and improve the soil health, crop yield and quality. Applications of the products have increased the yield of several economically important crops. A simple sand–live soil assay method for large-scale screening of the rhizosphere-competent bacteria, effective in suppressing plant pathogens, has been developed. Novel methods have been developed and patented, which enable fast screening of a large number of soil samples to identify and isolate useful rhizosphere-competent plant growth-promoting microbes. Screening for chickpea rhizosphere competitive bacteria having biological control properties was conducted at three different stages: (i) development of a screening method for large-scale initial selection of bacteria isolates from chickpea rhizosphere; (ii) testing of biocontrol activity under *in vitro* conditions; and (iii) screening of antibiotic-resistant mutants for rhizosphere competence in non-sterile field soil, which has disclosed *Pseudomonas fluorescens* NBRI 1303 (ATCC 55939) to be effective in suppressing plant pathogens, including *Fusarium oxysporum* f. sp. *ciceri*, *Rhizoctonia bataticola* and *Pythium* sp. in chickpea. The purified bacterial strain can be used as active agent for biocontrol compositions and can also be used for enhancement of chickpea plant growth and yield, as well as for the production of antibiotics directed towards phytopathogenic fungal diseases.[1] A composition useful for qualitative screening of phosphate solubilizing microorganisms has been developed, which was further improved as a synergistic composition for qualitative screening of phosphate-solubilizing microorganisms for its utilization for visual observation-based characterization.[2] Another synergistic composition useful as a microbiological growth medium and a high-throughput screening method has been developed for screening of phosphate-accumulating microorganisms.[3] A synergistic composition useful as bio-inoculant comprises bacterial strains of accession numbers NRRL B-30486, NRRL B-30487 and NRRL B-30488, individually or in all possible combinations, and optionally a carrier, with each of the strains showing plant promoter activity, phytopathogenic fungi-controlling activity, abiotic stress conditions tolerating capability and phosphate solubilization capability under abiotic stress conditions; further, a method of producing the composition and, in addition, a method of isolating said bacterial strains has been developed from milk of the cow 'Sahiwal'.[4] Another synergistic composition proved useful as plant and soil health enhancer, comprising bovine urine, neem and garlic, individually or in all possible combinations, with the treatment showing it has the ability to stimulate accumulation of nutrients in the plant biomass, proliferation of plant growth-promoting, phosphate solubilizing, abiotic stress tolerant and antagonists towards plant pathogenic fungi, control of phytopathogenic fungi in the rhizosphere of plants, and enhances the total phenolic contents of the plants.[5] A cost-effective method of producing high density *Trichoderma*-based formulation has been developed by preparing a talc-based fungal biocontrol preparation, which is 100 times as concentrated as compared to conventional products. The production procedure provides a check over quality at every step and ensures a longer shelf life. In the present process a commercially viable, cost effective and efficient technology has been presented for the preparation of biofungicide comprising *Trichoderma*.[6] A bio-inoculant composition comprising *Trichoderma* protoplast fusant relates to a novel recombinant fungal strain of *Trichoderma* sp. MTCC 5659 useful for enhancing the nutritional value and growth of plants. The invention further relates to a formulation useful as bio-inoculant, wherein the said formulation comprises MTCC 5659 optionally along with a carrier. The claimed strain has been developed via the protoplast fusion technique of two parent *Trichoderma* strains and is useful for stimulating the content of amino acids, trace elements, chlorophyll and plant growth and yield-attributing characters.[7]

7.7 Conclusion

Environmental stresses are becoming a major problem and productivity is declining at an unprecedented rate. Our dependence on chemical fertilizers and pesticides has encouraged the thriving of industries that are producing life-threatening chemicals and that are not only hazardous for human consumption but can also disturb the ecological balance. Biofertilizers can help to solve the problem of feeding an increasing global population at a time when agriculture is facing various environmental stresses. It is important to realize the useful aspects of biofertilizers and implement their application to modern agricultural practices. The new technology developed using the powerful tool of molecular biotechnology can enhance the biological pathways of production of phytohormones. If identified and transferred to the useful PGPR, these technologies can help provide relief from environmental stresses. However, the lack of awareness regarding improved protocols of biofertilizer applications to the field is one of the few reasons why many useful PGPR are still beyond the knowledge of ecologists and agriculturists. Nevertheless, the recent progress in technologies related to microbial science, plant–pathogen interactions and genomics will help to optimize the required protocols. The success of the science related to biofertilizers depends on inventions of innovative strategies related to the functions of PGPR and their proper application to the field of agriculture. The major challenges in this area of research lie with the identification of various strains of PGPR and it is essential to dissect the actual mechanism of their functioning for efficacy towards exploitation in sustainable agriculture.

Both biopesticides and biofertilizers require intensive research in order to achieve food security for the growing population and restore soil fertility. Nature has provided countless avenues for research in these fields, which need to be explored. The development of new biopesticides with multiple mode of action against pests and biofertilizers with multi-crop growth-promoting activities are most important for sustainable global agriculture. These two needs are to be prioritized in agricultural research by universities, research organizations and R&D wings of manufacturers for technology development to the farming community. The technologies so developed need to be transferred worldwide to achieve maximum benefits to society. At present, the patent scenario of biofertilizer/biopesticide is not very encouraging. Most of the patents in India have originated from public-funded organizations and the number of applications is also very low. There is need for investment from private organizations for giving a boost to the technology area. Further research will lead to newer combinations of products which are more efficacious for improving crop productivity without compromising quality of soil and environment.

7.8 Future Challenges

Microbial inoculants, which can fulfil diverse functions in plants, lead to promising solutions for a sustainable, environmentally friendly agriculture. While inoculants for plant growth promotion and biocontrol already exist, in the future, stress-protecting agents (stress conditions like those generated by salinity, drought, waterlogging, heavy metals and pathogenicity) will be of emerging importance not only due to climate change. Furthermore, to improve food quality by microbials is an important task. For all kinds of inoculants, microbial mixtures as multitasking inoculants are one alternative to overcome inconsistent *in vivo* effects. Using molecular technologies and exploiting new bioresources, new efficient BCAs can be detected. For example, endophytes with a specific intimate interaction are promising bioresources (Troufflard *et al*., 2010). In addition, naturally occurring antagonists on non-crop plants from natural vegetation are of interest. Initial investigations show high potential for microorganisms associated with mosses or with the natural vegetation of Tenerife (Amtmann *et al*., 2006). Another interesting point is the molecular analysis of the mode of interaction with the host plant as well as with pathogens, including its regulation, e.g. by '-omics' technologies. These results help to stabilize the effect or can result in new targets for the inoculants. In addition, it is necessary to study the ecological behaviour of microbes and the interaction with other eukaryotes. The possibility of such interactions must be considered in risk assessment studies before commercialization. Extensive research is also required to improve the scale-up and bioprocess development for microbial inoculants. Further optimization is required for fermentation and formulation processes. The lack of such improved protocols is among the reasons why many promising microbes, especially gram-negatives, are still not included in practical approaches. The recent progress achieved with genomic technologies will certainly help to optimize these processes. Beyond these technical questions, it is also necessary to revise the legislative processes to efficiently commercialize microbial inoculants. In this context, new and standardized methods have to be established for appropriate risk evaluation of microbial inoculants. Clearly, the future success of the industry producing microbial inoculants will depend on innovative business management, product marketing, extension education and progress in research. The field of plant growth will contribute substantially to making the 21st century the age of biotechnology by the development of innovative biological strategies in agriculture.

Notes

1 Nautiyal, C.S., 17 December 2002. *A biologically pure culture of bacteria which suppresses diseases caused by pathogens in chickpea crops and a culture of bacteria comprising a strain of* Pseudomonas fluorescens. US6495362 B1.

2 Nautiyal, C.S., Mehta, S. and Pushpangadan, P., 28 October 2003. *Composition for qualitative screening of phosphate solubilizing microorganisms and a qualitative method for screening microorganisms*. US6638730 B2; 19 December 2008. *A process for preparation of a synergistic composition useful as microbiological growth medium for screening of phosphate solubilising micro-organism*. IN 226577.

3 Nautiyal, C.S. and Chaudhry, V., 2 February 2012. *Synergistic composition useful as microbiological growth medium for rapid screening of phosphate accumulating microorganisms*. US20120028295 A1.

4 Nautiyal, C.S., Mehta, S., Singh, H.B. and Pushpangadan, P., 2003–2009. *Method for screening useful stress tolerant bacteria from cow and application thereof for promoting plant growth*. 2009, TT TT/T/2009/00056; 2008, AU 2002345299; 2006, AT 1423011; 2006, DE 1423011; 2006, DK 1423011; 2006, EP 1423011; 2006, ES 1423011; 2006, FI 1423011; 2006, FR 1423011; 2006, GB 1423011; 2006, IT 1423011; 2006, NL 1423011; 2006, SE 1423011; 2006, US 7097830; 2004, ZA 2003/2288; 2004, ZW 11/2003; 2003, CN 1479577.

5 Nautiyal, C.S., Mehta, S., Singh, H.B., Mansinghka, S.B., Dawle, S.H., Rajhans, N.E. and Pushpangadhan, P. 2003–2012. *Plant and soil health enhancer using cow urine and application thereof for promoting plant growth and controlling plant pathogenic fungi*. 2012, CA 2521037; 2010, AU 2004226117; 2009, IN 227512; 2009, JP 4422144; 2008, ZA 2005/07887; 2007, US 7297659; 2006, DE 1608606; 2006, EP 1608606; 2006, FR 1608606; 2006, GB 1608606; 2006, IT 1608606; 2006, NL 1608606; 2003, ZW 29/05.

6 Singh, P.C. and Nautiyal, C.S., 2011. *A cost effective method of producing high density* Trichoderma *based formulation*. 0070NF2011/IN.

7 Mishra, A. and Nautiyal, C.S., 2011–2013. *A novel recombinant strain of* Trichoderma *useful for enhancing nutritional value and growth of plants*. 2015, EP2658961B1; 2015, US9068189; 2014, US201403087482013; 2013, PCT/IN2012/000863; 2013, CA2825670A1; 2013, EP2658961A1.

Acknowledgement

The study was partially supported by New Initiative (as a Cross Flow Technology project) 'Root Biology and its Correlation to Sustainable Plant Development and Soil Fertility' from the Council of Scientific and Industrial Research (CSIR), New Delhi, India.

References

Abril, A., Zurdo-Pineiro, J.L., Peix, A., Rivas, R. and Velazquez, E. (2007) Solubilization of phosphate by a strain of *Rhizobium leguminosarum* bv. *trifolii* isolated from *Phaseolus vulgaris* in El Chaco Arido soil (Argentina). In: Velazquez, E. and Rodriguez-Berrueco, C. (eds) *Developments in Plant and Soil Sciences*. Springer, Dordrecht, the Netherlands, pp. 135–138.

Adesemoye, A.O., Torbert, H.A. and Kloepper, J.W. (2008) Enhanced plant nutrient use efficiency with PGPR and AMF in an integrated nutrient management system. *Canadian Journal of Microbiology* 54, 876–886.

Adesemoye, A., Torbert, H. and Kloepper, J. (2009) Plant growth-promoting Rhizobacteria allow reduced application rates of chemical fertilizers. *Microbial Ecology* 58(Suppl. 4), 921–929.

Akhtar, M.S. and Siddiqui, Z.A. (2009) Use of plant growth-promoting rhizobacteria for the biocontrol of root-rot disease complex of chickpea. *Australian Plant Pathology* 38(Suppl. 1), 44–50.

Alagawadi, A.R. and Gaur, A.C. (1988) Associative effect of *Rhizobium* and phosphate solubilizing bacteria on the yield and nutrient uptake of chickpea. *Plant Soil* 105, 241–246.

Alikhani, H.A., Saleh-Rastin, N. and Antoun, H. (2006) Phosphate solubilization activity of rhizobia native to Iranian soils. *Plant and Soil* 287(Suppl. 1–2), 35–41.

Ambrosini, A., Beneduzi, A., Stefanski, T., Pinheiro, F., Vargas, L. and Passaglia, L. (2012) Screening of plant growth promoting Rhizobacteria isolated from sunflower *Helianthus annuus* L. *Plant & Soil* 356, 245–264.

Amtmann, A., Hammond, J.P., Armengaud, P. and White, P.J. (2006) Nutrient sensing and signaling in plants: potassium and phosphorus. *Advances in Botanical Research* 43, 209–257.

Amtmann, A., Troufflard, S. and Armengaud, P. (2008) The effect of potassium nutrition on pest and disease resistance in plants. *Physiologia Plantarum* 133, 682–691.

Antoun, H. and Kloepper, J.W. (2001) Plant growth promoting rhizobacteria (PGPR). In: Brenner, S. and Miller, J.H. (eds) *Encyclopedia of Genetics*. Academic Press, New York, pp. 1477–1480.

Argelis, D.T., Gonzala, D.A., Vizcaino, C. and Gartia, M.T. (1993) Biochemical mechanism of stone alteration carried out by filamentous fungi living in monuments. *Biogeochemistry* 19, 129–147.

Armengaud, P., Breitling, R. and Amtmann, A. (2010) Coronatine-intensive 1 (COII) mediates transcriptional responses of *Arabidopsis thaliana* to external potassium supply. *Molecular Plant* 3(2), 390–405.

Ashbolt, N.J. and Inkerman, P.A. (1990) Acetic acid bacterial biota of the pink sugarcane mealy bug, *Saccharococcus sacchari*, and its environs. *Applied and Environmental Microbiology* 56(Suppl. 3), 707–712.

Ashrafuzzaman, M., Hossen, F.A., Ismail, M.R., Hoque, M.A., Islam, M.Z., Shahidullah, S.M. and Meon, S. (2009) Efficiency of plant growth promoting Rhizobacteria (PGPR) for the enhancement of rice growth. *African Journal of Biotechnology* 8(Suppl. 7), 1247–1252.

Banasco, P. (1998) Fluorescent *Pseudomonas* spp. as biocontrol agents against forage legume root pathogenic fungi. *Soil Biology and Biochemistry* 10(Suppl. 10–11), 1317–1323.

Bardi, L. and Malusà, E. (2012) Drought and nutritional stresses in plant: alleviating role of rhizospheric microorganisms. In: Haryana, N. and Punj, S. (eds) *Abiotic Stress: New Research*. Nova Science Publishers, Hauppauge, New York, pp. 1–57.

Barea, J.M., Azcon, R. and Azcon-Aguilar, C. (2002) Mycorrhizosphere interactions to improve plant fitness and soil quality. *Antonie Van Leeuwenhoek* 81, 343–351.

Barraquio, W.L., Segubre, E.M., Gonzalez, M.S., Verma, S.C., James, E.K., Ladha, J.K. and Tripathi, A.K. (2000) Diazotrophic enterobacteria: what is their role in the rhizosphere? In: Ladha, J.K. and Reddy, P.M. (eds) *The Quest for Nitrogen Fixation in Rice*. IRRI, Manila, pp. 93–118.

Barriuso, J. and Solano, B.R. (2008) Ecology, genetic diversity and screening strategies of plant growth promoting rhizobacteria (PGPR). *Journal of Plant Nutrition* 1–17.

Barriuso, J., Solano, B.R., Lucas, J.A., Lobo, A.P., Villaraco, A.G. and Mañero, F.J.G. (2008) Ecology, genetic diversity and screening strategies of plant growth promoting rhizobacteria (PGPR). In: Ahmad, I., Pichtel, J. and Hayat, S. (eds) *Plant-Bacteria Interactions: Strategies and Techniques to Promote Plant Growth.* Wiley, Weinheim, Germany, pp. 1–17.

Basak, B.B. and Biswas, D.R. (2009) Influence of potassium solubilizing microorganism (*Bacillus mucilaginosus*) and waste mica on potassium uptake dynamics by sudangrass (*Sorghum vulgare* Pers.) grown under two Alfisols. *Plant Soil* 317, 235–255.

Bashan, Y. and Holguin, G. (1997) *Azospirillum*–plant relations: environmental and physiological advances (1990–1996). *Canadian Journal of Microbiology* 43, 103–121.

Bashan, Y., Holguin, G. and de-Bashan, L.E. (2004) *Azospirillum*-plant relationships: physiological, molecular, agricultural, and environmental advances (1997–2003). *Canadian Journal of Microbiology* 50(Suppl. 8), 521–577.

Bashan, Y., de-Bashan, L.E., Prabhu, S.R. and Hernandez, J.P. (2014) Advances in plant growth-promoting bacterial inoculant technology: formulations and practical perspectives (1998–2013). *Plant Soil* 378, 1–33.

Berman-Frank, I., Lundgren, P. and Falkowski, P. (2003) Nitrogen fixation and photosynthetic oxygen evolution in Cyanobacteria. *Research in Microbiology* 154, 157–164.

Bhattacharyya, P.N. and Jha, D.K. (2012) Plant growth-promoting rhizobacteria (PGPR): emergence in agriculture. *World Journal of Microbiology and Biotechnology* 28, 1327–1350.

Bisen, K., Keswani, C., Mishra, S., Saxena, A., Rakshit, A. and Singh, H.B. (2015) Unrealized potential of seed biopriming for versatile agriculture. In: Rakshit, A., Singh, H.B. and Sen, A. (eds) *Nutrient Use Efficiency: from Basics to Advances*. Springer, New Delhi, India, pp. 193–206.

Biswas, B.C., Tewatia, R.C., Prasad, N. and Das, S. (1994) *Biofertilizers in Indian Agriculture*. Fertilizer Association of India, New Delhi, pp. 1–43.

Bodker, L., Kjoller, R. and Rosendahl, S. (1998) Effect of phosphate and Arbuscular mycorrizhal fungus *Glomus intraradices* on disease severity of root rot of peas (*Pisum sativum*) caused by *Aphanomyces euteiches*. *Mycorrhiza* 8, 169–174.

Brown, M.E. (1974) Seed and root bacterization. *Annual Review of Phytopathology* 12, 181–197.

Burr, T.J., Schroth, M.N. and Suslow, T. (1978) Increased potato yields by treatment of seed pieces with specific strains of *Pseudomonas fluorescens* and *P. putida*. *Journal of Phytopathology* 68, 1377–1383.

Cakmakc, R., Donmez, M.F. and Erdogan, U. (2007) The effect of plant growth promoting rhizobacteria on barley seedling growth, nutrient uptake, some soil properties, and bacterial counts. *Turkish Journal of Agriculture and Forestry* 31(Suppl. 3), 189–199.

Chakraborty, U., Chakraborty, B. and Basnet, M. (2006) Plant growth promotion and induction of resistance in *Camellia sinensis* by *Bacillus megaterium*. *Journal of Basic Microbiology* 46(Suppl. 3), 186–195.

Charest, M.H., Beauchamp, C.J. and Antoun, H. (2005) Effects of the humic substances of de- inking paper sludge on the antagonism between two compost bacteria and *Pythium ultimum*. *FEMS Microbiology Ecology* 52(Suppl. 2), 219–227.

Chen, Y.P., Rekha, P.D., Arun, A.B., Shen, F.T., Lai, W.A. and Young, C.C. (2006) Phosphate solubilizing bacteria from subtropical soil and their tricalcium phosphate solubilizing abilities. *Applied Soil Ecology* 34, 33–41.

Chung, H., Park, M., Madhaiyan, M., Seshadri, S., Song, J., Cho, H. and Sa, T. (2005) Isolation and characterization of phosphate solubilizing bacteria from the rhizosphere of crop plants of Korea. *Soil Biology and Biochemistry* 37, 1970–1974.

Cleyet-Marcel, J.C., Larcher, M., Bertrand, H., Rapior, S. and Pinochet, X. (2001) Plant growth enhancement by rhizobacteria. In: Morot Gaudry, J.F. (ed.) *Nitrogen Assimilation by Plants: Physiological, Biochemical and Molecular Aspects*. Science Publishers, Plymouth, UK, pp. 185–197.

Contreras-Cornejo, H.A., Macias-Rodriguez, L., Cortes-Penagos, C. and Lopez-Bucio, J. (2009) *Trichoderma virens*, a plant beneficial fungus, enhances biomass production and promotes lateral root growth through an auxin-dependent mechanism in *Arabidopsis*. *Plant Physiology* 149, 1579–1592.

Curl, E.A. and Truelove, B. (1986) *The Rhizosphere*. Springer, Berlin-Heidelberg, 288 pp.

Das Gupta, S.M., Khan, N. and Nautiyal, C.S. (2006) Biologic control ability of plant growth-promoting *Paenibacillus lentimorbus* NRRL B-30488 isolated from milk. *Current Microbiology* 53, 502–505.

Dashti, N., Zhang, F., Hynes, R. and Smith, D.L. (1998) Plant growth promoting rhizobacteria accelerate nodulation and increase nitrogen fixation activity by field grown soyabean (*Glycine Max* (L) Merr) under short season conditions. *Plant and Soil* 200(Suppl. 2), 205–213.

Dastager, S.G., Deepa, C.K. and Pandey, A. (2010) Isolation and characterization of novel plant growth promoting *Micrococcus* sp. NII-0909 and its interaction with cowpea. *Plant Physiology and Biochemistry* 48, 987–992.

De Freitas, J.R., Banerjee, M.R. and Germida, J.J. (1997) Phosphate-solubilizing rhizobacteria enhance the growth and yield but not phosphorus uptake of canola (*Brassica napus* L.). *Biology and Fertility of Soils* 24, 358–364.

Dekhil, S.B., Cahill, M., Stackebrandt, E. and Li, S. (1997) Transfer of *Conglomeromonas largomobilis* subsp. *largomobilis* to the genus *Azospirillum* as *Azopirillum largomobile* comb. nov., and elevation of *Conglomeromonas largomobilis* subsp. *parooensis* to the new type species of *Conglomeromonas*, *Conglomeromonas parooensis* sp. nov. *Systematic and Applied Microbiology* 20, 72–77.

Dewan, G.I. and Subba Rao, N.S. (1979) Seed inoculation with *Azospirillum brasilense* and *Azotobacter chroococcum* and the root biomass of rice (*Oryza sativa* L). *Plant and Soil* 53(Suppl. 3), 295–302.

Eckert, B., Weber, O.B., Kirchof, G., Halbritter, A., Stoffels, M. and Hartmann, A. (2001) *Azospirillum doebereinerae* sp. nov., a nitrogen-fixing bacterium associated with the C4-grass *Miscanthus*. *International Journal of Systematic Evolutionary Microbiology* 51(Suppl. 1), 17–26.

Egamberdiyeva, D. (2005) Plant-growth-promoting rhizobacteria isolated from a Calcisol in a semi-arid region of Uzbekistan: biochemical characterization and effectiveness. *Journal of Plant Nutrition and Soil Science* 168(Suppl. 1), 94–99.

El-Yazeid, A.A., Abou-Aly, H.A., Mady, M.A. and Moussa, S.A.M. (2007) Enhancing growth, productivity and quality of squash plants using phosphate dissolving microorganisms (bio phosphor) combined with boron foliar spray. *Research Journal of Agriculture and Biological Sciences* 3(4), 274–286.

Fallik, E., Sarig, S. and Okon, Y. (1994) Morphology and physiology of plant roots associated with *Azospirillum*. In: Okon, Y. (ed.) *Azospirillum–Plant Associations*. CRC Press, Boca Raton, Florida, pp. 77–84.

Gangwar, M. and Kaur, G. (2009) Isolation and characterization of endophytic bacteria from endorhizosphere of sugarcane and ryegrass. *The Internet Journal of Microbiology* 7(Suppl. 1).

García, J.A.L., Probanza, A., Ramos, B. and Mañero, F.J.G. (2001) Ecology, genetic diversity and screening strategies of plant growth promoting rhizobacteria. *Journal of Plant Nutrition and Soil Sciences* 164, 1–7.

García, J.A.L., Probanza, A., Ramos, B., Palomino, M.R. and Mañero, F.J.G. (2004) Effect of inoculation of *Bacillus licheniformis* on tomato and pepper. *Agronomy for Sustainable Development* 24(Suppl. 4), 169–176.

Gholami, A., Shahsavani, S. and Nezarat, S. (2009) The effect of plant growth promoting rhizobacteria (PGPR) on germination, seedling growth and yield of maize. *International Journal of Biological Life Sciences* 1(Suppl. 1), 35–40.

Graham, P.H. (1999) Biological dinitrogen fixation: symbiotic. In: Sylvia, D., Fuhrmann, J., Hartel, P. and Zuberer, D. (eds) *Principles and Applications of Soil Microbiology*. Prentice Hall, Upper Saddle River, New Jersey, pp. 322–368.

Gunnell, D., Eddleston, M., Phillips, M.R. and Konradsen, F. (2007) The global distribution of fatal pesticide self-poisoning: systematic review. *BMC Public Health* 21(7), 357.

Gupta, A.K. (2004) *The Complete Technology Book on Biofertilizers and Organic Farming*. National Institute of Industrial Research Press, India.

Hafeez, F.Y., Yasmin, S., Ariani, D., Mehboob-ur-Rahman Zafar, Y. and Malik, K.A. (2006) Plant growth-promoting bacteria as biofertilizer. *Agronomy for Sustainable Development* 26, 143–150.

Hallmann, J., Quandt-Hallmann, A., Mahaffee, W.F. and Kloepper, J.W. (1997) Bacterial endophytes in agricultural crops. *Canadian Journal of Microbiology* 43(Suppl. 10), 895–914.

Han, H.S. and Lee, K.D. (2005) Phosphate and potassium solubilizing bacteria effect on mineral uptake, soil availability and growth of eggplant. *Research Journal of Agricultural and Biological Sciences* 1, 176–180.

Han, H.S., Supanjani and Lee, K.D. (2006) Effect of co-inoculation with phosphate and potassium solubilizing bacteria on mineral uptake and growth of pepper and cucumber. *Plant Soil and Environment* 52 (Suppl. 3), 130–136.

Harman, G.E., Howell, C.R., Viterbo, A., Chet, I. and Lorito, M. (2004) *Trichoderma* species – opportunistic, avirulent plant symbionts. *Nature Reviews Microbiology* 2, 43–56.

Harrier, L.A. and Waston, C.A. (2004) The potential role of arbuscular mycorrhiza (AM) fungi in the bioprotection of plants against soil-borne pathogens in organic and/or other sustainable farming systems. *Pest Management Science* 60(Suppl. 2), 149–157.

Hiltner, L. (1904) Uber neuere erfahrungen und probleme auf dem gebiet der boden bakteriologie und unter besonderer berucksichtigung det grundungung und branche. *Arb. Deut. Landw. Ges.* 98, 59–78.

Jaizme-Vega, M.D.C., Rodríguez-Romero, A.S. and Guerra, M.S.P. (2004) Potential use of rhizobacteria from the *Bacillus* genus to stimulate the plant growth of micropropagated bananas. *Fruits* 59(Suppl. 2), 83–90.

Kaymak, H.C., Yarali, F., Guvenc, I. and Donmez, M.F. (2008) The effect of inoculation with plant growth rhizobacteria (PGPR) on root formation of mint (*Mentha piperita* L.) cuttings. *African Journal of Biotechnology* 7(Suppl. 24), 4479–4483.

Keswani, C., Mishra, S., Sarma, B.K., Singh, S.P. and Singh, H.B. (2014) Unraveling the efficient applications of secondary metabolites of various *Trichoderma* spp. *Applied Microbiology and Biotechnology* 98, 533–544.

Khammas, K.M., Ageron, E., Grimont, P.A. and Kaiser, P. (1989) *Azospirillum irakense* sp. nov., a nitrogen-fixing bacterium associated with rice roots and rhizosphere soil. *Research in Microbiology* 140(Suppl. 9), 679–693.

Khan, A.G. (2005) Role of soil microbes in the rhizosphere of plants growing on trace metal contaminated soils in phytoremediation. *Journal of Trace Elements in Medicine and Biology* 18, 355–364.

Khan, M.S., Zaidi, A., Ahemad, M., Oves, M. and Wani, P.A. (2010) Plant growth promotion by phosphate solubilizing fungi – current perspective. *Archives of Agronomy and Soil Science* 56, 73–98.

Kloepper, J.W. and Beauchamp, C.J. (1992) A review of issues related to measuring of plant roots by bacteria. *Canadian Journal of Microbiology* 38, 1219–1232.

Kloepper, J.W. and Schroth, M.N. (1978) Plant growth promoting rhizobacteria on radishes. In: *Proceedings of the Fourth International Conference on Plant Pathogen Bacteria*, vol. 2. Station de Pathologie Vegetale et Phytobacteriologie, INRA, Angers, France, pp. 879–882.

Kloepper, J.W., Lifshitz, R. and Zablotowicz, R.M. (1989) Free-living bacterial inocula for enhancing crop productivity. *Trends in Biotechnology* 7, 39–43.

Kloepper, J.W., Reddy, M.S., Rodríguez-Kabana, R., Kenney, D.S., Kokalis-Burelle, N., Martinez-Ochoa, N. and Vavrina, C.S. (2004) Application of rhizobacteria in transplant production and yield enhancement. *Acta Horticulturae* 631, 217–229.

Koumoutsi, A., Chen, X.-H., Henne, A., Liesegang, H., Hitzeroth, G. and Franke, P. (2004) Structural and functional characterization of gene clusters directing non ribosomal synthesis of bioactive cyclic lipopeptides in *Bacillus amyloliquefaciens* strain FZB42. *Journal of Bacteriology* 186, 1084–1096.

Kraus, J. and Loper, J. (1995) Characterization of genomic region required for production of antibiotic pyoluteorin by the biological control agent *Pseudomonas fluorescens* Pf-5. *Applied and Environmental Microbiology* 61(Suppl. 3), 849–854.

Kremer, R.J. and Kennedy, A.C. (1996) Rhizobacteria as biocontrol agents of weeds. *Weed Technology* 10(Suppl. 3), 601–609.

Kumar, R. and Chandra, R. (2008) Influence of PGPR and PSB on *Rhizobium leguminosarum* Bv. viciae strain competition and symbiotic performance in lentil. *World Journal of Agricultural Sciences* 4(Suppl. 3), 297–301.

Kumar, V., Behl, R.K. and Narula, N. (2001) Establishment of phosphate-solubilizing strains of *Azotobacter chroococcum* in the rhizosphere and their effect on wheat cultivars under greenhouse conditions. *Microbiological Research* 156, 87–93.

Lata, S.A.K. and Tilak, K.V.B.R. (2002) Biofertilizers to augment soil fertility and crop production. In: Krishna, K.R. (ed.) *Soil Fertility and Crop Production*. Science Publishers, Enfield, New Hampshire, pp. 279–312.

Laville, J., Blumer, C., Schroetter, C.V., Gaia, V., Défago, G., Keel, C. and Haas, D. (1998) Characterization of the hcnABC gene cluster encoding hydrogen cyanide synthase and anaerobic regulation by ANR in the strictly aerobic biocontrol agent *Pseudomonas fluorescens* CHA0. *The Journal of Bacteriology* 180 (Suppl. 12), 3187–3196.

Leach, A.W. and Mumford, J.D. (2008) Pesticide environmental accounting: a method for assessing the external costs of individual pesticide applications. *Environmental Pollution* 151, 139–147.

Lian, B., Fu, P.Q., Mo, D.M. and Liu, C.Q. (2002) A comprehensive review of the mechanism of potassium release by silicate bacteria. *Acta Mineral Sinica* 22, 179.

Lisette, J., Xavier, C. and Germida, J.J. (2003) Selective interactions between arbuscular mycorrhizal fungi and *Rhizobium leguminosarum* bv. viceae enhance pea yield and nutrition. *Biology and Fertility of Soils* 37, 261–267.

Liu, D., Lian, B. and Dong, H. (2012) Isolation of *Paenibacillus* sp. and assessment of its potential for enhancing mineral weathering. *Geomicrobiology Journal* 29(5), 413–421.

Lucy, M., Reed, E. and Glick, B.R. (2004) Application of free living plant growth-promoting rhizobacteria. *Antonie Van Leeuwenhoek* 86, 1–25.

Lugtenberg, B.J.J. and Dekkers, L.C. (1999) What makes *Pseudomonas* bacteria rhizosphere competent? *Environmental Microbiology* 1(Suppl. 1), 9–13.

Lynch, J.M. (1990) *The Rhizosphere*. Wiley, Chichester, UK, 458 pp.

Mäder, P., Fließbach, A., Dubois, D., Gunst, L., Fried, P. and Niggli, U. (2002) Soil fertility and biodiversity in organic farming. *Science* 296, 1694–1697.

Magalhães, F.M., Baldani, J.I., Souto, S.M., Kuykendall, J.R. and Döbereiner, J.A. (1983) New acid-tolerant *Azospirillum* species. *Anais da Academia Brasileira de Ciências* 55(Suppl. 4), 417–430.

Malusá, E. and Vassilev, N. (2014) A contribution to set a legal framework for biofertilizers. *Applied Microbiology and Biotechnology* 98, 6599–6607.

Malusá, E., Sas-Paszt, L., Zurawicz, E. and Popinska, W. (2007) The effect of a mycorrhiza-bacteria substrate and foliar fertilization on growth response and rhizosphere pH of three strawberry cultivars. *International Journal of Fruit Science* 6, 25–41.

Marilley, L. and Aragno, M. (1999) Phytogenetic diversity of bacterial communities differing in degree of proximity of *Lolium perenne* and *Trifolium repens* roots. *Applied Soil Ecology* 13, 127–136.

Mauch, F., Mauch-Mani, B. and Boller, T. (1988) Antifungal hydrolases in pea tissue. II. Inhibition of fungal growth by combinations of chitinase and beta-1, 3-glucanase. *Plant Physiology* 88(Suppl. 3), 936–942.

McAfee, J. (2008) *Potassium, a Key Nutrient for Plant Growth*. Department of Soil and Crop Sciences. Available at: http://jimmcafee.tamu.edu/files/potassium (accessed December 2015).

McCully, M. (2005) The rhizosphere: the key functional unit in plant/soil/microbial interactions in the field. Implications for the understanding of allelopathic effects. In: Harper, J., An, M., Wu, H. and Kent, J. (eds) *Proceedings of the 4th World Congress on Allelopathy: 21–26 August 2005*. International Allelopathy Society, Charles Sturt University, Wagga Wagga, New South Wales, Australia.

Mehnaz, S., Weselowski, B. and Lazarovits, G. (2007) *Azospirillum canadense* sp. nov., a nitrogen-fixing bacterium isolated from corn rhizosphere. *International Journal of Systematic and Evolutionary Microbiology* 57, 620–624.

Mehnaz, S., Weselowski, B., Aftab, F., Zahid, S., Lazarovits, G. and Iqbal, J. (2009) Isolation, characterization, and effect of fluorescent *Pseudomonads* on micropropagated sugarcane. *Canadian Journal of Microbiology* 55(Suppl. 8), 1007–1011.

Miller, S.A., Beed, F.D. and Harmon, C.L. (2009) Plant disease diagnostic capabilities and networks. *Annual Review of Phytopathology* 47, 15–38.

Mohammadi, K. and Sohrabi, Y. (2012) Bacterial biofertilizers for sustainable crop production: a review. *Journal of Agricultural and Biological Science* 7, 307–316.

Nagórska, K., Bikowski, M. and Obuchowski, M. (2007) Multicellular behavior and production of a wide variety of toxic substances support usage of *Bacillus subtilis* as a powerful biocontrol agent. *Acta Biochimica Polonica* 54(Suppl. 3), 495–508.

Negi, Y.K., Garg, S.K. and Kumar, J. (2005) Cold-tolerant fluorescent *Pseudomonas* isolates from Garhwal

Himalayas as potential plant growth promoting and biocontrol agents in pea. *Current Sciences* 89(Suppl. 12), 25.

NIIR (2012) *The Complete Technology Book on Biofertilizer and Organic Farming*, 2nd revd edn. NIIR Project Consultancy Services, New Delhi.

Nowak-Thompson, B., Gould, S.J., Kraus, J. and Loper, J.E. (1994) Production of 2,4-diacetylphloroglucinol by the biocontrol agent *Pseudomonas fluorescens* Pf-5. *Canadian Journal of Microbiology* 40(Suppl. 12), 1064–1066.

Ogbo, F.C. (2010) Conversion of cassava wastes for biofertilizer production using phosphate solubilizing fungi. *Bioresource Technology* 101, 4120–4124.

Okon, Y. (1985) *Azospirillum* as a potential inoculant for agriculture. *Trends in Biotechnology* 3(Suppl. 9), 223–228.

Okon, Y. and Kapulnik, Y. (1986) Development and function of *Azospirillum* inoculated roots. *Plant and Soil* 90(Suppl. 1–3), 3–16.

Orhan, E., Esitken, A., Ercisli, S., Turan, M. and Sahin, F. (2006) Effects of plant growth promoting rhizobacteria (PGPR) on yield, growth and nutrient contents in organically growing raspberry. *Scientia Horticulturae* 111(Suppl. 1), 38–43.

Owen, D., Williams, A.P., Griffith, G.W. and Withers, P.J.A. (2015) Use of commercial bio-inoculants to increase agricultural production through improved phosphorus acquisition. *Applied Soil Ecology* 86, 41–54.

Pal, K.K., Tilak, K.V.B.R., Saxena, A.K., Dey, R. and Singh, C.S. (2000) Antifungal characteristics of a fluorescent *Pseudomonas* strain involved in the biological control of *Rhizoctonia solani. Research in Microbiology* 155(Suppl. 3), 233–242.

Park, J., Bolan, N., Megharaj, M. and Naidu, R. (2010) Isolation of phosphate-solubilizing bacteria and characterization of their effects on lead immobilization. *Pedologist* 53, 67–75.

Parmar, P. and Sindhu, S.S. (2013) Potassium solubilization by rhizosphere bacteria: influence of nutritional and environmental conditions. *Journal of Microbiology Research* 3(1), 25–31.

Paul, D., Dinesh Kumar, N. and Nair, S. (2006) Proteomics of a plant growth-promoting rhizobacterium, *Pseudomonas fluorescens* MSP-393, subjected to salt shock. *World Journal of Microbiology and Biotechnology* 22 (Suppl. 4), 369–374.

Peng, G., Wang, H., Zhang, G., Hou, W., Liu, Y., Wang, E.T. and Tan, Z. (2006) *Azospirillum melinis* sp. nov., a group of diazotrophs isolated from tropical molasses grass. *International Journal of Systematic and Evolutionary Microbiology* 56(Suppl. 6), 1263–1271.

Pindi, P.K. and Satyanarayana, S.D.V. (2012) Liquid microbial consortium – a potential tool for sustainable soil health. *Journal of Biofertilizers and Biopesticides* 3, 4.

Polenko, D.R., Scher, F.M., Kloepper, J.W., Singleton, C.A., Laliberte, M. and Zaleska, I. (1987) Effects of root colonizing bacteria on nodulation of soybean roots by *Bradyrhizobium japonicum*. *Canadian Journal of Microbiology* 33, 498–503.

Prajapati, K., Sharma, M.C. and Modi, H.A. (2013) Growth promoting effect of potassium solubilizing microorganisms on *Abelmoscus esculantus*. *International Journal of Agriculture Sciences* 3(1), 181–188.

Prasad, H. and Chandra, R. (2003) Growth pattern of urdbean *Rhizobium* sp. with PSB and PGPR in consortia. *Journal of the Indian Society of Soil Science* 51, 76–78.

Probanza, A., Lucas García, J.A., Ruiz Palomino, M., Ramos, B. and Gutiérrez Mañero, F.J. (2002) *Pinus pinea* L seedling growth and bacterial rhizosphere structure after inoculation with PGPR *Bacillus* (*B. licheniformis* CECT 5106 and *B. pumilus* CECT 5105). *Applied Soil Ecology* 20(Suppl. 2), 75–84.

Raj, S.A. (2004) Solubilization of silicate and concurrent release of phosphorus and potassium in rice ecosystem. In: *Conference Paper Biofertilizers Technology*. Coimbatore, India, p. 372.

Rajawat, M.V.S., Singh, S., Singh, G. and Saxena, A.K. (2012) Isolation and characterization of K-solubilizing bacteria isolated from different rhizospheric soil. In: *Proceeding of 53rd Annual Conference of Association of Microbiologists of India,* p. 124.

Reinhold, B., Hurek, T., Fendrik, I., Pot, B., Gillis, M., Kersters, K., Thielemans, S. and de Ley, J. (1987) *Azospirillum halopraeferens* sp. nov., a nitrogen-fixing organism associated with roots of kallar grass (*Leptochloa fusca* (L.) Kunth). *International Journal of Systematic Bacteriology* 37, 43–51.

Rewari, R.B. and Tilak, K.V.B.R. (1988) Microbiology of pulses. In: Baldev, B., Ramanujam, S. and Jain, H.K. (eds) *Pulse Crops*. Oxford & IBH, New Delhi, pp. 373–411.

Rokhzadi, A., Asgharzadeh, A., Darvish, F., Nourmohammadi, G. and Majidi, E. (2008) Influence of plant growth-promoting rhizobacteria on dry matter accumulation and yield of chickpea (*Cicer arietinum* L.) under field condition. *American-Eurasian Journal of Agricultural and Environmental Sciences* 3(2), 253–257.

Saber, K., Nahla, L.D. and Chedly, A. (2005) Effect of P on nodule formation and N fixation in bean. *Agronomy for Sustainable Development* 25, 389–393.

Saikia, N. and Brezbaruah, B. (1995) Iron-dependent plant pathogen inhibition through *Azotobacter* RRLJ 203 isolated from iron-rich acid soils. *Indian Journal of Experimental Biology* 33, 571–575.

Sangeeth, K.P., Bhai, R.S. and Srinivasan, V. (2012) *Paenibacillus glucanolyticus*, a promising potassium solubilizing bacterium isolated from black pepper (*Piper nigrum* L.) rhizosphere. *Journal of Spices and Aromatic Crops* 21(2), 118–124.

Santi, C., Bogusz, D. and Franche, C. (2013) Biological nitrogen fixation in non-legume plants. *Annals of Botany* 111, 743–767.

Saxena, A.K. and Tilak, K.V.B.R. (1998) Free-living nitrogen fixers: its role in crop production. In: Verma, A.K. (ed.) *Microbes for Health, Wealth and Sustainable Environment*. Malhotra Publication Co., New Delhi, 25–64.

Saxena, A.K. and Tilak, K.V.B.R. (1999) Potentials and prospects of *Rhizobium* biofertilizer. In: Jha, M.N. (ed.) *Current Trends in Life Sciences,* Agromicrobes. Today & Tomorrow Printers & Publishers, New Delhi, pp. 51–78.

Schwintzer, R. and Tjepkema, J.D. (1990) *The Biology of Frankia and Actinorrhizal Plants*. Academic Press, San Diego, California, 99 pp.

Seshadri, S., Ignacimuthu, S., Vadivelu, M. and Lakshminarasimhan, C. (2007) Inorganic phosphate solubilization by two insect pathogenic *Bacillus* sp. *Developments in Plant and Soil Sciences* 102, 351–355.

Shaukat, K., Affrasayab, S. and Hasnain, S. (2006) Growth responses of *Helianthus annus* to plant growth promoting rhizobacteria used as a biofertilizer. *Journal of Agricultural Research* 1(Suppl. 6), 573–581.

Sheng, X.F. (2005) Growth promotion and increased potassium uptake of cotton and rape by a potassium releasing strain of *Bacillus edaphicus*. *Soil Biology and Biochemistry* 37, 1918–1922.

Sheng, X.F. and He, L.Y. (2006) Solubilization of potassium bearing minerals by a wild type strain of *Bacillus edaphicus* and its mutants and increased potassium uptake by wheat. *Canadian Journal of Microbiology* 52, 66–72.

Shoebitz, M., Ribaudo, C.M., Pardo, M.A., Cantore, M.L., Ciampi, L. and Curá, J.A. (2007) Plant growth promoting properties of a strain of *Enterobacter ludwigii* isolated from *Lolium perenne* rhizosphere. *Biochemical Journal* 41(Suppl. 9), 1768–1774.

Siddiqui, Z. (2006) PGPR: prospective biocontrol agents of plant pathogens. In: Siddiqui, Z. (ed.) *PGPR: Biocontrol and Biofertilization*. Springer, Dordrecht, the Netherlands, pp. 111–142.

Singh, J.S., Pandey, V.C. and Singh, D.P. (2011) Efficient soil microorganisms: a new dimension for sustainable agriculture and environmental development. *Agriculture, Ecosystems & Environment* 140, 339–353.

Singh, R. and Adholeya, A. (2003) Interactions between arbuscular mycorrhizal fungi and plant-growth promoting rhizobacteria. *Mycorrhiza News* 15, 16–17.

Sinha, R.K., Valani, D., Chauhan, K. and Agarwal, S. (2010) Embarking on a second green revolution for sustainable agriculture by vermiculture biotechnology using earthworms: reviving the dreams of Sir Charles Darwin. *Journal of Agricultural Biotechnology and Sustainable Development* 2(7), 113–128.

Sivaramaiah, N., Malik, D.K. and Sindhu, S.S. (2007) Improvement in symbiotic efficiency of chickpea (*Cicer arietinum*) by coinoculation of *Bacillus* strains with *Mesorhizobium* sp. *Cicer. Indian Journal of Microbiology* 47, 51–56.

Sofo, A., Tataranni, G., Xiloyannis, C., Dichio, B. and Scopa, A. (2004) Direct effects of *Trichoderma harzianum* strain T-22 on micropropagated shoots of GiSeLa6 (R) (*Prunus cerasus* × *Prunus canescens*) rootstock. *Environmental and Experimental Botany* 76, 33–38.

Son, T.T.N., Diep, C.N. and Giang, T.T.M. (2006) Effect of Bradyrhizobia and phosphate solubilizing bacteria application on soybean in rotational system in the Mekong delta. *Omonrice* 14, 48–57.

Sridevi, M., Mallaiah, K.V. and Yadav, N.C.S. (2007) Phosphate solubilization by *Rhizobium* isolates from *Crotalaria* species. *Journal of Plant Sciences* 2, 635–639.

Supanjani, Han, H.S., Jung, J.S. and Lee, K.D. (2006) Rock phosphate-potassium and rock-solubilising bacteria as alternative, sustainable fertilizers. *Agronomy for Sustainable Development* 26(Suppl. 4), 233–240.

Tarrand, J.J., Krieg, N.R. and Döbereiner, J. (1978) A taxonomic study of the *Spirillum lipoferum* group with description of a new genus, *Azospirillum* gen. nov. and two species, *Azospirillum lipoferum* (Beijerinck) comb. nov. and *Azospirillum brasilense* nov. *Canadian Journal of Microbiology* 24(Suppl. 8), 967–980.

Tchan, Y.T. (1984) Family II. Azotobacteriaceae. In: Krieg, N.R. and Holt, J.G. (eds) *Bergey's Manual of Systematic Bacteriology*. Williams and Wilkins, Baltimore, Maryland, pp. 1–219.

Tchan, Y.T. and New, P.T. (1984) Genus I. *Azotobacter beijerinck*. In: Krieg, N.R. and Holt, J.G. (eds) *Bergey's Manual of Systematic Bacteriology.* Williams and Wilkins, Baltimore, Maryland, pp. 220–229.

Tilak, K.V.B.R. (1993) *Bacterial Fertilizers*. Indian Council of Agricultural Research, New Delhi, pp. 4–33.

Tilak, K.V.B.R. and Subba Rao, N.S. (1987) Association of *Azospirillum brasilense* with pearlmillet (*Pennisetum americanum* (L) Leeke. *Biology and Fertility of Soils* 4(Suppl. 1–2), 97–102.

Tilak, K.V.B.R., Ranganayaki, N., Pal, K.K., De, R., Saxena, A.K., Nautiyal, C.S., Mittal, S., Tripathi, A.K. and Johri, B.N. (2005) Diversity of plant growth and soil health supporting bacteria. *Current Science* 89(Suppl. 1), 136–150.

Toro, M., Azcón, R. and Barea, J.M. (1998) The use of isotopic dilution techniques to evaluate the interactive effects of *Rhizobium* genotype, mycorrhizal fungi, phosphate solubilizing rhizobacteria and rock phosphate on nitrogen and phosphorus acquisition by *Medicago sativa*. *New Phytologist* 138, 265–273.

Troufflard, S., Mullen, W., Larson, T.R., Graham, I.A., Crozier, A., Amtmann, A. and Armengaud, P. (2010) Potassium deficiency induced the biosynthesis of oxylipins and glucosinolates in *Arabidopsis thaliana*. *Plant Biology* 10(1), 172.

Tucci, M., Ruocco, M., De Masi, L., De Palma, M. and Lorito, M. (2011) The beneficial effect of *Trichoderma* spp. on tomato is modulated by the plant genotype. *Molecular Plant Pathology* 12(4), 341–354.

Uroz, S., Calvaruso, C., Turpault, M.P., Pierrat, J.C., Mustin, C. and Frey-Klett, P. (2007) Effect of the mycorrhizosphere on the genotypic and metabolic diversity of the bacterial communities involved in mineral weathering in a forest soil. *Applied and Environmental Microbiology* 73, 3019–3027.

Vassilev, N., Medina, A., Azcon, R. and Vassilev, M. (2006) Microbial solubilization of rock phosphate as media containing agro industrial wastes and effect of the resulting products on plant growth and phosphorus uptake. *Plant Soil* 287, 77–84.

Vassilev, N., Martos, E., Mendes, G., Flor-Peregrin, E., Martos, V. and Vassileva, M. (2013) Biochar of animal origin: a sustainable solution of the highgrade rock phosphate scarcity. *Journal of the Science of Food and Agriculture* 93, 1799–1804.

Vazquez, P., Holguin, G., Puente, M., Lopez-Cortes, A. and Bashan, Y. (2000) Phosphate solubilizing microorganisms associated with the rhizosphere of mangroves in a semi-arid coastal lagoon. *Biology and Fertility of Soils* 30, 460–468.

Vessey, J.K. (2003) Plant growth promoting rhizobacteria as biofertilizers. *Plant and Soil* 255(Suppl. 2), 571–586.

Vitale, A., Cirvilleri, G., Castello, I., Aiello, D. and Polizzi, G. (2012) Evaluation of *Trichoderma harzianum* strain T22 as biological control agent of *Calonectria pauciramosa*. *Bio-Control* 57(5), 687–696.

Wang, X., Pan, Q., Chen, F., Yan, X. and Liao, H. (2011) Effects of co-inoculation with arbuscular mycorrhizal fungi and rhizobia on soybean growth as related to root architecture and availability of N and P. *Mycorrhiza* 21, 173–181.

Wani, P.A., Zaidi, A., Khan, A.A. and Khan, M.S. (2005) Effect of phorate on phosphate solubilization and indole acetic acid (IAA) releasing potentials of rhizospheric microorganisms. *Annals of Plant Protection Sciences* 13, 139–144.

Weller, D.M. (2007) *Pseudomonas* biocontrol agents of soilborne pathogens: looking back over 30 years. *Phytopathology* 97(Suppl. 2), 253.

Wheeler, C.T. and Miller, J.M. (1990) Current and potential uses of actinorrhizal plants in Europe. In: Schwintzer, C.R. and Tjepkema, J.D. (eds) *The Biology of Frankia and Actinorrhizal Plants*. Academic Press, San Diego, California, pp. 365–389.

White, P.J. and Karley, A.J. (2010) Potassium. In: Hell, R. and Mendel, R.R. (eds) *Cell Biology of Metals and Nutrients, Plant Cell Monographs*, vol. 17. Springer, Berlin, pp. 199–224.

WIPO (World Intellectual Property Organization) (2015) *Guidelines for Preparing Patent Landscape Reports.* Available at: http://www.wipo.int/edocs/pubdocs/en/wipo_pub_946.pdf (accessed December 2015).

Wu, S.C., Cao, Z.H., Li, Z.G., Cheung, K.C. and Wong, M.H. (2005) Effects of biofertilizer containing N-fixer, P and K solubilizers and AM fungi on maize growth: a greenhouse trial. *Geoderma* 125, 155–166.

Xie, C.H. and Yokota, A. (2005) *Azospirillum oryzae* sp. nov., a nitrogen-fixing bacterium isolated from the roots of the rice plant *Oryza sativa*. *International Journal of Systematic and Evolutionary Microbiology* 55(Suppl. 4), 1435–1438.

Young, J.P.W. (1992) Phylogenetic classification of nitrogen-fixing organisms. In: Stacey, G., Burris, R.H. and Evans, H.J. (eds) *Biological Nitrogen Fixation.* Chapman and Hall, New York, pp. 43–86.

Zaidi, A., Khan, M.S., Ahemad, M., Oves, M. and Wani, P.A. (2009) Recent advances in plant growth promotion by phosphate-solubilizing microbes. In: Khan, M.S. (ed.) *Microbial Strategies for Crop Improvement*. Springer, Berlin-Heidelberg, pp. 23–50.

Zarjani, J.K., Aliasgharzad, N., Oustan, S., Emadi, M. and Ahmadi, A. (2013) Isolation and characterization of potassium solubilizing bacteria in some Iranian soils. *Archives of Agronomy and Soil Science* 77, 7569.

Zhang, F., Dashti, N., Hynes, R.K. and Smith, D.L. (1996) Plant growth promoting rhizobacteria and soybean (*Glycine Max* (L.) Merr) nodulation and nitrogen fixation at suboptimal root zone temperatures. *Annals of Botany* 77(Suppl. 5), 453–459.

Zhao, F., Sheng, X., Huang, Z. and He, L. (2008) Isolation of mineral potassium-solubilizing bacterial strains from agricultural soils in Shandong Province. *Biodiversity Science* 16, 593–600.

Zhu, F., Qu, L., Hong, X. and Sun, X. (2011) Isolation and characterization of a phosphate-solubilizing halophilic bacterium *Kushneria* sp. YCWA18 from Daqiao Saltern on the coast of Yellow Sea of China. *Evidence-Based Complementary and Alternative Medicine Article ID* 615032, 1–6.

8 Seeds of Change: Genetically Modified Crops, Canada's Agricultural Growth Act and the Erosion of Farmers' Privilege

Elisabeth Abergel*

Department of Sociology and Institut des Sciences de l'Environnement, Université du Québec à Montréal, Montréal, Canada

8.1 Introduction

We live in a historical moment where societies face the prospect of having to balance food security needs and environmental sustainability. Large questions concerning the types of agricultural production and the amount of land devoted to agriculture as well as how much of a rural population shall remain productive on farms are all being decided directly or indirectly by governments, corporations and civil society groups in a context of global climate change that includes: land degradation, rapid urbanization, diminishing biodiversity, limited natural resources and water supplies. Issues of global inequality and rising poverty make searching for sustainable and equitable solutions to food production, the rural poor and agricultural development central aspects of the international political agenda. The challenge of what is known as sustainable intensification describes the pressure to intensify food production while using sustainable and socially appropriate means.

Agricultural biotechnology presents itself in a context of a catastrophic future, as a solution to many of our world's environmental and economic ills. Discourse around the promises of this technology becomes framed in a context of technological progress. Its introduction into the environment and our food system is presented as an inevitable solution that will rectify the past mistakes of agricultural industrialization and productivist excesses, reducing the potential ecological and health impacts of these crops to issues of risk management and better science. In this context, most countries where GM (genetically modified) agriculture has been legalized, regulatory tools were designed to minimize risks while maximizing benefits. GM crops are often framed as being unavoidable in the face of mounting environmental degradation and economic disparity between people and countries. Since we are told there is no such thing as zero risk, proponents of GM crops often argue that, despite uncertainties concerning their long-term impacts, their use will simplify farming and provide economic advantages over conventional and organic production systems. Some call agricultural biotechnology the second green revolution, the difference being according to McMichael (2012) that the first green revolution was 'largely a public initiative geared towards national markets for staples, its successor is a private initiative increasingly geared to global markets with dietary inequalities'. Since their introduction, corporations and states have framed GM crops as environmentally sustainable and economically desirable, leaving citizens unsure about their public benefit. It is claimed that GM plants were designed to alleviate world poverty and hunger. However, whilst defending a 'pro-poor' position, the biotechnology industry has been pursuing research and development on specific

*Corresponding author, e-mail: abergel.elisabeth@uqam.ca

crop varieties that are aimed at large-scale, monocultural, industrialized farming systems normally found in the global north, largely ignoring the needs of the developing world and the rural poor (Glover, 2010). In addition, the crops that have been genetically modified and commercialized around the world, i.e. maize, soybeans, cotton and canola, rank amongst the world's top agricultural commodities. GM crops tend to be 'global' crops geared towards exports and are not specifically adapted to local growing conditions or local needs. As a result, the rate and scale at which they have been commercially planted across the world since the mid-1990s means that they are likely to disrupt ecosystems, distort local and international food markets and the agricultural economy. The largest increases in GM hectarage have occurred in developing countries in recent years accounting for 52% of global biotech crops in 2012, representing a 100-fold increase since their introduction in 1996 (James, 2012). According to Clive James (2014): 'The 20 developing and eight industrial countries where biotech crops are produced represent more than 60 percent of the world's population.' Consequently, these crops outcompete the local, traditional crop varieties people rely on for nutrition; they create new power dynamics between farmers as well as increasing dependency on corporate actors and their state-sponsors for expensive technologically intensive inputs.

Introduced in the early 1990s, the first transgenic crop varieties to come out of the laboratory, greenhouses and experimental field plots passed through a set of newly designed regulations in order to enable their commercial scale releases. Because existing legislation and regulatory frameworks did not deal specifically with GM seed and crops, these had to be adapted to reflect the pending large-scale environmental and market releases of transgenic varieties domestically and globally. From the beginning efforts were made to keep regulatory burdens to a minimum, allowing this fledgling industry to flourish with full state support. From then on, with the support of corporate actors, the promise of biotechnology became the main focus of policy makers and legislators, eager to capitalize on this new technology through trade liberalization. The inclusion of Article 27 of the World Trade Organization (WTO), the Agreement on Trade-Related Aspects of Intellectual Property Rights (TRIPS), mandates that member countries offer patent protection for inventions. While exceptions are allowed for plants and animals (including genetically engineered varieties), the Article stipulates that member countries 'shall provide for the protection of plant varieties by patents or by an effective *sui generis* system or by any combination thereof' (Torrance, 2007). Most developed countries have adopted patent protection for GMOs except for Canada, which does not allow the patenting of GM plants and animals.

Even though Canada has been actively promoting agricultural biotechnology in the international arena, the legal positions it adopts globally are inconsistent with domestic patent law and legal experts seem to think that a transition towards patent protection for GM plants and animals is inevitable (Torrance, 2007). In spite of this inconsistency, Canada has a well-enforced intellectual property (IP) system for GM agriculture and for protecting the rights of the developers of GMOs at the expense of non-GM farmers. Several court decisions in trials opposing organic and conventional farmers and biotech corporations over genetic contamination issues have upheld the monopoly rights of corporations over the rights of farmers or to put it simply, they have upheld patent rights over farmers' rights. In particular, the decision by the Canadian Supreme Court in the *Schmeiser* v. *Monsanto* case, where Percy Schmeiser was found guilty of violating patent law when volunteer herbicide-tolerant (HT) canola plants were found growing in his fields, strengthened monopoly control over seed plants in Canada. Another landmark case involved farmers of the Saskatchewan Organic Directorate (SOD) who challenged Monsanto Co. and Bayer CropScience in a class action suit for the damages caused by transgene escape, mostly from herbicide-tolerant canola varieties. The failure to certify organic farmers as a 'class' by both the Saskatchewan Lower Court (2005) and the Canadian Supreme Court (2007) means that the only recourse that farmers hurt by genetic drift have to challenge GM corporations is an individual case basis only. These examples illustrate how the responsibilities of patent holders for GM crops involved in transgene contamination events have been largely avoided, instead placing the burden of contamination on farming communities (Abergel, 2012).

Countries with active biotechnology R&D programmes designed regulatory frameworks to accommodate GMOs while devising international agreements in order to enable the global circulation and commercialization of GM crop varieties. From the outset, there were distinct philosophical differences between North American authorities and EU countries regarding how best to regulate these crop varieties. While US and Canadian regulators defended a system based on the idea that genetic engineering poses no new risks to the environment and human/animal health and that it should be considered along an evolutionary continuum of traditional breeding, the EU contended that the process of genetic alteration was in itself risky and could introduce new, unforeseen hazards. Today, each system operates on a different set of scientific and sociocultural assumptions about the nature of risk despite pressures to reconcile these differences. The North American model regulates the final product only, claiming that it is familiar and substantially equivalent to its conventional counterpart, while the EU takes into account the process by which the product was generated during its safety assessments, implying that unknown risks might be introduced during the genetic modification phase. To simplify, Canadian regulators downplay the way in which the plant variety was produced, claiming that GM varieties are essentially the same as their conventional counterparts; they focus instead on the newly introduced trait in combination with the known plant variety, which means that the GM variety may or may not be subject to risk assessment regulations. The claim stems from the idea that conventional breeding and genetic engineering techniques exist on a continuum of genetic modification even though scientists have long disputed this.

A series of amendments to existing Seed Laws, new variety registration rules and new regulatory regimes enacted in different countries coinciding with the commercial introduction of new crop varieties resulted from the use of genetic engineering in the 1990s. This period also marks the rise of IP regimes applied to living organisms (bacteria, plants and animals) and their parts (seeds, genes, cellular components and other biological materials) as well as the processes used to create new transgenic varieties. Since the mid-1990s, when transgenic crop varieties were introduced in several countries, laws and regulations have been amended in order to make room for questions of uncertainty and risk, not foreseen at the time of regulatory and legislative development. Despite a stated reliance on risk management, many rules and regulations designed to ensure environmental and health 'safety' failed to foresee what some scientists had been warning against for a long time: irreversible impacts on biodiversity (contamination risks of transgenic traits to related varieties, impacts on pollinating insects, toxicity to soil microbiota and micro-fauna) and an accrued dependence on chemical pesticides, most notably due to the introduction of pesticide- and herbicide-tolerant crops, which leads to the complex management of plants and insects resisting several chemical pesticides and herbicides at once. Recently, scientific reports have cast doubt on the safety of glyphosate, the main ingredient in Monsanto's RoundUp herbicide, which is one of the most widely used chemical herbicides and which has been described as an 'environmental toxin' because of the residues found on food capable of disrupting mammalian health (Samsel and Seneff, 2013). The World Health Organization's International Agency for Research on Cancer (IARC) raised global concerns over RoundUp when it classified it as 'probably carcinogenic to humans' in March 2015 (IARC, 2015). Other scientific reports point to the toxicity of the mixture of ingredients and inert compounds found in RoundUp as the potential cause for cancer and embryonic cell death (Bénachour and Séralini, 2009).

Measures to eradicate so-called superweeds and 'volunteer' plants that resist multiple herbicides are proving costly and dangerous for farmers and the environment, as cocktails of toxic chemicals become necessary. The next generation of GM crops that tolerate multiple herbicides, stacked varieties, might actually exacerbate this process by creating what scientists call a 'herbicide treadmill' whereby increased herbicide use gives rise to evermore resistant 'superweeds' which limit chemical options and render weed management very difficult. Similarly, insect-tolerant plants like Bt maize also contribute to insect resistance to the Bt toxin, originally derived from the bacteria *Bacillus thuringiensis*, creating problems for small-scale and/or organic farmers who rely on this product for their crops. Mitigation measures such as distance between conventional and GM fields as well as the creation of insect refuges, whereby Bt crops are planted with alternating rows of non-Bt crops creating opportunities for Bt-resistant insects to mate with wild-type insects thus giving rise to non-resistant insect progeny, have not proven very effective and have led to problems associated with secondary pests. For instance, Chinese farmers have observed decreases in bollworm but report increases in spider mite infestations in Bt cotton crops (Wang *et al.*, 2008; Ho *et al.*, 2009). Engineered maize varieties such as SmartStax and SmartStax+ commercialized by Monsanto and Dow AgroScience contain six types of Bt toxins in addition to two or four herbicide tolerance traits stacked in order to decrease insect resistance. However, even though these maize varieties obtained regulatory approval in 2009 and were commercially released in 2010 in the USA and Canada, they have not been extensively tested for their environmental and health impacts nor have they been found to efficiently slow down the development of insect resistance. The complex nature of these new GM varieties, which stack or pyramid a large number of engineered resistance traits, introduces high levels of scientific uncertainty raising questions about the robustness of biosafety approvals.

The widespread introduction and global commercialization of GM crops gave rise to a variety of popular movements that oppose their large-scale dissemination in different parts of the world. The fight revolves around contending scientific interpretations of safety and whose science should prevail, but mostly, the struggle is largely focused on the perils of the 'corporate control' of food, seed, land and genes. While it would be difficult to generalize the various campaigns and particular objections that mobilize organized civil society groups as well as citizens and consumers in general, environmental concerns as well as human and animal health issues feature highly among them. Of course, the further corporatization of food and farming, the criminalization of small-scale and organic farmers accused of violating patent rights, as well as threats to their livelihoods due to widespread contamination by GM crops are among the many issues concerning citizens and consumers everywhere. There are also many groups opposed to GM for religious or more philosophical reasons having to do with the genetic alteration of living organisms and their genes using techniques not previously available. Throughout the debate surrounding GM foods and since their introduction, there has been a 'disconnect' between the enthusiastic and ideological support of the biotech industry by governments and the public's lack of confidence in biotech food and in the state's ability to regulate and protect the food supply. It seems that dismissing public concerns as 'ignorance' or 'lack of scientific literacy' has resulted in raising consumers' concerns about animal welfare, as well as the quality, origin and production methods of the foods they eat. The growing demand for organic products in developed countries can be directly correlated with the food safety perception of consumers.

In light of the above discussion, the purpose of this chapter is to discuss the Canadian context by focusing on the recently adopted Act to amend certain Acts relating to Agriculture and Agri-food, also known as the Agricultural Growth Act (AGA), a new legislation with far-reaching implications for GM and non-GM farmers, which was passed in February 2015. The AGA was adopted by parliament despite much debate and resistance on the part of civil society groups, farming and trade groups as well as opposition political parties. Little consideration for proposed amendments was given in spite of various testimonials by prominent national food and farming organizations as well as international NGOs before the Senate Agriculture and Forestry Committee during its examination of Bill C-18.[1] The AGA, an omnibus bill containing many provisions concerning various agricultural sectors, was meant to modernize current legislation by streamlining specific regulatory processes by amending several pieces of existing legislation. Among proposed changes, it gives greater protection for crop insurance and grants the minister of agriculture new powers regarding the registration of new plant varieties as well as the safety of imported materials. Most importantly, for our purposes we will be discussing the changes brought to the Seeds Act and the Plant Breeders' Rights Act, as they are highly contentious and carry potentially damaging provisions for farmers. What this new piece of legislation illustrates is how on a national and international level, the adoption of intellectual property rights (IPR) for GM plants based on UPOV 1991 in combination with harmonized international regulatory regimes are pushing the expansion of markets for these crops. The International Union for the Protection of New Varieties of Plants (UPOV) is an intergovernmental organization established in 1961 by the International Convention for the Protection of New Varieties of Plants to increase plant breeders' rights. We are witnessing dual but complementary processes aimed at forcing GM crops on to existing and new markets in emerging economies: the international harmonization of IPR rules in line with UPOV 1991 and the systematic erosion and streamlining of regulatory requirements for safety approvals across jurisdictions make it impossible to resist the transnational movement of GM crops. On the one hand the adoption of IPRs for plants and seeds, as a function of trade, has a disciplining effect on global agricultural markets to endorse GM crops whether or not specific countries might want to exclude them. Severely restricting farmers' privilege, the ability to save, store, exchange and reuse seed, creates a movement towards further consolidation of the seed industry in the hands of large corporations. It also encourages the monopoly control of breeding programmes via the private ownership and legalized acquisition of global germplasm, as the new law facilitates the transnational/cross-boundary movement of new GM varieties. On the other hand, harmonizing regulatory requirements for new biotechnology agricultural products and GM varieties across jurisdictions, including accelerated approval times and reduced regulatory burdens for safety approvals among trading partners, constitutes the systematic deregulation of GM crops on a global level. During the last decade the hectarage of GM crops has more than doubled around the world. While still concentrated in certain parts of the world, the biotech industry is poised to capitalize on increased demand for agricultural inputs and food from emerging economies. The enforcement of IPR on seed and the systematic deregulation of GM crops in the context of liberalized trade by the world's largest producers work in tandem to further expand the global reach of GM crop varieties, *de facto* forcing 'public acceptance' of GM foods and eliminating the potential for resisting their presence in our food systems.

8.2 Canada's Agricultural Sector at a Glance

Canadian agriculture has benefited from rising global economic growth and higher incomes in large emerging economies (Agriculture and Agri-Food Canada, 2015). As a result, it is projected that Canada will supply these markets in the near future contributing to higher income levels for Canadian farmers. Since 2000, when agricultural commodity prices were trending downwards due to increased global supplies, prospects for this country's agricultural exports have changed. In the current economic climate it seems that food prices have been increasing due to higher demand globally around key commodities such as cereals, meat, dairy, vegetable oil and sugar (Agriculture and Agri-Food Canada, 2015). Both food and non-food commodities have helped to set the trend for higher demand of agricultural goods. According to the latest figures contained in the Canadian Agriculture Outlook 2015 document, it appears that Canadian farmers have been the direct beneficiaries of global demand for agri-food products. The report states: 'Global demand for agri-food products is linked to economic growth rates of key emerging economies with large population bases, rapid urbanization and rising incomes.' These economies continue to drive global growth in demand for primary agricultural products, meat, dairy and feed, as well as processed food products. Growth in non-food demand from biofuels may be less than what was observed in the past ten years as mandates are met and expansion is driven by underlying economic conditions. Although a decrease in economic growth is projected in the BRIC (Brazil, Russia, India and China) economies, it is expected that in the near future, relatively high growth rates will be sustained. Other emerging markets, referred to as the N11 (Egypt, Bangladesh, Indonesia, Iran, Mexico, Nigeria, Pakistan, the Philippines, Turkey, South Korea and Vietnam), will also contribute to global growth (Agriculture and Agri-Food Canada, 2015, p. 3). Success for Canadian farmers means that access to these emerging markets will need to expand as commodity prices and demand are expected to level off in countries with high levels of disposable incomes. This focus on exports towards emerging economies has meant that Canada has to compete with other agricultural countries in the global marketplace.

Canada was the fifth largest agricultural and agri-food exporter in the world in 2012. Canadian agriculture accounted for 3.5% of the total value of world agricultural and agri-food exports in that same year. In 2014, Canada was the fifth largest producer of GM crops after the USA, Brazil, Argentina and India. Approximately 11 million hectares of GM crops were planted and consisted mainly of canola, maize and soybeans. The value of the Canadian seed market was estimated in 2012 to be around US$2.1 billion. The US seed market is the largest, worth approximately US$12 billion, followed by that of China, France, Brazil and India (ISF, 2012). In contrast, in many emerging economies, the formal seed business accounts for about 10–20% of total market value while 80–90% of seed is supplied by non-commercial or the informal market mainly through farmer-saved seed. From a Canadian perspective there is great growth potential for the commercial introduction of new 'improved' plant varieties, which depends on the implementation of IP regulations, particularly in markets where the informal seed business dominates. The American Seed Trade Association (ASTA) has identified five priority areas that would enhance the US seed trade. These recommendations echo the Canadian situation with regards to recent legislative changes. Among these, as they relate to IP protection, ASTA mentions the establishment of laws, enforcement mechanisms, seed legislation, science-based phytosanitary requirements, adventitious presence (AP) and low-level presence (LLP), parallel systems of plant variety protection (PVP) and utility patents, co-existence of farming systems, promotion of quality seed, movement of germplasm (International Treaty on Plant Genetic Resources for Food and Agriculture or ITPGRFA), acceptance of technology and seed applied technology (American Seed Trade Association, 2015). The case of low-level presence of GMOs will be used to illustrate how the AGA works in practice.

8.3 Canada's 'Novel' Approach to GM Regulations

From the outset Canadian regulators decided that no new laws would specifically be created in order to regulate GM crops. Hence, GM crops fall under the descriptor 'Plants with Novel Traits' or PNTs and are considered like any other novel plant variety produced through traditional breeding and/or genetic modification. The Canadian Food Inspection Agency (CFIA) was created in 1997 to consolidate the activities of four lead ministries: Agriculture and Agri-Food Canada (AAFC), Health Canada, Fisheries and Oceans Canada and Environment Canada in all matters relating to food oversight including GM foods. In 2013, it was decided that the CFIA would report to the Health minister and would no longer be under the responsibility of Agriculture and Agri-Food Canada. The CFIA is responsible for safety inspections and quarantine programmes for all food products (plants and animals). CFIA is responsible for food labelling taking into account the safety of food products (presence of potential allergens) as opposed to Health Canada, which oversees nutritional labels. The CFIA enforces policies and agricultural federal acts and regulations, including 13 federal statutes and 38 sets of regulations related to food safety. It is the main regulatory body when it comes to Canada's animal health and welfare and plant protection programmes, overseeing the scientific risk assessment of new GM plant varieties for environmental releases, for animal feed and for field trials. Health Canada regulates GM foods destined for human consumption. While there is no clear regulatory mechanism for GM animals, it is believed that transgenic animals such as GM fish would fall under the mandate of the Canadian Environmental Protection Act (CEPA), which contains a section dealing with 'Animate Products of Biotechnology' entering the environment and does not cover all living organisms, including transgenic animals. CEPA, which is a wide ranging set of laws and regulations relating to environmental matters, is enforced by Environment Canada, and acts as a 'safety net' because it deals with all animate products of biotechnology not covered by other federal legislation as in the example of transgenic animals. The CFIA coordinates activities with Environment Canada when issues of environmental safety arise. Agriculture and Agri-Food Canada does not regulate products of biotechnology, its role is to provide scientific advice in safety decisions as well as to assist in the development of biotechnology regulatory policy.

In Canada, the federal government introduced a regulatory framework for determining the relative safety of GM crops in 1994. The development of this framework was the result of close consultations between Canadian and US regulators and was based on a set of 'strategic regulations' that followed 'logical reasoning rather than empirical methodologies' according to one US regulator (Abergel, 2007). As such, scientific methodologies to determine the safety of GM crop varieties were not specified but rather a set of broadly defined guiding principles were introduced that would enable 'logical reasoning'. For a complete history and detailed account of the Canadian approach to regulating GM crops, refer to Barrett and Abergel (2000, 2002) and Abergel and Barrett (2002). Basically, GM plant varieties are considered 'Plants with Novel Traits' or PNTs, defined as 'a new variety of a species that has one or more traits that are novel to that species in Canada' (CFIA). Novelty is also characterized by a plant variety that is both stable and new relative to plant species cultivated in Canada and its potential to have an environmental effect. As such the category of PNTs includes plant varieties developed using techniques that include genetic engineering, mutagenesis, cell fusion and traditional breeding. PNTs do not refer specifically to GM varieties, which mean that the Canadian system does not make specific reference to genetic engineering when it assesses the final product. PNTs fall under the provisions of the Feeds Act, the Food and Drug Act and the Seeds Act depending on their final use. As a result, new varieties undergo mandatory reviews only by relevant regulatory authorities.

Developers of new GM crop varieties must obtain official authorization before they can grow a PNT on Canadian soil from the CFIA's Plant Biosafety Office (PBO), which takes the lead responsibility for regulating the environmental release of plants developed through biotechnology. Once it has evaluated the safety risks of PNTs, the PBO publishes non-confidential biosafety decision documents about unconfined releases through its website. It also provides information to provincial governments about the location of approved field trials involving PNTs. It is important to emphasize those regulators in the Plant Biosafety Office focus their safety assessment of PNTs solely on the newly introduced trait regardless of the process used to create the GM plant. According to the Seeds Regulations, the novel trait has to be 'intentionally selected, created or introduced into a distinct, stable population of cultivated seed of the same species,' and it must not be substantially equivalent in terms of safety and use of cultivated seed of the same species in Canada, particularly as it pertains to its environmental and biodiversity impacts.

The determination of a variety's novelty is made by proponents or in conjunction with regulatory authorities if unclear. Novelty acts as a regulatory trigger, a determination of novelty means that the variety is not substantially equivalent to its counterparts and that it must undergo risk assessment for determining its impacts on the environment and human health. However, if a plant is found not to be novel, 'based on scientific rationale', it is considered substantially equivalent to its counterpart and may be exempt from undergoing a thorough safety assessment under the Seeds Regulations. The PBO determines novelty and authorizes PNTs based on a safety assessment conducted by the Plant Biotechnology Risk Assessment Unit (PBRA), both under the authority of the CFIA. Assessors consider five criteria when a variety falls under the definition of a PNT: weediness potential, gene flow and invasiveness, plant-pest potential, impact on non-target species and impacts on biodiversity. The safety assessment is based on comparing the biology of the modified species with its natural counterpart as well as an analysis of the data provided by proponents collected from field trials as well as specific details about the genetic modification process. In cases of new varieties that fall under PNTs but are 'retransformation' or 'remutation' (Shearer, 2014), that is plants that use traits already approved in other varieties of the same species for the same use or plants that cannot be transformed using traditional breeding methods, no full risk assessment may be required as the trait has now become 'familiar'.

The principles that underpin the Canadian regulatory system are 'novelty' and 'substantial equivalence'. Each principle is meant to provide a sufficiently adequate assessment of the environmental and health risks posed by these crops using a step-wise approach that combines knowledge about: the plant variety, the trait being introduced, the environment, the interactions between the newly introduced trait, the plant and the release environment used to authorize small-scale field trials. The results of small-scale field trials then provide the basis for authorizing large-scale unconfined releases. Focusing on 'logical reasoning' and prior knowledge about the PNT, the novel trait and the environment are studied as independent factors. Moreover, empirical observations consist in observing interactions between these factors over limited time and space without clear indications of the quantity and quality of data needed to provide significant knowledge. As such, the Canadian system opts for an approach that fails to anticipate risk, instead generating ecological risk by enabling incomplete studies and restricted experimental design to pass for assurances of safety.

The term 'trade science' was used earlier to describe the type of scientific principles devised by Canadian regulators (in line with their US counterparts) to assess the impacts of GMOs on the environment and in human/animal health. These 'scientific' principles were specifically created to facilitate the trade and transboundary movement of GM products (Abergel, 2007). The lack of comprehensive scientific assessment of GM varieties and the decision to adopt the concept of substantial equivalence for determining environmental safety assessments kept the Canadian requirements of extensive safety data to a minimum. The data requirements necessary for safety assessments, whilst not clearly defined experimentally, have been severely critiqued for their insufficient ability to observe and/or predict adverse ecological effects. Since commercial crop developers are responsible for demonstrating the lack of 'novelty' of their products, i.e. their substantial equivalence to naturally occurring counterparts, this essentially guarantees very few findings of novelty when it comes to GM crop varieties. Clearly, these principles were largely defined to stimulate the biotechnology industry as no independent review of findings is mandated by regulations. Not only do the data generated by proponents remain unverified, they are to this day unverifiable as the information is considered confidential business information. As more PNTs passed through the system, these served as baselines of hazard acceptability (Abergel, 2007) during the assessment of subsequent varieties obtained through genetic engineering technologies. These regulatory procedures became operationalized and, with time, effectively posing as 'sound science' even though severely limited in scope. By insisting on the scientific robustness of its regulatory system in key international arenas, Canadian regulators and agricultural industry actors hope to convince trading partners to not only import Canadian goods but also adopt a similar attitude towards GMOs. Continual reference to the scientific basis of Canada's decision-making system around GM varieties is no assurance of safety or scientific consensus or even validity. The Canadian system works because it adopts what some critics have called a 'don't look, don't see' approach to scientific testing.

Since the elaboration of the Canadian regulatory framework, which essentially views GM varieties as safe until proven otherwise, scientific data have been accumulating over the last two decades that show the adverse effects of these crops. While it would be far too complex to discuss their implications here, suffice it to say that no other issue has had such a polarizing effect on the conduct of science, its ethos and the dissemination of its results. I am referring here to the lack of unbiased, independent research, the suppression of key studies and data, and the high-level of political, informational and economic interference that corporate actors have engaged in. Since the early 2000s more than 100 plants with novel traits have been deregulated by CFIA, including several GM plant varieties.

8.4 The Agricultural Growth Act

Prior to the introduction of the Agricultural Growth Act by the Conservative government of Canada, the Standing Senate Committee on Agriculture and Forestry was given the mandate in 2011 to study the status of agricultural research and innovation in Canada. Specifically, the committee examined R&D in the context of 'developing new markets domestically and internationally; enhancing agricultural sustainability; improving food diversity and security' (Canada, 2014). The publication of its report entitled *Innovation in Agriculture: The Key to Feeding a Growing Population* consulted 170 witnesses from different sectors of agriculture and gathered information from fact-finding missions throughout Canada and the USA. The recommendations made in this report mirror the changes contained in the AGA. The following section deals with the most relevant recommendations.

One of the key findings of the Senate Standing Committee on Agriculture and Forestry report was the need to 'establish a legislative and regulatory framework that facilitates quick and predictable approval and commercialization of new products and processes. Laws and regulations should also be modern and flexible in order to adapt to rapidly changing technologies and manufacturing processes.' The authors of the report placed emphasis on Canadian agriculture as a key sector of the economy, therefore advocating for 'laws and regulations [need] to be harmonized with those of Canada's major trading partners and be based on scientific evidence' (Canada, 2014, p. 34). This finding led to the belief that the crop variety registration system under the Seeds Regulations needed to be improved. At that time, seed required no registration as a variety in order to be legally sold in Canada, meaning that farmers could use traditional seed and retain their customary use of traditional crop varieties. Spurred by the Canadian Seed Trade Association (CSTA), it was noted that the system for registering new varieties was inflexible and needed to be adapted to new realities. The proposed system for new plant varieties (GM or conventional) suggests that only registered seed varieties be sold for human consumption, severely restricting the use of non-registered crop varieties and favouring large commercial developers.

The Canadian seed registration system includes a merit assessment prior to variety registration, followed by, in the case of PNTs, approval for unconfined release. A new seed variety can go through one of these three routes:

Part I, which is similar to what is done now, requires that the crop kind or variety be recommended by an official committee and undergo up to three years of trials, be grown out in fields and be compared to check varieties. The second route, or Part II, does not require such merit assessment but does require a recommendation from a committee. The third route, or Part III of the new framework, allows application to be made directly to the CFIA, which is responsible for the seed variety programme. Recommendation from an official committee or merit assessment is therefore not required. In order to be registered faster, crops can be transferred from Part I to Part II to Part III on a priority basis, depending on the merit of the justification and the degree of consensus within the crop value chain. However, this new framework cannot be implemented without amending the *Seeds Regulations* (Canada, 2014: 35)

This means that under the 'simplified' system a new variety could be registered without undergoing any merit assessment. Merit refers to the quality assessment of new varieties compared to reference varieties. New varieties must demonstrate equal or superior quality for the specific characteristics being assessed for that crop in order to be registered by a recommending committee. Assessments take disease resistance, yield and other agronomic as well as market parameters into account, which they measure against a commercial registered reference variety. However, crops such as maize are exempt from variety registration in Canada because in 1996 growers felt that the time needed to assess new hybrid maize varieties, which could take to 2–3 years, would place Canadian growers at a disadvantage when compared to US farmers. After a crop has been recommended for registration, developers must apply to the CFIA's Variety Registration Office (VRO) for seed certification by providing extensive information. If the plant is a PNT, the novel engineered traits must be approved for unconfined release.

Along with this recommendation, the Senate Standing Committee on Agriculture and Forestry highlights the need to 'simplify' application procedures for assessing the safety of PNTs. Currently, new crop developers need to submit safety data to the CFIA for food and feed safety as well as environmental safety and to Health Canada for the evaluation of these plants on human and animal health. Companies find it onerous to have to submit the same data to different agencies that work on different timelines (Canada, 2014: 36). During public consultations, Monsanto representatives also testified that Canada should accelerate its approval process so that 'biotechnology-derived plants can be commercialized faster outside Canada which can affect the competitiveness of domestic producers' (Canada, 2014: 36). Other industry actors asked for more predictability in order to better 'plan investments'. Based on the testimonies of Monsanto, CSTA and Dow Agro Sciences Canada, the committee recommended that Health Canada and the CFIA alter the Safety Assessment Process for Plants with Novel Traits to 'centralize information and data collection, accelerate the review process, and improve predictable timeframes' (Canada, 2014: 36). The report also recommends that the safety assessment system be modified to accommodate the commercialization of PNTs for molecular farming (the use of agricultural crops to produce products other than food, feed or fibre, such as pharmaceuticals or industrial biomaterials).

The Standing Committee report produced a series of recommendations, which formed the basis for altering seed legislation in Canada. The section dealing with IP provides the rationale for changing the current legislation and accelerating the patent application system. Witnesses, mostly from industry or commodity groups, requested an extension to the duration of patent protection to allow innovators to recoup the costs of developing their product. Currently, patent protection lasts 20 years in Canada. In the case of plant varieties, the Senate Standing Committee on Agriculture and Forestry recommends that the Plant Breeders' Rights Act (1990) be made current to the standards of UPOV1991, which expands the rights of breeders. Witnesses claim that under the UPOV 1978 system, companies were discouraged from selling new international varieties or acquiring new genetic materials as it afforded them limited IPR protection. In terms of research and innovation, IPRs were thought to stimulate R&D and encourage private investment in agricultural research, although problems such as limits on knowledge sharing, industrial concentration and higher prices from patented products and technologies were also identified.

Another important recommendation dealt with the international harmonization of the regulatory framework for approving novel products. It is remarked that accelerating scientific reviews for new products and varieties that have been approved in other countries would increase Canada's competitiveness. Standardized review times of 180 days as well as mutual recognition of scientific data can also speed up approvals. Hence, products already approved in Canada could be approved in the USA or elsewhere. Joint scientific reviews across jurisdictions among trading partners could make it possible for an applicant to submit a single application for approval as data requirements could be harmonized. In fact, regulatory harmonization, it was suggested, could be integrated into new trade agreements. Canada has signed a number of multilateral and bilateral trade agreements over the last decades and government negotiators have continually emphasized the removal of tariff and non-tariff barriers for agricultural goods. These trade agreements have clearly increased export markets for Canadian products. While this strategy has been successful, Canada's advocacy around its science-based regulations has also been the focus of negotiations. While Canadian trade negotiators promote Canada's regulatory system for GM plants as scientifically based and burden-free, it is clearly a bargaining tool to open new markets for Canadian products as well as for allowing entry of foreign products of biotechnology into the country. However, as seen earlier, differences in regulatory cultures and GM risk perception mean that tensions are often expressed between trading partners. This is particularly evident in the case of biotechnology-derived foods where the EU stipulates that seed should not contain any GM materials, thus creating a non-tariff barrier and sanitary and phytosanitary measures affecting trade relations. Canada recently concluded two key international trade deals, which were not yet ratified at the time of writing this article: the Trans-Pacific Partnership Agreement (TPP), which involves 12 countries around the Pacific, and the Comprehensive Economic Trade Agreement (CETA) signed with the EU. In both cases, these deals contain important provisions with potentially profound impacts on the agricultural sector, placing pressure on Canada's IPR system to allow the patenting of higher life forms such as plants or animals. In addition both deals contain new IPR enforcement mechanisms. In light of these developments, the following section deals with recent amendments to the seed legislation, signalling Canada's willingness to comply with trading partners.

8.5 Amendments to the Seeds Act and Plant Breeders' Rights

On 25 February 2015, Bill C-18 also known as the Agricultural Growth Act (AGA) became law. It was created in order to amend several acts that affect various measures relating to agriculture. The fact that the Canadian economy posted some of the best results among G7 countries following the global financial crisis, in terms of both production and job creation, is due in large part to agriculture leading the Conservative government to propose legislation to further strengthen the industry. The AGA reinforces and advocates modern science and technology, innovation, and international practice in the agriculture industry. It is believed that the proposed amendments will stimulate investment in developing new crop varieties and agricultural innovation. Most notably, it amends several key aspects of the Plant Breeders' Rights Act, including the duration and scope of those rights and conditions for the protection of those rights. It also provides for exceptions to the application of those rights. In addition, the AGA also modifies The Feeds Act, The Fertilizers Act, The Seeds Act, The Health of Animals Act and The Plant Protection Act.

The AGA also alters the Agriculture and Agri-Food Administrative Monetary Penalties Act to, among other things, increase the maximum limits of penalties that may be imposed for certain violations. Other amendments include changes to the Agricultural Marketing Programs Act so that advanced payment programmes to farmers might be better accessed, managed and delivered so as to modernize this programme. Lastly, the AGA provides clarifications to the Farm Debt Mediation Act so that the Minister of Agriculture and Agri-Food might participate in the mediation process when that minister is a guarantor of a farmer's debt.

As was mentioned earlier, the Standing Senate Committee on Agriculture and Forestry recommends the need for Agriculture and Agri-food Canada and the Canadian Food Inspection Agency to bring the Plant Breeders' Rights Act up to the standards of the 1991 Act of the International Convention for the Protection of New Varieties of Plants. It is this recommendation in particular that opposition members objected to and the National Farmers Union (NFU), Canada's second largest farming group, proposed instead an alternative Farmers Seeds Act in line with the UN International

Treaty on Plant Genetic Resources for Food and Agriculture, which Canada ratified in 2002.

Before discussing changes to the Plant Breeders' Act, it is important to understand the gradual streamlining of plant variety registration in Canada because it sets the context for the adoption of UPOV '91. As was mentioned earlier, most seeds are subject to mandatory variety registration in Canada in order to ensure their safety, quality, stability standards and the identification and traceability of PNTs (including GM varieties). In 2009, regulations were amended to create a 'flexible variety registration system', which would reduce the regulatory burden for developers. Crops necessitating variety registration were divided in three groups, each group stipulating a different level of variety registration requirements. Crops in Part I required pre-registration testing that includes field trials and in some cases laboratory testing as well as merit assessment. Part II required that a new variety undergo pre-registration testing but no merit assessment. Finally, there were no requirements for pre-registration testing or merit assessment for crop kinds included in Part III and proponents could directly apply to the CFIA for variety registration. The introduction of this tiered system was aimed at reducing the regulatory burden by moving crop kinds from Part I, to Part II and eventually Part III depending on the level of consensus and rationale established within each crop value chain that includes variety development, distribution, production and end-use (Regulations amending the Seeds Regulations, 2014). The 2009 amendment was intended to reduce the costs of variety registration for small local breeders as well as large multinational companies. In 2014, another amendment was introduced that further reduces the variety registration requirements for oilseeds, soybeans and forages by moving them from Part I to Part III. According to Boehm (2009), former president of the NFU, this is significant because it encourages all crop kinds to eventually be moved to Part III, accelerating the registration and commercialization of new seed varieties. Suggesting that the possibility that GM varieties will now be registered faster with minimum regulatory requirements, Boehm fears that this severely threatens organic agriculture. Of particular concern is the fact that Part III contains crop kinds very likely to be GM varieties, the 2014 amendment concerning soybeans and forages making this a reality.

The new Agricultural Growth Act contains provisions that strengthen IPRs that see the adoption of and implementation of UPOV 91 by plant breeders. Prior to the AGA, Canadian legislation met the requirements for UPOV 78, which underwent several changes over time to keep up with the UPOV requirements for plant breeders' rights protections. Until the AGA became law, Canada was one of only two UPOV members from developed countries with legislation that did not comply with UPOV 91. Innovations in agriculture and in plant variety technology prompted this legislative change, because, according to the federal government, UPOV 78 placed Canadian breeders and farmers at a competitive disadvantage. Aligning Canadian regulations with UPOV 91 is believed to 'level the playing field for [our] producers, but it is also expected to encourage foreign breeders to release their varieties in Canada' giving Canadian farmers access to new varieties used by competitors (Sanford, 2014).

Bill C-18 also allows for the recognition of foreign data and reviews for new feed, seed and fertilizer registrations. According to lawmakers, streamlining the registration process for new products coming into Canada should not compromise safety, even though foreign registration systems may have different requirements. The AGA is thought to promote access to new Canadian and foreign varieties and expand the protection of new Canadian varieties abroad, both seen as important aspects of competitiveness. Arguments supporting the new legislation suggest that strengthening IPRs for plant breeders means boosting investment in R&D for new crop development, most notably focusing on the development of new varieties with higher crop yields. Under the new amendments, IP protection for plant breeders goes from 18 to 20 years. In addition, the inclusion of the provision of 'essentially derived' materials means that plant breeders can claim royalties if new varieties can be proven to be simple derivations of old varieties.

Section 5 of the AGA (Agricultural Growth Act, 2014) deals with changes to the Plant Breeders Act and stipulates that the holder of the plant variety has the exclusive right:

- (a) To produce and reproduce propagating material of the variety;
- (b) To condition propagating material of the variety for the purposes of propagating the variety;
- (c) To sell propagating material;

- (d) To export or import propagating material of the variety;
- (e) To make repeated use of propagating material of the variety to produce commercially another plant variety if the repetition is necessary for that purpose;
- (f) In the case of a variety to which ornamental plants belong, if those plants are normally marketed for purposes other than propagation, to use any such plants or parts of those plants as propagating material for the production of ornamental plants or cut flowers;
- (g) To stock propagating material of the variety for the purpose of doing any act described in any of paragraphs (a) to (f); and
- (h) To authorize, conditionally or unconditionally, the doing of any prescribed act described in paragraphs (a) to (g).

What this signifies in practice is that, according to Boehm, plant breeders can now hold both patent and plant breeders' rights on the same variety. More importantly, they also have the right to collect royalties anywhere in the seed and food system and can potentially collect royalties on a farmer's whole crop (Boehm, 2015). He argues that what is known as farmers' privilege can be removed and that the issue of rights is inverted in favour of breeders while farmers retain limited and conditional 'privilege' to save or reuse their own seed. In fact, a farmer's privilege is severely limited by 'reasonable limits', which can be described as restrictions that can be imposed on farmer's saved seed depending on 'the type of variety, the number of cycles of reproduction, remuneration proportion of harvested material, etc' (Boehm, 2015). The AGA now allows the Minister of Agriculture to authorize or revoke on a case-by-case basis a farmer's privilege. Effectively, this means that the minister controls the fate of seed instead of farmers who have long traditionally held that role. During parliamentary debates, an opposition member stated:

I am very interested in the issue of farmers' privilege, where the government's idea of representing farmers is to take away rights from them and allow the minister to decide whether or not they have those rights. In this process it is very easy for the large corporate seed lobbyists to just call up the minister, have a meeting, take his staff out for drinks and, lo and behold, the privileges that farmers have enjoyed year after year, decade after decade, can be annulled without any process (Angus, 2014).

Changes brought about by the AGA illustrate how the expansion of PBR alters the scope of farmers' privilege, making its future uncertain.

8.6 The Implications of AGA for Intellectual Property Rights, Seed Ownership and Agriculture

IPRs are monopoly rights granted for a prescribed period to developers of new products or processes that cover patents, designs, trademarks, copyrights and trade secrets. In agriculture these include Plant Variety Rights and Plant Breeders' Rights. The idea is to promote innovation and open markets for new products. In 2001, the US Council on Foreign Relations stated that trade disputes between countries were caused by differences in planting of GMOs and in regulatory approaches. Trade frictions between the EU and Canada and the USA over GM crops heated up when the EU declared a moratorium on the importation and growing of GM crops in 2000. According to the Council on Foreign Relations:

> in principle the WTO's agreement on Sanitary and Phytosanitary regulations (the 'SPS Agreement') should govern these issues. However, in practice it cannot settle trade disputes, as the WTO cannot rule on the legitimacy of a nation's food safety laws. Accordingly, trade differences can be dealt with through intellectual property as the WTO's agreement on Trade-Related Intellectual Property Rights (TRIPs) requires countries to adopt greater protection of intellectual property, including protection of private property rights in agriculture.

Since most IPRs in agriculture are privately owned, the commercialization of GM crops is directly linked to a country's trade policy. During the 'green revolution' in agriculture beginning in the 1950s, key innovations were publicly owned and advanced by public- and foundation-funded agricultural research programmes.

One of the most contentious issues contained in this legislative amendment is the provision for Canada to adopt UPOV 91. UPOV 91 constitutes a fundamental shift from UPOV 78. First, it restricts the rights of farmers to freely use their seeds or other propagation materials for further cultivation, although there is an optional exception in Article 15.2 of UPOV 91 which allows farmers to 'use for propagating purposes, on their own holdings, the product of the harvest which they have obtained by planting, on their own holdings, the protected variety' (Association for Plant Breeding for the Benefit of Society (APREBES), 2015). However, this provision is optional and it is limited because it excludes propagation material which is not the result of the harvest (fruits or berries) and prevents exchange and selling of protected materials. This exception means that breeders can still demand royalties even where seed saving is permitted. The AGA acknowledges this optional provision but reserves the right to prevent farmers from using their 'privilege' on a case-by-case basis, as decided by the Minister of Agriculture. It is unclear how the notion of privilege will be interpreted in disputes involving farmers accused of infringing PBR rules. The question is will the gradual enforcement and strengthening of breeders' rights eventually eliminate farmers' privilege?

Second, plants and varieties in some countries can be patented under UPOV 91 in addition to PVP. In previous UPOV Acts, double protection was not allowed. Third, the harvest belongs to the breeder when unauthorized material has been propagated, as their rights extend to harvested material. Breeders can theoretically claim ownership of the harvest and products made from the harvest if farmers use protected plant materials without paying royalty fees. Finally, the use of protected varieties for breeding purposes is restricted, as newly produced varieties need to be significantly different from the parent cultivars. Plant varieties that do not constitute 'true innovation' will be considered as an 'essentially derived' variety and will be the property of the first breeder as previously discussed.

Until this new amendment Canada was still operating under the provisions of UPOV 78. In short, this controversial amendment as it pertains to the Seed Act and to the Plant Breeders' Rights Act leaves open many questions about the future of farming in Canada, as many farmers will no longer be able to rely on their saved crop seed for planting. Some locally produced seed varieties might be supplanted by 'new and innovative' global varieties developed mostly through genetic engineering. According to the NFU, 'this legislation will fundamentally restructure agriculture in ways that will profoundly affect farmers, rural communities and the food system they supply' (Slomp, 2014). By 'removing the regulatory burden' on large seed producers, new crop variety developers and more generally the whole input industry this new law inherently leaves farmers vulnerable to increased corporate control of agriculture, by generating revenue on a farmer's entire production and not just purchased seed (Boehm, 2015). The reinforcement of IPRs over seed and crop varieties in line with UPOV 91 means that, as Boehm suggests, 'the plant breeder/seed corporation would have total control of seed, including exclusive control of conditioning (cleaning and treating), stocking (bagging or storage), importing, exporting, and sales of seed. This upsets the thousands of years of normal agricultural practice whereby farmers always saved seed for their next crop'. These changes increase dependence on external seed suppliers and open the door for farmers having to pay to reuse their own seed (Boehm, 2015). It clearly jeopardizes farmers' autonomy.

8.7 Low-Level Presence: An Example of Genetically Modified Crops Deregulation

In a related but indirect way the AGA opens the door to the low level presence of previously unauthorized GMOs into the country. Canada currently does not allow GM foods or feeds unapproved by Health Canada to enter the agricultural system. In an effort to address 'trade disruptions', the current government is working on a plan that would allow low levels of contaminated crops into the country without the requirement for risk assessment by the Health Department. This is in line with the AGA's streamlining of foreign data requirement for seed registration and for allowing the unconfined release of PNTs. The amount of low-level presence of GM crops into imported food and feed shipments that would bypass risk assessment would be set at 0.2%. This initiative was prompted by the realization that, as the amount of GM crops planted commercially increases globally, there is a greater chance of unintended contamination of both authorized and unauthorized GM materials. The adoption of thresholds for unauthorized GM varieties makes it clear that, in the eyes of Canadian authorities, GM contamination is inevitable and cannot be stopped. Allowing low-level presence of unauthorized GM varieties and materials makes it impossible to prevent them from entering our food system. Yet Canadian authorities do not address issues of public interest, focusing instead on trade issues. Questions remain about the legal aspects of this policy since unauthorized varieties should simply not be permitted to enter Canada. This development undermines the credibility of the Canadian regulatory system at a time when citizens are asking for more transparency and accountability from authorities concerning food safety. According to an article published in the *Epoch Times*, when questioned about this latest initiative officials declared that: 'the government has been taking action to seek global solutions to prevent unnecessary trade disruptions, ensure open and predictable trade that is based on science, and make the agriculture sector more competitive' (Baker-Voakes, 2015). Hence, it is obvious that the same logic prevails as in the AGA and that the Canadian government is concentrating efforts to allow the presence of GMOs at home and abroad using a variety of means. By enabling unauthorized GM varieties to enter Canada 'through the backdoor', low-level presence is but one of the many consequences of the amendments contained in the AGA converging to reduce and eventually eliminate the need for regulations.

8.8 Conclusion

The AGA combines the usual blatant pro-trade pro-GM stance: unquestioned support for large industry players, farmers and private research needs. In addition, under the guise of regulatory harmonization, this piece of legislation ignores the concerns of small-scale farmers and disregards the need for research into less costly non-GM alternatives that stimulate innovation and spur local food security. In some ways this legislation represents a new type of law making as it further entrenches IP laws and discounts all other forms of agriculture and farming that do not fit 'the agricultural biotechnology narrative'. By expanding IPRs and plant breeders' rights and streamlining regulatory requirements to a minimum for the transboundary movement, including import and export of GMOs, it will increase the number of applications filed by breeders' of new crops for new plant varieties and it will extend them to more countries regardless of public opinion.

By taking seeds away from the control of farmers and access to rural communities, this further erodes the capacity to preserve biodiversity by privatizing and technologizing seed. It confirms agricultural policy to international trade commitments, focusing mostly on industrial objectives. Today, large seed firms dominate the market and the seed sector has been reshaped on a global scale as a result of substantial private investments and IPR protection (UNEP-CBD, 1996). The vertical integration of the seed sector causes less competition between seed firms and, as a result of IPR protection, higher prices for seed. A globally integrated seed sector has also focused breeding research on major crops with important commercial value (which may not be food) and has taken incentives from domestic breeding programmes. Besides restricting market access to specific crop varieties, seed companies control the food system and exert pressure on the costs of inputs (fertilizers, pesticides, herbicides, etc.) for maximum crop yield. Monopoly ownership over seed and seed varieties will mean limited choices for farmers who will depend on external suppliers for their crops. It will also likely decrease availability and access to certain kinds of crops, mostly non-patented or publicly owned varieties.

Second, the choice to grant the minister of agriculture new exclusive decision-making powers regarding food safety, new variety registrations and seed control more generally bypasses regulatory frameworks and legislative processes. Instead, the minister alone will now exert new powers on a 'case-by-case' basis regarding the ability of farmers to store and save their seed, whether some crop varieties and agricultural products should be exempt from regulatory approvals based on another country's safety assessment, etc. It will be up to the minister to decide whether new crop varieties need no further studies. This goes against calls by civil society groups and the Royal Society of Canada for a more transparent and impartial regulatory system for biotechnology crops. In its 2001 report, the Royal Society of Canada examined and critiqued the regulatory system providing insightful comments about the ambiguity of the science used to determine the safety of GM foods (Royal Society of Canada, 2001).

While the AGA appears to be a modernizing piece of legislation relating to 'competitiveness' and 'innovation' in terms of opening the door to private investments for new crop development and for opening markets to Canadian agricultural goods, in fact it actually deregulates several key areas of Canadian agriculture that protect farmers and public safety, further eroding local capacity for innovation and rural community development. It is a 'disciplining' legislation in that it removes rights away from farmers and farming communities and places undue power in the hands of the private seed developers and the global seed industry. By granting the Minister of Agriculture new decision and discretionary powers regarding specific situations, this legislation also bypasses regulatory and legislative processes that ensure a minimum of government accountability. CFIA is the agency responsible for enforcing standards and policies established by Health Canada relating to the safety and nutritional quality of food. It is also responsible for food inspection, animal health, plant protection and consumer protection. The CFIA appears to have conflicting mandates; in the area of agricultural biotechnology it both regulates and supports the industry while protecting the public interest. In addition, the agency evaluates the safety of GM foods and feeds under several legislative acts and manages plant breeders' rights. The AGA expands the powers of the CFIA and of the Minister of Agriculture. As a side effect of this legislation, global trade rules can be used to 'effectively narrow the menu of regulatory choices open to governments', focusing instead on keeping markets open for GM products and providing a kind of regulatory 'predictability' among trading partners (Abergel, 2007). IPRs not only provide the basis for opening markets for Canadian GM products but they also give incentives for crop developers and breeders to focus on specific types of commercial traits and plant varieties that can be genetically engineered.

The strong and unwavering support of the Canadian federal government for the biotech industry in recent decades makes this latest element of the AGA particularly questionable. All these factors contribute to what Nixon describes as a kind of slow violence: 'a violence that occurs gradually and out of sight, a violence that is dispersed over time and space, an attritional violence that is typically not viewed as violence of delayed destruction typically not viewed as violence at all' (Nixon, 2011). The restructuring of agriculture in the developed and developing world participates in slow violence, which is 'embedded in neoliberal order of austerity measures, structural adjustment and rampant deregulation, corporate mergers, etc.' as these contribute in widening the gap between rich and poor, North and South. These economic measures directly threaten the autonomy of farmers everywhere as they further erode away the conditions for environmental and economic sustainability already in crisis in many parts of the world.

From a farming point of view, one can ask, after two decades of transgenic crops, have conditions improved? How can the Canadian experience with GMOs serve as a lesson to other countries and in particular for farmers in the global south? As more and more countries amend existing seed legislation or adopt new seed laws to conform to international trade agreements, it is important to ask how much we want trade imperatives to determine not just how we eat and grow food but how much power, as democratic societies, we can exert over these vital choices.

Note

1 *Bill C-18 (Historical) House Debate: Agricultural Growth Act: An Act to Amend Certain Acts Relating to Agriculture and Agri-food*. Assented to 25 February, 2015. 2nd Session, 41st Parliament. Available at: https://openparliament.ca/bills/41-2/C-18 (accessed 15 October 2015).

References

Abergel, E. (2007) Trade, science, and Canada's regulatory framework for determining the environmental safety of GE crops. In: Taylor, I.P. (ed.) *Genetically Engineered Crops: Interim Policies, Uncertain Legislation*. The Haworth Press, New York, pp. 173–206.

Abergel, E. (2012) The paradox of governing through the courts: the Canadian GM contamination debate. In: Koç, M., Sumner, J. and Winson, A. (eds) *Critical Perspectives in Food Studies*. Oxford University Press, Don Mills, Ontario, pp. 260–275.

Abergel, E. and Barrett, K. (2002) Putting the cart before the horse: a review of biotechnology policy in Canada. *Journal of Canadian Studies* 37, 135–161.

Agriculture and Agri-Food Canada (2015) *2015* Canadian Agricultural Outlook. Available at: http://www.agr.gc.ca/eng/about-us/publications/economic-publications/alphabetical-listing/2015-canadian-agricultural-outlook/?id=1423751592650 (accessed 15 October 2015).

American Seed Trade Association (2015) *Key Issues in International Seed Trade.* Available at: http://www.amseed.org/issues/international/key-issues (accessed 15 October 2015).

Angus, C. (2014) House of Commons Debates, 41st Parliament, 2nd Session, Edited Hansard, No. 147, November 24, 2014. Available at: https://openparliament.ca/bills/41-2/C-18 (accessed 15 October 2015).

Association for Plant Breeding for the Benefit of Society (APREBES) (2015) *From UPOV 78 to UPOV 91.* Available at: http://www.apbrebes.org/content/upov-78-upov-91 (accessed 15 October 2015).

Baker-Voakes, K. (2015) Canada to allow unapproved GM crop contamination in imports. *The Epoch Times*, 18 July 2015.

Barrett, K. and Abergel, E. (2000) Breeding familiarity: environmental risk assessment for genetically engineered crops in Canada. *Science and Public Policy* 27, 2–12.

Barrett, K. and Abergel, E. (2002) Defining a safe genetically modified organism: boundaries of scientific risk assessment. *Science and Public Policy* 29, 47–58.
Bénachour, N. and Séralini, G.-E. (2009) Glyphosate formulations induce apoptosis and necrosis in human umbilical, embryonic, and placental cells. *Chemical Research in Toxicology* 22(1), 97–105.
Boehm, T. (2009) Government ignores concerns with variety registration change and plows ahead anyway. Available at: http://www.nfu.ca/story/government-ignores-concerns-variety-registration-change-and-plows-ahead-anyway (accessed 15 October 2015).
Boehm, T. (2015) Bill C-18 will take from farmers' pockets. *The Western Producer*. Available at: https://www.producer.com/2015/02/bill-c-18-will-take-from-farmers-pockets (accessed 15 October 2015).
Canada (2014) *Innovation in Agriculture: The Key to Feeding A Growing Population. Canada Parliament, Standing Senate Committee on Agriculture and Forestry, Report* 6. Available at: http://www.parl.gc.ca/Content/SEN/Committee/412/agfo/rep/rep06jun14-e.htm (accessed 15 October 2015).
Canada (2015) Seeds Act, Revised 27 February 2015. Government of Canada, Justice Laws Website. Available at: http://laws-lois.justice.gc.ca/eng/acts/s-8/ (accessed 15 October 2015).
Glover, D. (2010) The corporate shaping of GM crops as a technology for the poor. *The Journal of Peasant Studies* 37(1), 67–90.
Ho, P., Zhao, J.H. and Xue, D. (2009) Access and control of agro-biotechnology: Bt cotton, ecological change and risk in China. *The Journal of Peasant Studies* 36(2), 345–364.
International Agency for Research on Cancer (IARC) (2015) IARC Monographs Vol. 112: evaluation of five organophosphate insecticides and herbicides. Available at: https://www.iarc.fr/en/media-centre/iarc-news/pdf/MonographVolume112.pdf (accessed 15 October 2015).
International Seed Federation (ISF) (2012) *Seed exports* 2012. Available at: http://www.worldseed.org/isf/seed_statistics.html (accessed 15 October 2015).
James, C. (2012) *Global Status of Commercialized Biotech/GM Crops: 2012*. ISAAA Brief No. 44. ISAAA, Ithaca, New York.
James, C. (2014) *Global Status of Commercialized Biotech/GM Crops: 2014*. ISAAA Brief No. 49. ISAAA, Ithaca, New York.
McMichael, P. (2012) *Development and Social Change: A Global Perspective*, 5th edn. Sage Publications Inc., Thousand Oaks, California.
National Farmers Union (NFU) (2015) *About the National Farmers Union.* Available at: http://www.nfu.ca/about/about-national-farmers-union (accessed 15 October 2015).
Nixon, R. (2011) *Slow Violence and the Environmentalism of the Poor*. Harvard University Press, Cambridge, Massachusetts.
Royal Society of Canada (2001) *Elements of Precaution: Recommendations for the Regulation of Food Biotechnology in Canada*. Government of Canada, Ottawa.
Samsel, A. and Seneff, S. (2013) Glyphosate's suppression of cytochrome P450 enzymes and amino acid biosynthesis by the gut microbiome: pathways to modern diseases. *Entropy* 15, 1416–1463.
Sanford, G. (2014) House of Commons Debates, 41st Parliament, 2nd Session, Edited Hansard, No. 147, November 24, 2014. Available at: https://openparliament.ca/bills/41-2/C-18 (accessed 15 October 2015).
Shearer, H. (2014) Regulation of Plants with Novel Traits: Canadian Perspectives on the 'Novelty' Trigger. In: Eaglesham, A. and Hardy, R.W.F. (eds) *North American Biotechnology Council (NABC) Report 26: New DNA-Editing Approaches: Methods, Applications and Policy for Agriculture*, pp. 193–200. Available at: http://nabc.cals.cornell.edu/Publications/Reports/nabc_26/26_5_2_Shearer.pdf (accessed).
Slomp, J. (2014) Opinion: Bill C-18, the Agricultural Growth Act – Growth for whom? *Farm Focus of Atlantic Canada*, 21 January 2014. Available at: http://www.atlanticfarmfocus.ca/Opinion/2014-01-21/article-3585069/Opinion%3A-Bill-C-18,-the-Agricultural-Growth-Act%26mdash%3BGrowth-for-whom%3F/1 (accessed 15 October 2015).
Torrance, A.W. (2007) Intellectual property as the third dimension of GMO regulation. *Kansas Journal of Law & Public Policy* 16(3), 257–285.
UNEP-CBD (1996) *The Impact of Intellectual Property Rights Systems on the Conservation and Sustainable Use of Biological Diversity and on the Equitable Sharing of Benefits from Its Use.* Available at: http://www.iisd.ca/biodiv/cop3/3_22_vfinal.htm (accessed 15 October 2015).
Wang, S., Just, D.R. and Pinstrup-Andersen, P. (2008) Bt-cotton and secondary pests. *International Journal of Biotechnology* 10(2/3), 113–121.

9 Recent Innovations in Agricultural Biotechnology: Challenging the Status Quo

Kathleen L. Hefferon*

Department of Food Sciences, Cornell University, Ithaca, New York

9.1 Introduction

Over the past two decades, the patent environment for agricultural biotechnology has been geared toward the enhanced production of crops in order to both increase yield and lower costs. The innovations that arose from this momentum were in response to the need for farmers to produce food more efficiently to sustain a soon to be burgeoning global population, under increasingly difficult environmental conditions resulting from climate change. With climate change are expected increases in both biotic (insects, viruses, fungi and other pests) as well as abiotic (extreme temperatures, drought, flooding) stresses. Unfortunately, patent agricultural innovation has been increasingly politicized, and this new social influence has not served well to stimulate innovation.

As technologies for the genetic engineering of crop plants advance and improve, some effort has been made to utilize these innovations in order to treat or prevent human disease. This can take place either through the use of plants as platforms to produce pharmaceuticals or through the enhancement of phytochemicals, which have health benefits already made by plants through metabolic engineering. This review discusses recent patents regarding aspects of plant biotechnology pertaining to methodologies involved in the generation of transgenic crops, as well as innovations that may improve human health (Table 9.1). With regard to methodologies, procedures that can hasten the regulatory process for genetically modified crops, as well as technologies that facilitate plant biotechnology for a wide variety of uses in general are presented. With respect to human health, a number of innovations using plant virus expression vectors to generate vaccines and other biopharmaceutical proteins in plant tissues are discussed. Genetic engineering approaches to enhance the production of bioactive compounds that improve human health as well as reduce harmful or unwanted compounds are also described. The chapter concludes with a discussion of the future directions for patents in agricultural biotechnology in today's political climate.

9.2 Plant Biotechnology Methodologies

9.2.1 Method and system for data collection and analysis to assist in facilitating regulatory approval of a product

One of the biggest obstacles to the release of transgenic crops can be found in the navigation of the regulatory process. Genetically engineered and other speciality crops require approval from the United States Department of Agriculture (USDA), the Food and Drug Administration (FDA) and/or the Environmental Protection Agency (EPA). It is required that the crop is established as substantially equivalent to its conventional crop counterpart. This procedure is costly, time taking and requires multiple steps, including the collection of multiple data sets and analysis. As a result, small market crop developers are discouraged from commercializing novel or specialized crops due to the significant

*Corresponding author, e-mail: klh22@cornell.edu

Table 9.1. Patent details discussed in this chapter.

Topic Covered	Patent number	Title	Inventors/Institute	Date
Plant biotechnology	US20110224933	Method and System for Data Collection and Analysis to Assist in Facilitating Regulatory Approval of a Product	Elizabeth Hood *et al.*, Infinite Enzymes, LLC State University of Arkansas	15 September 2011
	EP20110155205	Use of Non-Agrobacterium Bacterial Species for Plant Transformation	Xudong Ye *et al.*, Monsanto Technology LLC	6 June 2014
Glycosylation engineering	US20120283420 A1	Production of Multi-Antennary N-Glycan Structures in Plants	Nagels and Weterings, University of Ghent, Bayer Crop Science	30 July 2012
Plastid biotechnology	61593555	Synthetic Brassica-Derived Chloroplast Transit Peptides	Justin M. Lira *et al.*, Dow AgriSciences	1 February 2013
	48904120	Synthetic Chloroplast Transit Peptides	Justin M. Lira *et al.*, Dow AgriSciences	1 February 2013
	WO2012078279 A3	Method of Microalgal Chloroplast Transformation Using Functional Selection in the Absence of Antibiotics	Anastasios Melis, Hsu-Ching Chen Wintz, The Regents of the University of California	30 July 2013
Plant-derived pharmaceuticals	US 13/060,414	DNA Replicon System for High-Level Rapid Production of Vaccines and Monoclonal Antibody Therapeutics in Plants	Hugh Mason *et al.*, Arizona State University	17 July 2013
	US 20120315323 A1	Influenza Hemagglutinin Antibodies, Compositions and Related Methods	Vidadi Yusibov *et al.*, Fraunhofer USA, Inc.	24 July 2014
	US 20120015899 A1	Modified Plant Virus Particles and Uses therefor	George Lomonossoff *et al.*, Plant Bioscience, Limited, Aura Biosciences (John Innes Center, University of Warwick)	12 September 2011
	US 20140287507 A1	Incorporation of Plant Virus Particles and Polymers as 2D and 3D Scaffolds to Manipulate Cellular Behaviors	Qian Wang and Lim Andrew Lee, University of South Carolina	24 October 2012
	WO 2013121405 A1	Oral Unit Dosage Forms and Uses of Same for the Treatment of Gaucher Disease	Yoseph Shaaltiel and Salit Tzaban, Protaliz, Ltd.	19 August 2013
Bioactive compounds in plants	WO 2014141147 A1	Modifying Soybean Oil Composition Through Targeted Knockout of the Fad2-1a/1b Genes	Mathis *et al.*, Cellectis	13 March 2014

Continued

Table 9.1. Continued.

Topic Covered	Patent number	Title	Inventors/Institute	Date
	US20140007305 A1	Reducing Levels of Nicotinic Alkaloids in Plants	Takashi Hashimoto and Akira Kato, 22nd Century Limited, LLC	28 July 2014
	WO/2010/148338	Production of Alkaloids by Liliaceae Cell Culture	Venkatesh Srinivasan *et al.*, Phyton Holdings, LLC	29 August 2014
	WO/2013/146555	Method for Producing Steviol Glycoside	Ono; Eiichiro Suntory Holdings	22 September 2014
	US8624083 B2	Nucleic Acid Sequences Encoding Transcription Factors Regulating Alkaloid Biosynthesis and their Use in Modifying Plant Metabolism	Page *et al.*, National Research Council of Canada	24 April 2014

hurdles to reach regulatory approval status. In patent US20110224933, a computational method for collecting comparative data for use in petitions for regulatory approval of a crop or crop product is provided for (Hood *et al.*, 2011).[1] The computer program enables the search of literature databases that can be used to identify data relative to determining substantial equivalence for the product to provide a first data set. Furthermore, the computational methods described in this invention determine experimental data to be collected for the product based in part on the first data set. Experimental data for the product are collected to provide a second data set, and comparisons between the first and the second data set indicate substantial equivalence for the product. In addition to this, methods are provided to set up a service that assists in procuring regulatory approval of the crop or crop product, including the provision of a web site that can be stored on a web server with secure access for clients (Ding *et al.*, 2002; Hull *et al.*, 2008). The method also enables a document to be compiled comprising data for submission to a regulatory body using the comparative data (McDonald and Pereira, 2005) (Figs 9.1a, b).

9.2.2 Use of non-*Agrobacterium* bacterial species for plant transformation

Agrobacterium-mediated transformation remains one of the principal means to generate transgenic plants. A number of plant species are recalcitrant to *Agrobacterium*-mediated transformation, and efficiency is low in others. Additionally, since *A. tumefaciens* enters plant tissues at wound sites and does not naturally infect unwounded tissues, the use of certain tissues as transformation targets is not available (Shewry *et al.*, 2007). Ye *et al.* (2014)[2] have devised a method for producing transgenic plants and plant cells by using non-*Agrobacterium* bacterial species, which can in turn be widely used for plant transformation. Recently, reports have been published describing transformation of plants by bacteria other than *Agrobacterium* sp. In patent 38723193, the inventors utilized *Rhizobium*, a phylogenetically related but distinct species from *Agrobacterium*. In *Rhizobium*, the conjugal transfer system requires an oriT as the nicking site and TraA or Mob as a nicking enzyme, which is different from the conventional elements used in T-DNA mobilization (VirD2 and RB and LB sites, respectively) (Farrand *et al.*, 1996). Unlike VirD2, which was found to have plant NLS (nuclear localization signal) at its C-terminus for plant nuclear targeting, the TraA or Mob does not have an obvious NLS. The precise mechanism and site of integration of DNA in plants by TraA remains unclear. Members of the *Rhizobium* species are known to symbiotically associate with plant roots in specialized nitrogen-fixing nodules. In addition to host-specific nodulation of plant roots, such as legumes, some plant growth-promoting effects by members of the Rhizobiales are known in the absence of nodulation (Fraley *et al.*, 1985; Broothaerts *et al.*, 2005).

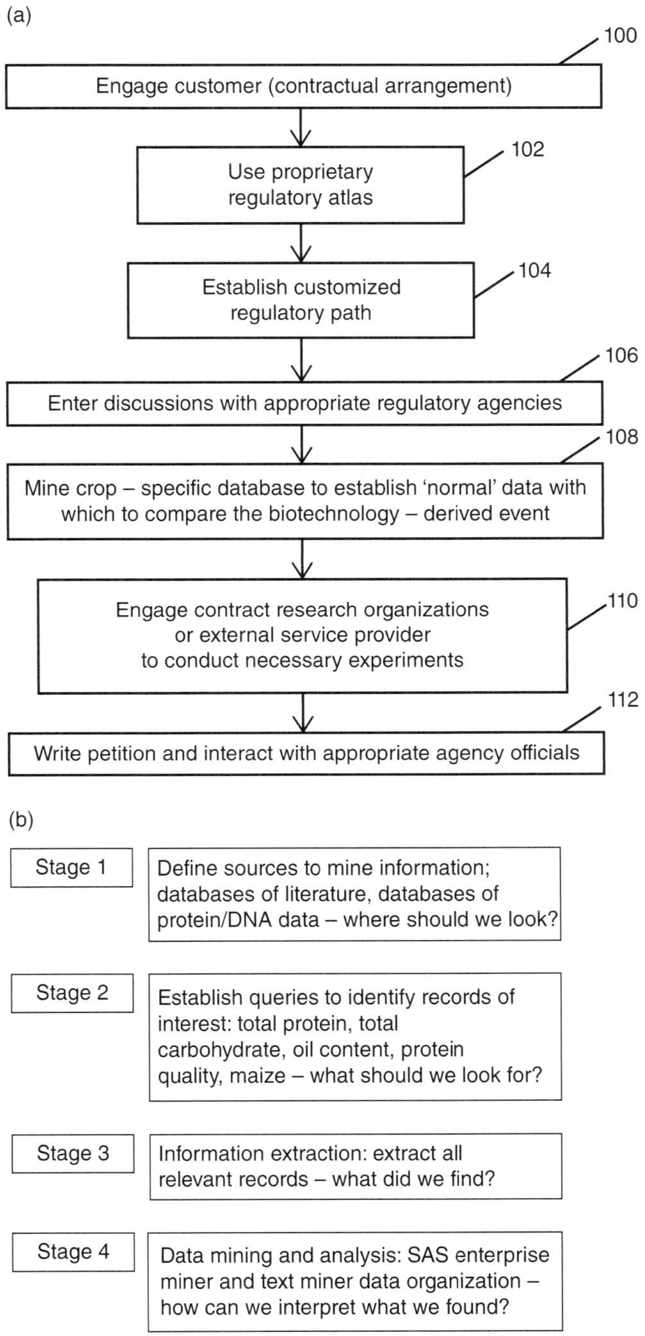

Fig. 9.1. (a) Flow chart indicating one example of a methodology and (b) various stages according to one example of a methodology for mining literature and databases for data describing the variation in composition of maize seed, target protein safety and growth parameters.

9.2.3 Synthetic *Brassica*-derived chloroplast transit peptides

Plastids are subcellular organelles found in plant cells that perform specialized functions within cells. In general, plastids such as chloroplasts are responsible for photosynthesis, as well as the synthesis and storage of certain chemical compounds. Other types of plastids include etioplasts, chromoplasts, gerontoplasts, leucoplasts, amyloplasts, elaioplasts and proteinoplasts. Plastids contain their own genome and machinery for protein synthesis, however, they interact closely with the nuclear genome of the plant cell as well. Many proteins that are encoded by the nuclear genome are destined for the chloroplast, and require transit peptides to be targeted to and function within these organelles (Bruce, 2000). The agricultural industry takes advantage of this fact, and several herbicides have been developed that target nuclear encoded genes whose proteins are synthesized in the chloroplast. For example, triazine-derived herbicides successfully inhibit photosynthesis by displacing a plastoquinone molecule from its binding site in the 32 kD polypeptide of the photosystem II. Mutant plants have been obtained that are resistant to triazine herbicides due to the inability of plastoquinone to be displaced by triazine herbicides. Glyphosate, as another example, inhibits the function of 5-enol pyruvyl-3-phosphoshikimate synthase (EPSPS), which is an enzyme involved in the synthesis of aromatic amino acids. The way by which precursor proteins are imported into the chloroplast can take place by several distinct mechanisms. These mechanisms can involve the use of chloroplast transit peptides (CTPs) found at the N-terminus of the precursor protein (Lee *et al.*, 2008). The patent described by Lira *et al.* (2013)[3] describes the use of chimeric polypeptides comprising an amino acid sequence that controls the transit of the chimeric polypeptides into chloroplasts.

9.2.4 Method of micro-algal chloroplast transformation using functional selection in the absence of antibiotics

A recent invention by Melis and Wintz[4] provides a novel method by which transgenic transformants can be selected in green microalgae. This 'recovery of function' method utilizes the concept that recovery of photosynthetic function in micro-algal chloroplasts can be used as a selectable marker without the requirement of antibiotic selection (Potvin and Zhang, 2010; Grundman *et al.*, 2012; Oey *et al.*, 2014). This method results in extremely low levels of false positives during screening, and directs the segregation of chloroplast DNA so that homoplasmy is achieved (Fig. 9.2). Since the transformed microalgae do not contain antibiotic resistance genes, public concerns as well as economic costs involved in the use of antibiotics on an industrial scale are eliminated.

9.3 Plant-Derived Pharmaceuticals

In recent years, plants have emerged as an attractive system for the production of therapeutic proteins for human health (Giddings, 2001). Plants are generally considered to have several advantages over other expression systems, including: (i) the lack of animal pathogens such as prions and viruses; (ii) their relatively low cost; and (iii) large-scale production of safe and biologically active valuable recombinant proteins. The next section describes several recent patents which pertain to the production of plant-made vaccines and pharmaceuticals.

9.4 Modified Plant Virus Particles and Uses Thereof

Many drugs cannot be effectively delivered by conventional means such as injection, inhalation or oral ingestion. Pharmaceuticals derived and delivered by edible plant tissue provide an alternative means for treating subjects that can provide better bioavailability, increase drug half-life and target the therapeutic protein of interest to specific organs such as the gut-associated lymphoid tissue (GALT), where it can be released in a sustained manner (Mardanova *et al.*, 2015).

The research lab of George Lomonossoff works with virus-like particles (VLPs) of cowpea mosaic virus (CPMV) (Sainsbury *et al.*, 2014). These VLPs contain viral coat proteins that have been modified to carry a heterologous agent. This chimeric VLP can thus carry therapeutic molecules to a particular target site in a human or animal subject, which can then be released at that target site. For example, if a coat protein of CPMV has been modified to carry an Arginine-Glycine-Aspartic (RGD) targeting motif so that the therapeutic agent can be carried to a particular site in the body, such as the plasma (Jobsri *et al.*, 2015). Therapeutic molecules delivered

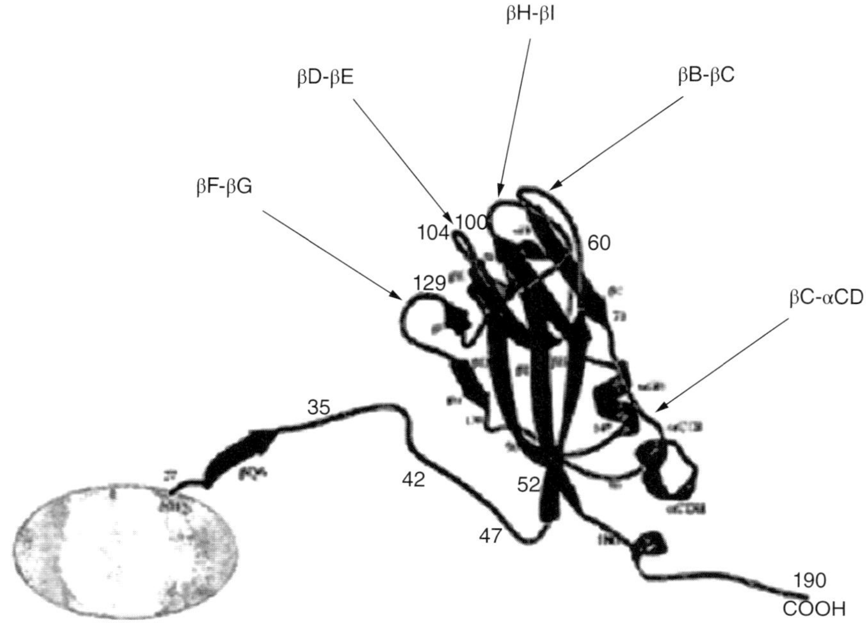

Fig. 9.2. Non-limiting illustration of the CCMV coat protein structure showing surface exposed loops (loops that were tested for targeting peptide insertions are in bold), core structure, C-terminus and non-resolved (highly mobile) N-terminal RNA binding domain.

using plant VLPs such as this can include a nucleic acid, polypeptide or other small molecule such as an anticancer drug (e.g. placitaxel).

9.5 DNA Replicon System for High-Level Rapid Production of Vaccines and Monoclonal Antibody Therapeutics in Plants

The geminivirus bean yellow dwarf virus has been engineered as an expression vector for the production of pharmaceutical proteins (Hefferon and Fan, 2004). Mason *et al.* (2013)[5] have been awarded a patent that further utilizes the geminivirus replicon expression system for the expression of heterologous proteins. In this case, the inventors have designed a geminivirus-based construct that can efficiently express heterologous proteins which consist of more than one subunit from a single vector. This vector contains a long intergenic region (LIR) of a geminivirus genome along with the promoter, the gene of interest, as well as a nucleic acid comprising of Rep/RepA protein of a geminivirus genome. The long intergenic region contains the cis–acting elements capable of mediating excision and replication by the Rep protein. Multi subunit proteins such as human monoclonal antibodies have been produced in plants using this technology (Chen *et al.*, 2011).

9.6 Incorporation of Plant Virus Particles and Polymers as 2D and 3D Scaffolds to Manipulate Cellular Behaviours

Another patent released in 2012 also describes the use of a plant virus expression vector in a more novel fashion. In this case, Wang and Lee (2012)[6] used tobacco mosaic virus (TMV) to enhance mesenchymal stem cell differentiation towards a bone-like phenotype. Cell-binding sequences (as long as 25 amino acids) can be inserted at the carboxyl terminus of the coat protein (CP) of TMV, and the assembled virus particle acts as a highly stabilized, macromolecular extracellular matrix. These chimeric virus particles thus can mimic natural extracellular matrix (ECM) proteins, which play a number of roles including cell adhesion, migration, proliferation and cell differentiation. TMV-ECM mimics are easy to produce and purify, unlike wild-type ECM proteins (Luckanagul *et al.*, 2012;

Sitasuwan *et al.*, 2012, 2014). Initial studies demonstrated that coating surfaces with virus particles were able to accelerate stem cell differentiation towards bone-line tissues from 21 to 14 days of cell culture.

9.7 Influenza Haemagglutinin Antibodies, Compositions and Related Methods

Yusibov *et al.* (2014)[7] have filed a patent regarding the making and use of influenza haemagglutinin antibodies. Specifically, the invention makes use of anti-H5N1 haemagglutinin monoclonal antibodies that can inhibit the haemagglutination of highly pathogenic avian influenza and thus protect against infection in animal challenge models (Canizares *et al.*, 2005). The mAbs can be used as a therapeutic tool in the treatment and/or prevention of human infection, as well as a diagnostic tool for identifying H5N1 human isolates. The antibodies are produced in plants using various techniques, including virus vector expression systems such as the TMV-based launch vector pGRD4-H5 HA (Fig. 9.3). The replication functions were provided *in trans* using the pSoup helper plasmid (Hellens *et al.*, 2000; Musiychuk *et al.*, 2007).

9.8 Oral Unit Dosage Forms and Uses of Same for the Treatment of Gaucher Disease

Shaaltiel and Tzaban[8] have filed a further patent for treating Gaucher disease by expressing recombinant glucocerebrosidase (GCD) in plant cells. Gaucher disease is caused by a recessive gene. Enzyme replacement therapy can be used to treat various lysosomal storage disorders such as Gaucher disease (Tekoah *et al.*, 2013, 2015; Shaaltiel *et al.*, 2015).

Gaucher disease is difficult to treat as the glycosylated protein is expensive to produce and has been delivered in the past by injection. Oral delivery of conventional GCD results in degradation under harsh conditions such as the stomach and upper gastrointestinal tract. As a result, the preparation of biologically active agents such as GCD is a complex and costly process.

Shaaltiel and Tzaban (2014) have previously used recombinant GCD (rGCD) expressed in carrot tissue culture cells to treat Gaucher disease patients. In the current patent, the authors were able to show that oral administration of plant-produced rGCD at levels corresponding to 1–1920 units/kg/14 days effectively treated Gaucher disease. This amount of rGCD is up to 16-fold higher than an amount in units of GCD administered by intravenous (i.v.) injection. The administration is performed preprandially or over a light meal such that the stomach pH is above 2.

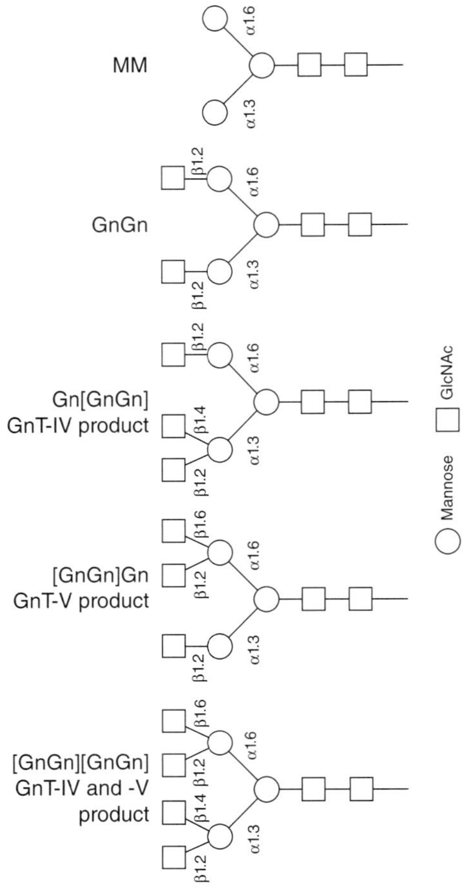

Fig. 9.3. *N*-Glycan structures, using the Consortium for Functional Glycomics symbols and their corresponding abbreviations (Nagels *et al.*, 2011).

9.9 Production of Multi-Antennary N-Glycan Structures in Plants

Over half of all human therapeutic proteins are glycosylated; this affects their function, location and activity. Glycosylation refers to the covalent binding of oligosaccharide chains to a specific site on a protein. For example, an oligosaccharide binding to the amide nitrogen of an asparagine (Asn) residue results in N-glycosylation. The glycosylation profiles of plant proteins have become increasingly easy to manipulate and have enormous implications, particularly in the field of molecular pharming. Plant-made proteins of pharmaceutical interest must have N-glycosylation patterns that resemble that of mammals (Saint-Jore-Dupas *et al.*, 2007). Much work has been accomplished in this field, but of particular interest has been the ability to manipulate multi-antennary N-glycan structures of plant-derived therapeutic proteins so that they more closely resemble their mammalian counterparts (Bakker *et al.*, 2006; Strasser *et al.*, 2008).

An important difference between mammalian- and plant N-glycan structures is that plants do not synthesize multi-antennary glycans. Tetra-antennary glycans, for example, can determine the bio-availability and the half-life of glycoproteins. This inability of plants to provide this form of glycosylation thus reduces its range of uses in pharmaceutical development. In the patent of Bieke Nagels and Koen Weterings,[9] methods and means are provided to produce multi-antennary N-glycosylation structures of glycoproteins in plants and plant cells. The human glycosylation machinery enzymes GnT-IV and GnT-V, which are required to produce multi-antennary N-glycan structures, have been introduced into plants (Nagels *et al.*, 2011). The plants possess mammalian N-acetylglucosaminyltransferase IV and/or N-acetylglucosaminyltransferase V, as well as reduced levels of β-(1,2)-xylosyltransferase and α-(1,3)-fucosyltransferase activity, preferably no detectable β-(1,2)-xylosyltransferase and no detectable α-(1,3)-fucosyltransferase activity.

9.10 Bioactive Compounds in Plants

9.10.1 Modifying soybean oil composition through targeted knockout of the *FAD2-1A/1B* genes

Bioactive compounds produced by plants have also been exploited for potential health benefits and as treatments for a variety of diseases. Technologies range from innovative extraction techniques to biological approaches, including gene editing of plant genomes. For example, an invention described by Mathis *et al.* (2014)[10] investigates the effect of modifying the oil composition of soybeans to increase the oleic acid content and reduce linoleic acid content. The result of this is a lowering of cholesterol and other improved health benefits. This was accomplished by reducing or eliminating expression of the fatty acid desaturase 2 (*FAD2*) *1A* and *1B* genes through gene editing (Pham *et al.*, 2010, 2011). Sequence-specific, rare cutting endonucleases such as a TAL effector endonuclease were used to introduce mutations into the coding regions of *FAO2-1A* and/or *FA2-1B* to knock out gene function (Mahfouz and Li, 2011). The resulting soybean crop would not be considered transgenic and thus could be marketable in Europe and similar-minded regions of the world.

9.10.2 Production of alkaloids by Liliaceae cell culture

Other bioactive compounds have been generated using cell culture technologies (Parekh *et al.*, 2008). Srinivasan *et al.* (2014)[11] were able to produce the alkaloid cyclopamine from a cell culture based on Liliaceae. Cyclopamine can block the Hedgehog (Hh) pathway that plays a role in human embryogenesis and tissue differentiation. Specifically, cyclopamine can block a protein known as 'Smoothened', which acts in the Hh pathway to promote cell growth. Cyclopamine has been shown to play a role in blocking the action of mutated genes that produce basal cell skin carcinomas, and further studies using mouse cell culture models have suggested that other cancers may be treated by cyclopamine such as medulloblastomas in the brain and rhabdomyosarcomas in muscle. The invention therefore describes an efficient method for production of cyclopamine and other alkaloids of the *Veratrum* genus of the Liliaceae family using undifferentiated cell culture. An objective of this invention is to obtain amounts of cyclopamine at yields that are commercially significant from large volume aerated fermenters. The plant cells used can be *Veratrum* or *Amianthium* cells.

9.10.3 Method for producing steviol glycoside

Steviol glycosides are a member of diterpenoids and elicit a sweet taste that is 300 times the sweetness of sugar (Brandle and Telmer, 2007). Derived from the leaves of *Stevia rebaudiana* plants, steviol glycosides are used as non-caloric sweeteners in the food industry. Ono (2014)[12] has filed a patent describing a method for producing steviol glycosides. Since steviol glycosides are naturally occurring, they are

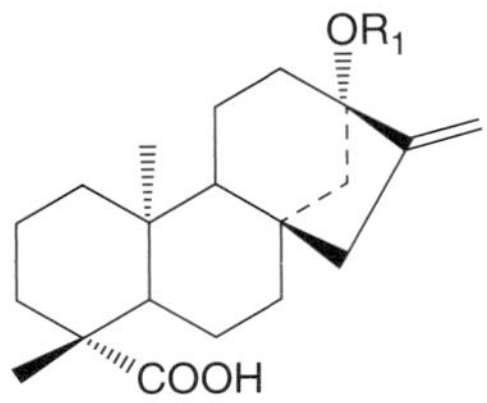

Fig. 9.4. Chemical structure of a steviol glycoside.

more likely to enjoy public acceptance than other, artificial sweeteners such as aspartame. In the leaf, steviol is modified with sugars into a glycoside called rebaudioside AS, with four glucose moieties attached (Fig. 9.4). Rebaudioside A and stevioside are the main components of sweetness in *Stevia* (Mohamed *et al.*, 2011; Madhav *et al.*, 2012).

9.10.4 Reducing levels of nicotinic alkaloids in plants

Many secondary metabolites including the alkaloid nicotine can be harmful to human health and as a result, its expression in plants is undesirable (Bush *et al.*, 1993). Two patents have recently been filed regarding the regulation of nicotine alkaloid biosynthesis in plants. Hashimoto and Kato (2014)[13] have filed a patent that relates to the down-regulation of alkaloid synthesis in plants, but not exclusively tobacco plants. The inventors were able to suppress the genes *A622* and *NBB1* to bring about a decrease in nicotinic alkaloid levels in tobacco plants. This can result in increasing the value of tobacco as a biomass resource, without the requirement of alkaloid extraction (tobacco in general contains 1–5% alkaloids or 800 lbs of alkaloids per acre of conventional tobacco biomass) (Beetham *et al.*, 1999; Chintapakorn and Hamill, 2003).

Similarly, Page and Todd (2014)[14] filed a patent which describes altering the nicotinic alkaloid content of tobacco plants by down-regulating the expression of transcription factors that regulate alkaloid biosynthesis. The invention includes the generation of a genetically engineered plant cell that comprises at least 21 consecutive nucleotides, in sense or antisense orientation, encoding a transcription factor that regulates nicotinic alkaloid biosynthesis (Helliwell and Waterhouse, 2005). Reduction of expression of this transcription factor by gene silencing results in a reduction of nicotine content in plants. The invention includes the selection of a target mutant reduced alkaloid plant and selective breeding of that plant to produce a population of plants with decreased expression of the transcription factor that positively regulates alkaloid biosynthesis (Sato *et al.*, 2001).

9.11 Conclusion

The innovations in agricultural biotechnology that are described in this chapter have largely focused on facilitating the genetic engineering of plants and developing plants with traits that have improved health benefits. While there is a political dimension these days to agricultural patent activity, the patents described in this chapter exemplify the importance of plant biotechnology in addressing various aspects of human health.

The private sector has filed and maintained a disproportionate amount of today's agricultural biotech patents. More specifically, over two-thirds of agricultural-based patents are held by only three multinational companies. As a result of this, there is concern for the direction in which agricultural biotechnology may be taken, so that priorities are made for the needs of industrial countries rather than for developing countries. For example, 'orphan crops' such as chickpea, sorghum, millet, pigeonpea, groundnut and cassava are often neglected by large multinational corporations who wish to address the needs of the wealthy regions. As a result, agricultural development is often perceived as being uneven across the globe. Moreover, a number of environmentalist NGOs such as Greenpeace have staunchly resisted the forward movement of these new technologies and, in particular, new patents that are based on seeds. Disputes over plant breeding methods have created an increasingly politicized environment (Vani and Doble, 2011; Blakeney, 2012; Ricroch *et al.*, 2015). The patents described in this chapter, however, present examples of how public research institutes can also play a significant role with respect to advancements made in agricultural biotechnology. Many of the patents covered here come from universities, publically funded research institutes and small biotechnology companies. It is noteworthy that the first patent described in this chapter could be beneficial to smaller entities such as startup companies to navigate better through the regulatory hurdles that currently surround agricultural biotechnology. The patents described in this chapter therefore can be viewed as a case in point that publically funded and small institutes have been and continue to play important roles in the advancement of agricultural biotechnology.

Notes

1–14 See Table 9.1 for details of patents.

References

Bakker, H., Rouwendal, G.J.A., Karnoup, A.S., Florack, D.E.A., Stoopen, G.M., Helsper, J.P.F.G., van Ree, R., van Die, I. and Bosch, D. (2006) An antibody produced in tobacco expressing a hybrid β-1,4-galactosyltransferase is essentially devoid of plant carbohydrate epitopes. *Proceedings of the National Academy of Sciences USA* 103, 7577–7582.

Beetham, P., Kipp, P.B., Sawycky, X.L., Arntzen, C.J. and May, G.D. (1999) A tool for functional plant genomics: Chimeric RNA/DNA oligonucleotides cause *in vivo* gene-specific mutations. *Proceedings of the National Academy of Sciences USA* 96, 8774–8778.

Blakeney, M. (2012) Patenting of plant varieties and plant breeding methods. *Journal of Experimental Botany* 63(3), 1069–1074.

Brandle, J.E. and Telmer, P.G. (2007) Steviol glycoside biosynthesis. *Phytochemistry* 68(14), 1855–1863.

Broothaerts, W., Mitchell, H.J., Weir, B., Kaines, S., Smith, L.M.A., Yang, W., Mayer, J.E., Roa-Rodríguez, C. and Jefferson, R.A. (2005) Gene transfer to plants by diverse species of bacteria. *Nature* 433(7026), 629–633.

Bruce, B.D. (2000) Chloroplast transit peptides: structure, function and evolution. *Trends in Cell Biology* 10(10), 440–447.

Bush, L.P., Fannin, F.F., Chelvarajan, R.L. and Burton, H.R. (1993) Biosynthesis and metabolism of nicotine and related alkaloids. In: Gorrod, J.W. and Wahren, J. (eds) *Nicotine and Related Alkaloids: Absorption, Distribution, Metabolism, and Excretion*. Springer-Verlag, New York, pp. 1–30.

Canizares, M.C., Nicholson, L. and Lomonossoff, G.P. (2005) Use of viral vectors for vaccine production in plants. *Immunology & Cell Biology* 83, 263–270.

Chen, Q., He, J., Phoolcharoen, W. and Mason, H.S. (2011) Geminiviral vectors based on bean yellow dwarf virus for production of vaccine antigens and monoclonal antibodies in plants. *Human Vaccines* 7(3), 331–338.

Chintapakorn, Y. and Hamill, J.D. (2003) Antisense-mediated down-regulation of putrescine N-methyltransferase activity in transgenic *Nicotiana tabacum* L. can lead to elevated levels of anatabine at the expense of nicotine. *Plant Molecular Biology* 53, 87–105.

Ding, J., Berleant, D., Nettleton, D. and Wurtele, E. (2002) Mining MEDLINE: abstracts, sentences, or phrases? *Pacific Symposium on Biocomputing* 2002, 326–337.

Farrand, S., Hwang, K.I. and Cook, D.M. (1996) The tra region of the nopaline-type Ti plasmid is a chimera with elements related to the transfer systems of RSF1010, RP4, and F. *Journal of Bacteriology* 178(14), 4233–4247.

Fraley, R.T., Rogers, S.G., Horsch, R.B., Eichholtz, D.A., Flick, J.S., Fink, C.L., Hoffmann, N.L. and Sanders, P.R. (1985) The SEV system: a new disarmed Ti plasmid vector system for plant transformation. *Nature Biotechnology* 3, 629–635.

Giddings, G. (2001) Transgenic plants as protein factories. *Current Opinion in Biotechnology* 12(5), 450–454.

Grundman, O., Khozin-Goldberg, I., Raveh, D., Cohen, Z., Vyazmensky, M., Boussiba, S. and Shapira, M. (2012) Cloning, mutagenesis, and characterization of the microalga *Parietochloris incisa* acetohydroxyacid synthase, and its possible use as an endogenous selection marker. *Biotechnology and Bioengineering* 109(9), 2340–2348.

Hefferon, K.L. and Fan, Y. (2004) Expression of a vaccine protein in a plant cell line using a geminivirus-based replicon system. *Vaccine* 23(3), 404–410.

Hellens, R.P., Edwards, E.A., Leyland, N.R., Bean, S. and Mullineaux, P.M. (2000) pGreen: a versatile and flexible binary Ti vector for Agrobacterium-mediated plant transformation. *Plant Molecular Biology* 42, 819–832.

Helliwell, C.A. and Waterhouse, P.M. (2005) Constructs and methods for hairpin RNA-mediated gene silencing in plants. *Methods in Enzymology* 392, 24–35.

Hull, D., Petifer, S.R. and Kell, D.B. (2008) Defrosting the digital library: bibliographic tools for the next generation web. *PLoS Computational Biology* 4, 10.

Jobsri, J., Allen, A., Rajagopal, D., Shipton, M., Kanyuka, K., Lomonossoff, G.P., Ottensmeier, C., Diebold, S.S., Stevenson, F.K. and Savelyeva, N. (2015) Plant virus particles carrying tumour antigen activate TLR7 and Induce high levels of protective antibody. *PLoS One* 10(2), e0118096. DOI: 10.1371/journal.pone.0118096. eCollection 2015.

Lee, D.-W., Kim, J.K., Lee, S., Choi, S., Kim, S. and Hwang, I. (2008) *Arabidopsis* nuclear-encoded plastid transit peptides contain multiple sequence subgroups with distinctive chloroplast-targeting sequence motifs. *The Plant Cell* 20(6), 1603–1622.

Luckanagul, J., Lee, L.A., Nguyen, Q.L., Sitasuwan, P., Yang, X., Shazly, T. and Wang, Q. (2012) Porous alginate hydrogel functionalized with virus as three-dimensional scaffolds for bone differentiation. *Biomacromolecules* 13(12), 3949–3958.

Madhav, H., Bhasker, S. and Chinnamma, M. (2012) Functional and structural variation of uridine diphosphate glycosyltransferase (UGT) gene of *Stevia rebaudiana*-UGTSr involved in the synthesis of rebaudioside A. *Plant Physiology and Biochemistry* 63, 245–253.

Mahfouz, M.M. and Li, L. (2011) TALE nucleases and next generation GM crops. *GM Crops* 2(2), 99–103.

Mardanova, E.S., Kotlyarov, R.Y., Kuprianov, V.V., Stepanova, L.A., Tsybalova, L.M., Lomonosoff, G.P. and Ravin, N.V. (2015) Rapid high-yield expression of a candidate influenza vaccine based on the ectodomain of M2 protein linked to flagellin in plants using viral vectors. *BMC Biotechnology* 15, 42.

McDonald, R. and Pereira, F. (2005) Identifying gene and protein mentions in text using conditional random fields. *BMC Bioinformatics* 6(S6). Available at: http://www.biomedcentral.com/1471-2105/6/S1/S6 (accessed 24 May 2005).

Mohamed, A.A., Ceunen, S., Geuns, J.M., Van den Ende, W. and De Ley, M. (2011) UDP-dependent glycosyltransferases involved in the biosynthesis of steviol glycosides. *Journal of Plant Physiology* 168(10), 1136–1141.

Musiychuk, K., Stephenson, N., Bi, H., Farrance, C.E., Orozovic, G., Brodelius, M., Brodelius, P., Horsey, A., Ugulava, N., Shamloul, A.M., Mett, V., Rabindran, S., Streatfield, S.J. and Yusibov, V. (2007) A launch vector for the production of vaccine antigens in plants. *Influenza and Other Respiratory Viruses* 1, 19–25.

Nagels, B., Els, J.M., Van Damme, E.J.M., Pabst, M., Callewaert, N. and Weterings, K. (2011) Production of complex multiantennary *N*-glycans in *Nicotiana benthamiana* plants. *Plant Physiology* 155(3), 1103–1112.

Oey, M., Ross, I.L. and Hankamer, B. (2014) Gateway-assisted vector construction to facilitate expression of foreign proteins in the chloroplast of single celled algae. *PLoS One* 9(2), e86841.

Parekh, S., Srinivasan, V. and Horn, M. (2008) Bioprocessing using novel cell culture systems. *Advances in Applied Microbiology* 63, 105–143.

Pham, A.-T., Lee, J.D., Shannon, J.G. and Bilyeu, K.D. (2010) Mutant alleles of FAD2-1A and FAD2-1B combine to produce soybeans with the high oleic acid seed oil trait. *BMC Plant Biology* 10, 195.

Pham, A.-T., Lee, J.D., Shannon, J.G. and Bilyeu, K.D. (2011) A novel-allele in a soybean plant introduction offers an alternate means to produce soybean seed oil with 85% oleic acid content. *Theoretical and Applied Genetics* 123(5), 793–802.

Potvin, G. and Zhang, Z. (2010) Strategies for high-level recombinant protein expression in transgenic microalgae: a review. *Biotechnology Advances* 28(6), 910–918.

Ricroch, A., Harwood, W., Svobodová, Z., Sági, L., Hundleby, P., Badea, E.M., Rosca, I., Cruz, G., Salema Fevereiro, M.P., Marfà Riera, V., Jansson, S., Morandini, P., *et al.* (2015) Challenges facing European agriculture and possible biotechnological solutions. *Critical Reviews in Biotechnology* 1, 1–9.

Sainsbury, F., Saxena, P., Aljabali, A.A., Saunders, K., Evans, D.J. and Lomonossoff, G.P. (2014) Genetic engineering and characterization of Cowpea mosaic virus empty virus-like particles. *Methods in Molecular Biology* 1108, 139–153.

Saint-Jore-Dupas, C., Faye, L., Gomord, V. (2007) From planta to pharma with glycosylation in the toolbox. *Trends in Biotechnology* 25(7), 317–323.

Sato, S., Hashimoto, T., Hachiya, A., Tamura, K., Choi, K.B., Morishige, T., Fujimoto, H. and Yamada, Y. (2001) Metabolic engineering of plant alkaloid biosynthesis. *Proceedings of the National Academy of Sciences USA* 98(1), 367–372.

Shaaltiel, Y., Gingis-Velitski, S., Tzaban, S., Fiks, N., Tekoah, Y. and Aviezer, D. (2015) Plant-based oral delivery of β-glucocerebrosidase as an enzyme replacement therapy for Gaucher's disease. *Plant Biotechnology Journal* 13(8), 1033–1040.

Shewry, P.R., Baudo, M., Lovegrove, A., Powers, S., Napier, A., Wad, J.L., Baker, J.M. and Beale, M.H. (2007) Are GM and conventionally bred cereals really different? *Trends in Food Science and Technology* 18, 201–209.

Sitasuwan, P., Lee, L.A., Bo, P., Davis, E.N., Lin, Y. and Wang, Q. (2012) A plant virus substrate induces early upregulation of BMP2 for rapid bone formation. *Integrative Biology (Cambridge)* 4(6), 651–660.

Sitasuwan, P., Lee, L.A., Li, K., Nguyen, H.G. and Wang, Q. (2014) RGD-conjugated rod-like viral nanoparticles on 2D scaffold improve bone differentiation of mesenchymal stem cells. *Frontiers in Chemistry* 2, 31.

Strasser, R., Stadlmann, J., Schähs, M., Stiegler, G., Quendler, H., Mach, L., Glössl, J., Weterings, K., Pabst, M. and Steinkellner, H. (2008) Generation of glyco-engineered *Nicotiana benthamiana* for the production of monoclonal antibodies with a homogeneous human-like N-glycan structure. *Plant Biotechnology Journal* 6, 392–402.

Tekoah, Y., Tzaban, S., Kizhner, T., Hainrichson, M., Gantman, A., Golembo, M., Aviezer, D. and Shaaltiel, Y. (2013) Glycosylation and functionality of recombinant β-glucocerebrosidase from various production systems. *Bioscience Reports* 33(5), pii: e00071.

Tekoah, Y., Shulman, A., Kizhner, T., Ruderfer, I., Fux, L., Nataf, Y., Bartfeld, D., Ariel, T., Gingis-Velitski, S., Hanania, U. and Shaaltiel, Y. (2015) Large-scale production of pharmaceutical proteins in plant cell culture-the protalix experience. *Plant Biotechnology Journal* 13(8), 1199–1208.

Vani, K.P. and Doble, M. (2011) Sustainable development in agriculture, food and nutrition – a patent analysis. *Recent Patents on Food, Nutrition & Agriculture* 3(2), 133–141.

10 Chinese Innovation System: The Case of Agricultural Knowledge Sharing

Liliana Mitkova[1]* and Xi Wang[2†]

[1]*Institut de Recherche en Gestion, University of Paris Est Marne-la-Vallée (UPEM) Champs-sur-Marne France;* [2]*Central University of Finance and Economics, Beijing, China*

10.1 Introduction

The open innovation model has been adopted so as to adapt to the rapid diffusion of knowledge, shortening of the life cycle of products and high international competitiveness. This model proposes to enhance a firm's innovative ability by acquiring knowledge from external sources, as well as benefit financially by using external paths to market for internally generated technologies (Chesbrough, 2003; Gassmann and Enkel, 2004). Enterprises can benefit from technological and market discontinuities into the open innovation model by sharing knowledge with other institutions and firms. In fact, knowledge sharing is a key component of open innovation (Islam, 2012) and Lichtenthaler (2011) underlines the link between these two concepts. The author identifies open innovation as 'systematically performing knowledge exploration, retention, and exploitation inside and outside an organization's boundaries throughout the innovation process'.

In knowledge management, knowledge sharing is analysed by various approaches that refer to how organizations create, retain and share knowledge (Bogers, 2012). Recently, the studies of knowledge sharing, which is the means by which an organization obtains access to its own and other organizations' knowledge, has emerged as a key research area from a broad and deep field of study on technology transfer and innovation, and more recently from the field of open innovation management (Cummings, 2003). Cummings (2003) suggests three types of researches concerning the knowledge-sharing activities: (i) analyses of the form and the location of the knowledge investigate how they affect the sharing processes; (ii) studies concerning the types of rules of engagement and managerial practices adopted by the parties for the flows of resources and knowledge; and (iii) researches of specific knowledge-sharing activities are important in that they are the means through which the parties seek to facilitate knowledge sharing. However, despite these studies, how effectively to manage knowledge sharing in an open innovation perspective is not yet fully understood (Enkel *et al.*, 2009). The main purpose of this chapter is focused on the second approach, offering an overview of the different measures and tools implemented in the Chinese innovation system in order to support the knowledge sharing under an open perspective.

A company's capacity to make use of knowledge sharing thus to adapt the open innovation model has been viewed as a source of sustainable competitive advantage of Chinese firms and their successful integration in the international technological

*Corresponding authors, e-mail: Liliana.Mitkova@u-pem.fr
†E-mail: miracle_wangxi@163.com

landscape. China is determined to enhance the global competitiveness of its firms and to be defined as an 'innovation-oriented' country by 2020 and a 'leading science power' by 2050 (Chen and Li-Hua, 2011). In general, the innovation policy has been promoted by strengthening the internal R&D capacities and recently by adopting a more open approach toward the international technologies in order to catch up and improve internal innovative performance (Chen and Qu, 2003). Recent literature debates the relative importance of different tools of the Chinese national policies as a major determinant of Chinese innovation system improvement (Hung, 2009; Liu, 2010). The academic works put the accent on the external factors to implement the open model in China (Savitskaya *et al.*, 2010), especially the development of the intellectual property (IP) system and government policy (Deng, 2009).

The first section of this chapter offers the theoretical background of knowledge sharing in an open perspective. The second part presents an overview of Chinese agricultural knowledge sharing, illustrating how this system supports innovation activities.

10.2 Knowledge Sharing and Open Innovation

Knowledge sharing, working as a major supporting process for knowledge management and open innovation (Llopis-Córcoles, 2011), is an activity through which knowledge (i.e. information, skills or expertise) is exchanged among friends, families, communities or organizations (Bukowitz and Williams, 1999). It is a two-way or dual process, including both the supply of new knowledge and the demand for new knowledge, enquiring and contributing to knowledge through activities such as learning-by-observation, listening and asking, sharing ideas, giving advice, recognizing cues and adopting patterns of behaviour (Bosua and Scheepers, 2007). Authors offer other definitions of knowledge sharing as the means by which an organization obtains access to its own and other organizations' knowledge from a more self-centred or inbound view, a key research sector of technology transfer, innovation and strategic management (Cummings, 2003). Or it refers to the provision of task information and know-how to help others and to collaborate to solve problems, develop new ideas or implement policies or procedures (Wang and Noe, 2010).

The practices and initiatives of knowledge sharing, in terms of organizational and individual learning, often form a key component of knowledge management programmes (Alavi and Leidner, 2001; Riege, 2005). According to Nonaka and Takeuchi (1995), knowledge sharing also acts as a critical stage in the knowledge transfer process. Some see knowledge management and knowledge transfer as processes that are undertaken largely for the purpose of creating a knowledge sharing culture, fostering collaboration and communication, and so in turn enhancing organizational innovation (Liebowitz, 2002). Therefore, the coordination of the process within the organization becomes the key success factor for the company.

10.2.1 Coordination mechanisms of knowledge sharing

According to Binz-Scharf (2003), the coordination mechanisms of a knowledge sharing system could be classified into two different levels: individual and organizational level coordination. Both coordination forms can be carried out by formal or informal mechanisms; the former type of mechanism is referred to as 'hierarchy' and the latter as 'social networks'.

The hierarchical structures are used in order to coordinate knowledge processes in complex organizations comprised of multiple specialized units (Tsai, 2002). The analysis of hierarchical structure as a coordination mechanism of communication has played an important role in organizational research (Powell, 1990). Tsai (2002) also discussed the role of centralization in hierarchical structure, and deemed that it might restrain knowledge sharing in different units unless they were required to do so by the higher authority.

Intra-organizational networks 'subsume relations between and among actors under a governance structure that handles conflict resolution and channel behaviour' (Fountain, 1999). Knowledge networks are characterized by fluidity of actors and their ties. This fluidity can be defined as 'all actors within an observable knowledge network have their own cognitive knowledge networks, which refer to their perceptions of the overall observable knowledge network' (Contractor *et al.*, 2000).

The organization and the coordination of knowledge sharing is also influenced by two different factors that we can refer to as institutional and organizational level factors.

10.2.2 Institutional and organizational factors of knowledge sharing

Knowledge sharing requires a focus on more than simply the transfer of the specific knowledge and needs to focus on structuring and implementing the arrangement in a way that bridges both existing and potential relationship issues, and examining the form and location of the knowledge to ensure its successful achievement (Cummings, 2003). Knowledge sharing can be supported by acting on certain contextual and organizational variables that influence knowledge flows (Binz-Scharf, 2003).

The early literature demonstrates knowledge sharing dilemma grounded on the analysis of institutional (macro-level) variables (Cummings, 2003), and many insights into how broad environmental factors may affect knowledge sharing activities (Rousseau, 1985). Researchers have contributed to identify the patterns of national, institutional and organizational factors that best support knowledge transfer or knowledge diffusion in general among firms (Mansfield, 1975). The four main factors affecting knowledge transfers, which include the activities and modes, organizational profiles of the parties, broad environmental factors and technological absorptive capacity of host country, emphasize the importance of national and organizational framework building (Contractor and Sagafi-Nejad, 1981). The contextual variables affecting knowledge flows are the environment, organizational structural characteristics and organizational cultural norms (Gupta and Govindarajan, 1991). The existence and richness of knowledge-sharing channels could determine the attitude toward knowledge-sharing behaviour (Kwok and Gao, 2005).

The organizational-level research highlighted lower economic costs (Mansfield *et al.*, 1979), time cost (Mansfield and Romeo, 1980), relationship ties (Hansen, 1999), business strategy (Grant, 1996), tacitness and embeddedness (Zander, 1991), the recipient's absorptive capacity and causal ambiguity (Szulanski, 1996), facilitating activities (Lall, 2000) and categorized subsidiaries (Gupta and Govindarajan, 1991) as key variables.

10.2.3 Integrating knowledge sharing into open innovation

Rothwell (1994) identifies five generations of innovation model since Schumpeter's linear process, from the chains that link markets and inventions (Kline and Rosenberg, 1986) to extensive networking. Open innovation has been proposed as a new paradigm that centres on opening up organizational boundaries in order to use and recombine internal and external knowledge to develop and commercialize valuable innovations (Chesbrough, 2003). At its core, open innovation is about 'the use of purposive inflows and outflows of knowledge to accelerate internal innovation, and expand the markets for external use of innovation, respectively'. Meanwhile, other researchers also provide some guidelines about open innovation and knowledge sharing (Islam, 2012). Kirschbaum (2005) argued that 'open innovation is finding and selecting innovative knowledge that is right for the organization'. Buganza and Verganti (2009) regarded that open innovation required enterprises to develop both their 'internal knowledge' and 'adaptive capacity' of using external knowledge. Lichtenthaler (2011) identified open innovation as 'systematically performing knowledge exploration, retention, and exploitation inside and outside an organization's boundaries throughout the innovation process.' A particular and increasingly important area within open innovation, which combines knowledge inflows and outflows, is the collaborative development of innovation by multiple organizations (Bogers and West, 2010). According to Enkel *et al.* (2009), this 'coupled process' of open innovation 'refers to co-creation mainly with complementary partners through alliances, cooperation, and joint ventures during which give and take are crucial for success.' As a coupled process, knowledge sharing combines the inbound and outbound processes of open innovation by allowing firms jointly to develop and commercialize innovation. The value of such collaborative innovation, both in terms of economic welfare and corporate competitive advantage, has grown tremendously in recent years (Verspagen and Duysters, 2004; West *et al.*, 2014).

Huysman and De Wit (2002) inspected the set of knowledge-sharing practices that facilitate and structure knowledge sharing for knowledge workers, and identified three primary motivations for sharing knowledge: knowledge acquisition, knowledge reuse and knowledge creation. Successful knowledge-sharing goals must centre on a culture and depend on the synergy of effective motivation, flat and open organizational structures, and modern technology-supported suitable sharing platforms

(Riege, 2005). According to Kim and Nelson (2000), knowledge sharing could occur through a dynamic learning process in the value chain, where organizations continually interact with customers and suppliers to innovate or creatively imitate.

To understand better the rationale and impediments of knowledge sharing in open collaborative innovation, the next section presents the Chinese experience in a practical perspective to derive the main opportunities and constraints of a knowledge sharing system with the example of the agriculture sector.

10.3 Development of China's Current Knowledge-Sharing System

In order to illustrate the development of the Chinese innovation system, this section is focused on the main practices at the institutional and organizational level of knowledge sharing.

10.3.1 From the institutional perspective

Academics, policy makers and practitioners are increasingly interested in the roles that the public sector is playing in the effective management of knowledge sharing (Currie and Suhomlinova, 2006). In pursuit of systematic improvement, public sector institutions transform themselves into intermediaries supporting the sharing process. In this vein, the Chinese government becomes an active participant in the national knowledge system, providing assistance of the sharing of the knowledge flow between government institutions, local players and firms. It includes: (i) the establishment of an enabling environment via appropriate laws, especially IP rights and regulations for knowledge sharing; (ii) the development of an internal 'horizontal' knowledge-sharing system between different governmental functional institutions at strategic level; and (iii) the founding of vertical sharing coordination between different levels of players within the governmental institutions. To implement these strategic aims the Chinese government has introduced various actions. For example, the government has promulgated and implemented the National Knowledge Management Standard GB/T23703. Also, in July 2009, the Chinese Ministry of Science and Technology (MOST) and another five ministries had jointly launched the National Technical Innovation Engineering General Plan. Moreover, in December 2009, they issued Building and Implementing Measures on Promoting the Development of Industry Technology Innovation Strategic Alliance (Trial), and, in January 2010, MOST issued Notice on Choice of a Batch of Industrial Technology Innovation Strategy Alliance Pilot Work.

At the local level, local institutions also take active roles in the process of building the knowledge-sharing structure. In this sense, they set up special funds for stimulating industry–university–research cooperation, mainly used for innovation programmes and R&D collaborations. For example, the subsidies and incentives for science and technology (S&T) intermediary service institutions (e.g. quality measuring, testing and certification services, and technology consulting and technology trade intermediary services) are established. Moreover, the subsidies for investment into industry–university–research cooperation programmes are put in place and the subsidies for S&T business incubator development are introduced.

Hence, China's knowledge sharing system from the institutional perspective is mainly in generalizing the government's role in the system by optimizing the external policy environment and coordination mechanism (Fig. 10.1). The main players in this knowledge system are enterprise, university and research institutions. The interactions between these main players are analysed in an organizational perspective.

10.3.2 From the organizational perspective

General view of knowledge-sharing modes in China

At the level of organizations the Chinese current knowledge sharing is implemented via a knowledge alliance, which takes three main forms: the project programme, the industry technology alliance and the scientific research alliance. Table 10.1 illustrates a general comparison between these three possibilities:

- The project cooperation programme is a temporary organized knowledge-sharing structure established by relevant parties to accomplish specific scientific research aims in a given time span. The goal is to create a technological consortium between the firms setting up the mechanism of sharing in the research and production activities. The organization has strong structure, procedures, pertinence, timeliness, etc. China's current project teams are usually aiming

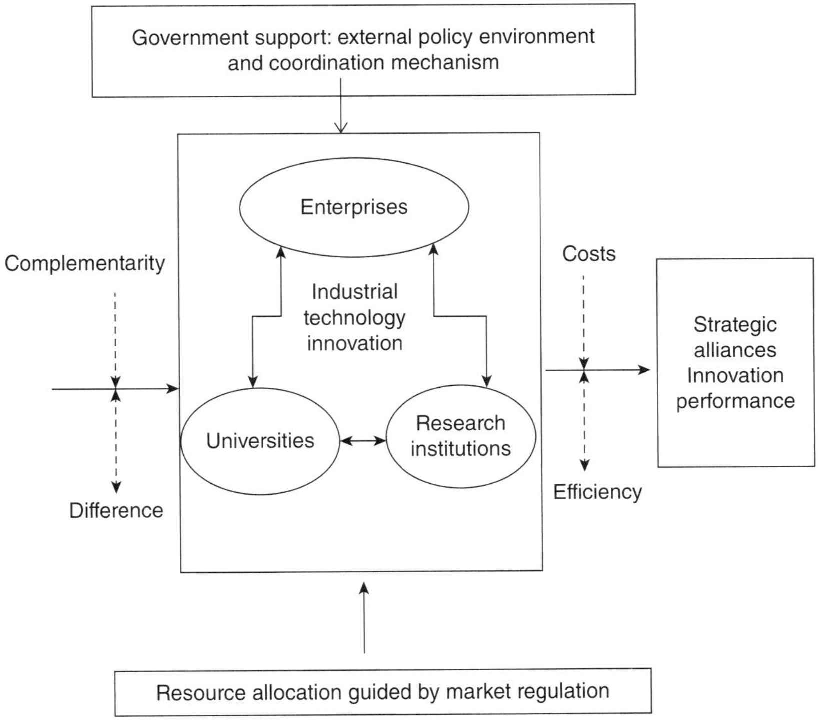

Fig. 10.1. Knowledge sharing mechanism: framework.

Table 10.1. Comparison of three organizational models.

Organizational model	Project cooperation programme	Industry technology alliance	Scientific research-based alliance
Members in organization	Enterprises, universities, research institutions		
Feature	Advantage: Targeted, time-sensitive; Shortcoming: consistent, persistent	Complementary advantages, benefit and risk sharing	Long-term stability
Shared method	Market-oriented mainly	Government subsidies and market-oriented combination	Market-oriented mainly
Shared degree	Lesser extent	Deeper level	Deepest

at a certain key technology or to tackling key problems, and the teams are always disbanded once the task is completed. This form of sharing lacks consistency and continuity. Therefore, this organization type is suitable for a specific period of time within specific areas of knowledge sharing, but is not conducive to real-time updates and further optimization.

- The industry technology alliance is established by enterprises, universities, research institutions or other organizations in order to promote inter-organizational knowledge sharing and industrial technology innovation capability. This form supports joint development, complementary advantages, benefit and risk sharing. The knowledge-sharing organization is based on common agreement. In general, when the technological alliance is based on government funding, the members in this alliance should share common technologies with each other or even with other organizations outside the alliance under certain conditions.

- The scientific research-based alliance is funded by the government or firms, with the main task of knowledge sharing and research activities. Generally, this form needs the joint investment of funds from enterprises, research institutions and universities to establish collaborative R&D institutions, laboratories or engineering and technology centres. The scientific research alliance is a long-term and stable organizational structure, which makes it quite suitable for knowledge sharing.

In addition, China currently also has established industry clusters such as the Pearl River Delta area of the home appliance industry with a relatively free organizational structure of knowledge sharing.

In China, the main problem of these knowledge sharing forms is how to balance between cooperation and competition. Generally speaking, there exist two typical knowledge sharing models in Chinese enterprises: the competition and cooperation model; and the orderly chaos model (Fig. 10.2).

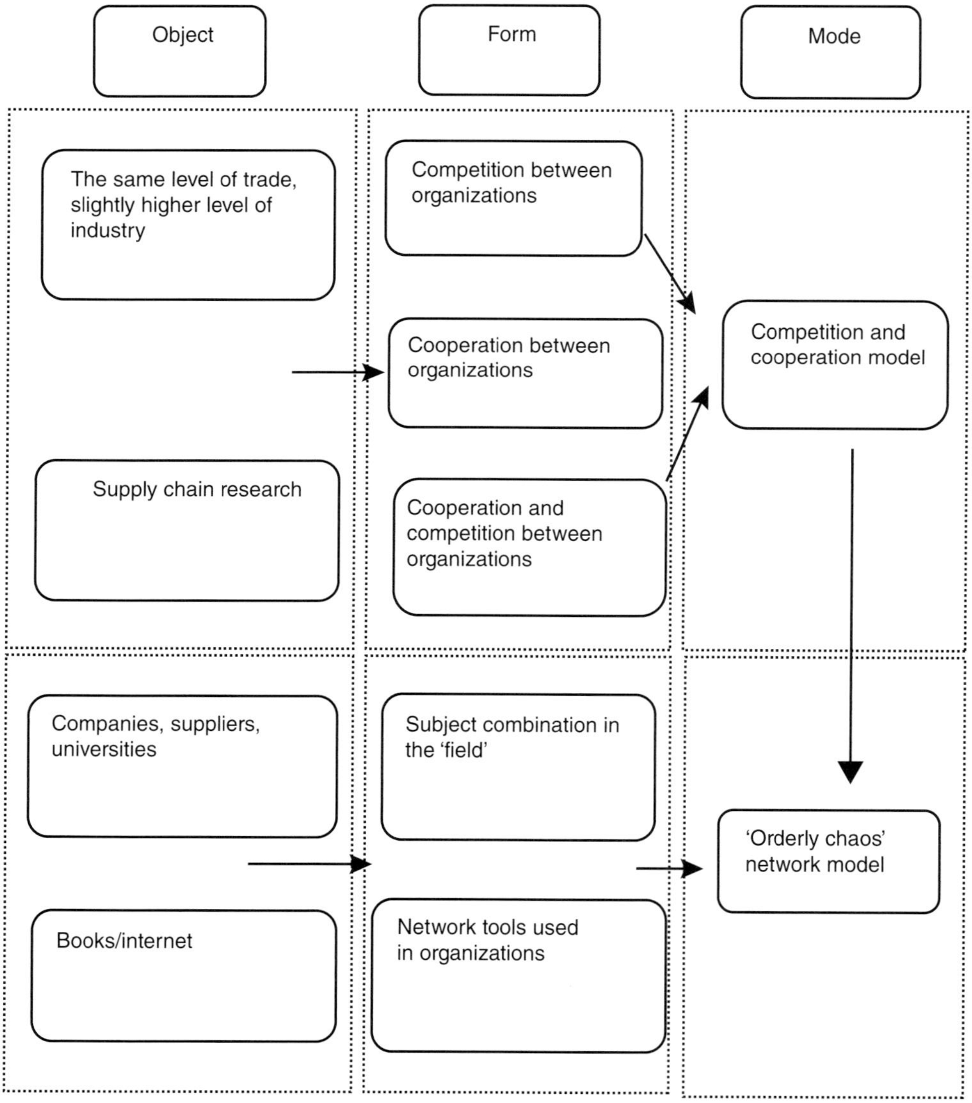

Fig. 10.2. Knowledge sharing models.

They are applicable for enterprises at different stages of their life cycle.

- The competition and cooperation model is suitable for high-tech enterprises to generate new knowledge in their infancy and development stage. The infancy stage is characterized as the intensive use of tacit knowledge, when the high-tech enterprise generally could not achieve the amount of new explicit and tacit knowledge it needs to generate new knowledge by merely relying on the internal R&D efforts. Thus it is quite common to use the competition and cooperation model to fulfil its need in different knowledge to increase the accessibility and possibility of new knowledge and ensure that the emergence of new knowledge is sooner than other competitors in the same industry.
- The 'orderly chaos' network model is more suitable for knowledge sharing in the mature stage. With the continuous expansion of R&D activities, the amount of knowledge in the enterprise becomes chaotic. In the chaos, tacit knowledge is easily overlooked while explicit knowledge is difficult to summarize. This 'orderly chaos' network model is used to find an orderly procedure in chaotic environments and to optimize organizational knowledge sharing.

Chinese agricultural knowledge sharing

Agricultural knowledge sharing refers to the process of the consistent realization of innovation by mutually sharing, learning and transforming the agricultural knowledge developed by agricultural producers, agricultural researchers and agricultural companies through some form of media. We can analyse the interaction of the three main bodies of agricultural knowledge sharing:

- Agricultural producers are the units or the individuals who are directly engaged in agricultural production, they are also the direct users and practitioners of agricultural knowledge. However, once in the face of disasters, they will be at a loss because of their own knowledge limitations with the difficulty to improve the agricultural planting system.
- Agricultural research. The theoretical research and experimental investigations are far from the real concerns of the producers. In China there have always been problems with the impossibility of implementing research results in the planting system.
- The agricultural corporation is an organized agricultural team focused on agricultural cultivation. The specificity of this structure is the development of inside knowledge exchanges within the team with the difficulty to obtain external comprehensive agricultural planting knowledge.

These three main players in the Chinese agricultural landscape organize different types of knowledge sharing in order to innovate and improve their practices:

- Knowledge sharing in a small range, which mainly refers to the mutual exchange of experience among individual farmers. This kind of knowledge sharing is usually carried on through face to face teaching between father and son, master and apprentice. The main shortcoming of this exchange is its small range and little influence on the innovation, besides of that with the risk to be lost. Nevertheless, this mode represents the main way of knowledge sharing in current Chinese farming.
- Knowledge sharing in mature organizations involves knowledge sharing within agricultural organizations and companies. These structures are more open for external experts specialized in agriculture.
- The building of an Internet sharing platform to promote the knowledge sharing among the three main bodies – agricultural producer, agricultural research and agricultural corporation – and to realize the maximum efficiency of the agricultural innovation system (Fig. 10.3).

A famous Confucius proverb considers that there must be one man good enough to act as a teacher as well as men that walk together to move forward in the knowledge. This proverb has been fully adopted in the building of the Chinese innovation system and the improvement of agricultural knowledge sharing. The Chinese concept of 'Internet + agriculture' offering an agricultural sharing platform has now become the main aim of the development of agricultural knowledge sharing in China.

10.4 Conclusion

The latest academic and managerial works highlight the major role of a knowledge-sharing system in the management of innovation activities,

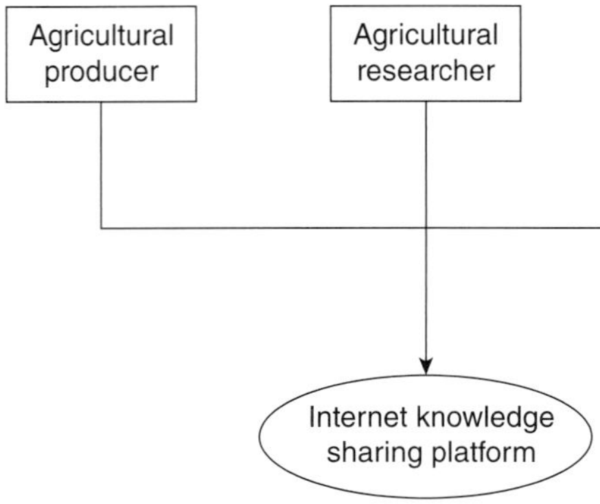

Fig. 10.3. The future agricultural knowledge sharing system.

especially in the case of an open innovation model. Traditionally, the debates dedicated to the Chinese innovation system focused on the barriers for an open model (Savitskaya *et al.*, 2010). Nevertheless, the implementation of the open model in China is dependent on the development of various measures supporting the knowledge-sharing system at institutional and organizational levels. At the institutional level the government aims to build a sustainable intermediary Internet platform for knowledge sharing. In this sense, the possible ways to support knowledge sharing in China in the agriculture sector are:

- To establish the government portal website providing public services and advice;
- To build the government knowledge-management centre, specializing in knowledge mining, discovery, application and innovation; and
- To establish a government knowledge feedback platform, fully adapted to outside opinion in order to test the efficiency of knowledge sharing.

Through the building of a knowledge platform at national level the government can expand the optimization of the information process and emphasize on knowledge mining, knowledge communication and knowledge transfer, as well as to develop the information service and improve the knowledge-management mechanism.

A successful knowledge-sharing system supporting an open innovation model requires a focus on more than simply transfer of the specific knowledge and practices. One of the main issues is how to structure and implement the arrangements in a way that bridges both existing and potential relationships, and examining the form and location of the knowledge to ensure its complete transfer. The participation in this process in order to support the open innovation model and efficient agriculture needs some improvement for various reasons.

1. The lack of proper technical support, effective computer network, communication system and technological platform in the enterprises facilitating the knowledge sharing;
2. The defect of the information system characterized by a slow speed of information transmission and a serious phenomenon of information recession and distortion. There is no direct contact and communication and coordination between players. Competitive relationship between them, the independence of the organizations and the difficulty of knowledge transfer across the hierarchical levels are the important factors that affect knowledge sharing.
3. Various managerial cultural restrictions are evident. The open channel of tacit knowledge communication is missed and the communication with the environment is not guaranteed.
4. The lack of an incentive mechanism supporting knowledge sharing. The interactions with external and internal players rely voluntarily on the employees. In fact, the firms do not develop specific mechanisms to promote the knowledge sharing culture.

References

Alavi, M. and Leidner, D. (2001) Knowledge management and knowledge management systems: Conceptual foundations and research issues. *MIS Quarterly* 25(1), 107–136.

Binz-Scharf, M.C. (2003) Exploration and exploitation: Toward a theory of knowledge sharing in digital government projects. Dissertation, Universität St. Gallen, Switzerland.

Bogers, M. (2012) Knowledge sharing in open innovation: an overview of theoretical perspectives on collaborative innovation. In: Heredero, C.P. and López, D. (eds) *Open Innovation at Firms and Public Administrations: Technologies for Value Creation.* IGI Global, Hershey, Pennsylvania, pp. 1–14.

Bogers, M. and West, J. (2010) Contrasting innovation creation and commercialization within open, user and cumulative innovation. User and Cumulative Innovation. Working paper. Mads Clausen Institute for Product Innovation, University of Southern Denmark, 13.

Bosua, R. and Scheepers, R. (2007) Towards a model to explain knowledge sharing in complex organizational environments. *Knowledge Management Research and Practice* 5(2), 93–109.

Buganza, T. and Verganti, R. (2009) Open innovation process to inbound knowledge: Collaboration with universities in four leading firms. *European Journal of Innovation Management* 12(3), 306–325.

Bukowitz, W.R. and Williams, R.L. (1999) *The Knowledge Management Fieldbook*. FT Press London.

Chen, D. and Li-Hua, R. (2011) Modes of technological leapfrogging: five case studies from China. *Journal of Engineering and Technology Management* 28(1), 93–108.

Chen, J. and Qu, W.G. (2003) A new technological learning in China. *Technovation* 23(11), 861–867.

Chesbrough, H.W. (2003) *Open Innovation: The new imperative for creating and profiting from technology.* Harvard Business Press, London.

Contractor, F.J. and Sagafi-Nejad, T. (1981) International technology transfer: Major issues and policy responses. *Journal of International Business Studies* 12(2), 113–135.

Contractor, N., Carley, K., Levitt, R., Monge, P.R., Wasserman, S., Bar, F. and Kunz, J. (2000) *Co-evolution of Knowledge Networks and 21st Century Organizational Forms: Computational Modelling and Empirical Testing*. University of Illinois, Urbana-Champaign, Illinois.

Cummings, J. (2003) *Knowledge Sharing: A Review of the Literature*. World Bank, Washington, DC.

Currie, G. and Suhomlinova, O. (2006) The impact of institutional forces upon knowledge sharing in the UK NHS: the triumph of professional power and the inconsistency of policy. *Public Administration* 84(1), 1–30.

Deng, P. (2009) Why do Chinese firms tend to acquire strategic assets in international expansion? *Journal of World Business* 44(1), 74–84.

Enkel, E., Gassmann, O. and Chesbrough, H. (2009) Open R&D and open innovation: exploring the phenomenon. *R&D Management* 39(4), 311–316.

Fountain, J.E. (1999) A note on the critical incident technique and its utility as a tool of public management research. Paper presented at the Annual Meeting of the Association of Public Policy and Management, Washington, DC, November 4–6, 1999.

Gassmann, O. and Enkel, E. (2004) Towards a theory of open innovation: three core process archetypes. In: *Proceedings of R&D Management Conference,* Lisbon, Portugal, 6–9 July.

Grant, R.M. (1996) Toward a knowledge-based theory of the firm. *Strategic Management Journal.* 17(S2), 109–122.

Gupta, A.K. and Govindarajan, V. (1991) Knowledge flows and the structure of control within multinational corporations. *Academy of Management Review* 16(4), 768–792.

Hansen, M.T. (1999) The search-transfer problem: The role of weak ties in sharing knowledge across organization subunits. *Administrative Science Quarterly* 44(1), 82–111.

Hung, S.W. (2009) Development and innovation in the IT industries of India and China. *Technology in Society* 31(1), 29–41.

Huysman, M.H. and De Wit, D. (2002) *Knowledge Sharing in Practice*. Springer Science and Business Media, Germany.

Islam, A.M. (2012) Methods of open innovation knowledge sharing risk reduction: a case study. *International Journal of e-Education, e-Business, e-Management and e-Learning* 2(4), 294–297.

Kim, L. and Nelson, R.R. (2000) *Technology, Learning, and Innovation: Experiences of Newly Industrializing Economies*. Cambridge University Press, Cambridge, UK.

Kirschbaum, R. (2005) Open innovation in practice. *Research-Technology Management* 48(4), 24–28.

Kline, S.J. and Rosenberg, N. (1986) *An Overview Of Innovation. The Positive Sum Strategy*: Harnessing Technology for Economic Growth. National Academy Press, Washington, DC.

Kwok, S.H. and Gao, S. (2005) Attitude towards knowledge sharing behaviour. *Journal of Computer Information Systems* 46(2), 45–51.

Lall, S. (2000) Technological change and industrialization in the Asian newly industrializing economies: achievements and challenges. In: Kim L. and Nelson R. (eds) *Technology, Learning & Innovation: Experiences of Newly Industrializing Economies*. Cambridge University Press, Cambridge, UK, pp. 13–68.

Lichtenthaler, U. (2011) Open innovation: Past research, current debates, and future directions. *The Academy of Management Perspectives* 25(1), 75–93.

Liebowitz, S. (2002) *Rethinking the Networked Economy: The True Forces Driving the Digital Marketplace*. AMACOM Division of the American Management Association, Dallas, Texas.

Liu, X. (2010) Can international acquisition be an effective way to boost innovation in developing countries?

Evidence from China's TFT-LCD industry. *Journal of Science and Technology Policy in China* 1(2), 116–134.
Llopis-Córcoles, Ó. (2011) Understanding Knowledge Sharing in Organizations: Multi-level Research Through a Social Cognitive Perspective. Paper presented at the DIME-DRUID ACADEMY Winter Conference 2011, Aalborg, Denmark, 20–22 January 2011.
Mansfield, E. (1975) International technology transfer: forms, resource requirements, and policies. *The American Economic Review* 65(2), 372–376.
Mansfield, E. and Romeo, A. (1980) Technology transfer to overseas subsidiaries by US-based firms. *The Quarterly Journal of Economics* 95(4), 737–750.
Mansfield, E., Romeo, A. and Wagner, S. (1979) Foreign trade and US research and development. *The Review of Economics and Statistics* 61(1) 49–57.
Nonaka, I. and Takeuchi, H. (1995) *The Knowledge-creating Company*. Oxford University Press, New York.
Powell, W. (1990) Neither market nor hierarchy: network forms of organization. *Research in Organizational Behavior* 12, 295–336.
Riege, A. (2005) Three-dozen knowledge-sharing barriers managers must consider. *Journal of Knowledge Management* 9(3), 18–35.
Rothwell, R. (1994) Towards the fifth-generation innovation process. *International Marketing Review* 11(1), 7–31.
Rousseau, D. (1985) Issues of level of organizational research: multi-level and cross-level perspective. In: Cumming, L. and Staw, B.M. (eds) *Research of Organizational Behaviour*. JAJ Press, Greenwich, Connecticut, pp. 1–37.
Savitskaya, I., Salmi, P. and Torkkeli, M. (2010) Barriers to open innovation: Case China. *Journal of Technology Management and Innovation* 5(4), 10–21.
Szulanski, G. (1996) Exploring internal stickiness: impediments to the transfer of best practice within the firm. *Strategic Management Journal* 17(S2), 27–43.
Tsai, W. (2002) Social structure of 'coopetition' within a multiunit organization: Coordination, competition and intra-organizational knowledge sharing. *Organization Science* 13(2), 179–190.
Verspagen, B. and Duysters, G. (2004) The small worlds of strategic technology alliances. *Technovation* 24(7), 563–571.
Wang, S. and Noe, R.A. (2010) Knowledge sharing: a review and directions for future research. *Human Resource Management Review* 20(2), 115–131.
West, J., Salter, A., Vanhaverbeke, W. and Chesbrough, H. (2014) Open innovation: The next decade. *Research Policy* 43(5), 805–811.
Zander, U. (1991) Exploiting a Technological Edge: Voluntary and Involuntary Dissemination of Technology. Doctoral dissertation, Institute of International Business, Stockholm.

11 Intellectual Property Rights Regime for Agricultural Biotechnology in India

Seweta Srivastava,[1*] Sujit Kumar Yadav,[1] Ravindra Kumar,[2] Ashwani Kumar,[1] Vinit Pratap Singh,[3] Asha Sinha[4] and Rajendra Kumar[1]

[1]*UP Council of Agricultural Research, Lucknow, Uttar Pradesh, India;* [2]*ICAR – Indian Agricultural Research Institute, Karnal, Haryana, India;* [3]*Lovely Professional University, Phagwara, Punjab, India;* [4]*Institute of Agricultural Sciences, Varanasi, Uttar Pradesh, India*

11.1 Introduction

The biotechnology sector in India, in particular its plant biotechnology branch, is uniquely poised for a major growth transformation. Inventions involving living organisms have formed one of the controversial issues of biotechnology that are often nurtured politically (Archana, 2013). As the country's agricultural biotechnology boundaries have advanced in the 1990s into r-DNA, transgenics and molecular marker-assisted plant breeding techniques, the Government of India needs to come up to date with a matching policy support and regulatory framework that was created to furnish the path of excellence in R&D, sustainability and biosafety. To a certain extent, improvements on the policy front have been flattered by a vivacious non-governmental sector that profoundly interceded the sensitivities of modern biotechnology. This chapter emphasizes on the matter of contention and proceeds to further the path of progress achieved by agricultural biotechnology companies in India in the field of commercialization of biotechnological services and products. The particular focal point of this chapter is on intellectual property rights (IPR) related to plant biotechnology.

Biotechnology covers a cluster of technologies where the phrase 'biological material' encompasses all inorganic and organic substances present in living systems. The Patents (Amendment) Act 2002 recognizes the grant of patent to 'A new product or process involving an inventive step and proficient of industrial application'. The Trade Related Intellectual Property Rights (TRIPs Article 27) makes patent available in all fields of technology provided they are new (novelty), have industrial utility and involve an inventive step (non-obviousness). Hence, biotechnological patents may include the process of production of genetically engineered organisms and genes, the preparation of chemical substances utilizing microorganisms, basic studies involving biochemical or physiological processes, studies on the structure and the role of molecules such as DNA, RNA, specific hormones etc. (Agnihotri, 2008).

To date, the public agricultural research effort from developing countries has made relatively little progress in developing and commercializing agricultural biotechnology innovations, with a few notable exceptions. Some innovations, such as virus elimination using *in vitro* propagation, marker-assisted

*Corresponding author e-mail: shalu.bhu2008@gmail.com

breeding and use of genetic cultivar recognition to enhance the performance of germplasm conservation, have been used for years in many countries. However, this chapter concentrates on the scientific know-how that has most attracted the interest of farmers, breeders and private firms, i.e. the evolution of new cultivars with transgenic traits (Wright and Pardey, 2006).

Under the Indian Patent Act, 1970, plants and animals in whole or any part thereof, except microorganisms and necessarily biological processes for production or propagation of plants and animals, are considered non-patentable; however, genetically modified (GM) organisms/recombinants can be protected. Some examples of patentable products and processes are recombinant protein products, vectors, transformed microorganisms, cell lines and a process for producing novel microorganisms/products. The Patent Act 1970 was amended in 2002 and again in 2005 to incorporate the provisions for granting product patent in all fields of technology. The inventions related to biotechnology, viz. strains/microorganisms as such existing in nature and discovered or a process for their preparation are not patentable, but a process for the preparation of the products utilizing microorganisms is patentable. However, several countries such as the USA and Australia allow patenting of microorganisms.

Two compelling aspects, i.e. the scientific and technical pool of human assets, and the wealthy biodiversity of the country, exist as the basis of India's biotechnology potential, which could be exploited for patenting technologies (Kumar and Das, 2010). In addition to the patent application and disclosure details, patenting of microorganisms involves the organism *per se*. Thus, in order to streamline this, the Budapest Treaty (1973) was set up to enable international recognition of the deposit of microorganisms for the determination of patent products. Accordingly it is essential to deposit the microorganisms in an International Depository Authority (IDA) and obtain an accession number before filing the patent application. There are 39 authorities given the status of IDA globally. In India, the Microbial Type Culture Collection (MTCC) has been established at Chandigarh for deposition and long-term conservation of microorganisms (http://mtcc.imtech.res.in). An exhaustive report on patenting of microorganisms has been prepared by the patent facilitating centre in India. For claim purposes, the correct description and definition of microbes detailing new substances and processes involved as well as microorganisms producing the substances are important.

Biotechnology can be categorized back to four different stages of its development.

1. Based on the traditional knowledge and know-how such as medical distillates, preparing fermented foods, etc. in various tribes.
2. The application of microorganisms started on an industrial scale during the Pasteur era, which elaborates fermentation of antibiotics, mass production of alcohol and the development of classical vaccines against various human diseases.
3. The third stage began in the 1970s with two essential techniques, viz. Hybridoma technology and recombinant DNA technology.
4. The fourth stage of biotechnology includes interdisciplinary techniques such as information technology and nano-technology, especially utilizing bioinformatics, which is the foundation of modern biotechnology (Raju, 2006).

11.2 The Implications of Intellectual Property Rights for Agricultural Biotechnology

IPRs are a set of laws that confer exclusive rights on inventors or products of inventors for a given period of time (Blakeney, 2011). The role of IPRs became prominent in the protection of plant varieties in the second half of the 20th century (Wright *et al.*, 2007). The practice of IPRs to plant protection first took place in the USA when the vegetative propagated plant was patented in 1930. Prior to this event, traditional farming was based on free exchange, saving, collecting and replanting of seeds among farmers. However, the initiation of IPRs, particularly for agricultural research tools and databases through various patenting systems, has led to the expression of concerns among different communities such as farmers, academia, scientists, industries and Governmental agencies especially in developing countries (Adenle *et al.*, 2012).

11.3 Forms of Intellectual Property Rights Used in Agriculture and Biotechnology

With the significant scientific upgrading of agricultural biotechnology, enhancement of food production is expected to improve the food security position of India (Science and Technology Policy, 2001). This scientific upgradation also enhances the productivity of land by reducing the pressure to extend the hectarage under cultivation, however anxiety has been increased by the risks involved (Sharma, 2001; Shiva, 2001). The formulation of a clear policy on the protection of biotechnological innovation has been hampered by the disparity of various interests. The formulation of a clear policy on the protection of biotechnological innovation has been hampered by the disparity of various interests. By accomplishing the 'new norms of intellectual property rights' the Government of India has sanctioned the following policy objectives:

- To assure the basic needs and food security of the people on a sustainable basis;
- To encourage a forward and enhancing livelihood by the elimination of hunger and malnutrition; and
- To create an IPR regime promptly having a strong, supportive and exhaustive policy environment, which boosts the incentives for the generation and protection of intellectual property by all category of inventors (Science and Technology Policy, 2003).

Legal mechanisms consist of patents, trademarks, trade secrecy rights, plant variety protection and plant breeders' rights, among others, while biological mechanisms are composed of genetic use restriction technologies (GURTs) and hybridization. These legal and biological mechanisms are summarized in Table 11.1.

11.4 Legal Intellectual Property Rights Mechanisms in India

The Indian Patent Act, 1970 was revised to grant patents for product, drugs, agro-chemicals and microorganisms over an ephemeral period. A product patent regime has been recognized in India since 1 January 2005. The new patent law increased the protection period for inventions from 14 to 20 years. *Sui-generis* legislation in the form of the Protection of Plant Varieties and Farmers Rights (PPVFR) Act in 2001 has been passed by the Parliament to confer protection of IPRs in respect of new plant varieties. These developments have created permissible authorized situations for international partnerships in biotechnology R&D. By determining the foreign direct investment (FDI) reserves in biotechnology, the Government of India has sent suitable signals to international investors. Further, these modifications have been in conformity with the WTO-TRIPs, which sealed superior creditability to these changes.

Meanwhile, the government has also brought in regulations for agricultural biotechnology. Apart from the bio-safety policies and their regulations that shape part of the Environment (Protection) Act 1986, the Indian Parliament endorsed the National Biodiversity Act in 2002 to bestow regulatory controls over admittance to biological resources in India.

11.5 The Role of the Patent in Agricultural Biotechnology

The earlier accessible IPRs applicable to agricultural biotechnology were complex (Safrin, 2002). The utilization of IPRs is the key feature influencing the success of any technological invention evolved

Table 11.1. Forms of IPRs used in agriculture (Fuglie *et al.*, 1996; Fernandez-Cornejo, 2004).

Form of IPR	Example	Focus crops/products	Year (country)
Technology (hybrids)[a]	Hybrids	Sexually propagated plants	1920 (USA)
Plant patents	US Plant Patent Act	Asexually propagated plants	1930 (USA)
Plant variety protection	Plant Variety Protection Act	Sexually propagated plants[b]	1970 (USA)
Utility patents	US Patent Act	Microorganisms	1980 (USA)
		Plants	1985 (USA)
		Animals	1987 (USA)
Trade secret law	US Uniform Trade Secrets Act	Parental lines of hybrids	1994 (USA)
Technology	Genetic Use Restriction Technology (GURT)	Genetic modification technology/crops	1998 (USA)

[a]Hybrids of maize, sorghum, and wheat were developed in 1920, 1952 and 1968, respectively.
[b]Hybrids became allowable for protection in the USA in 1985.

in the market or to be made, thereby propping up the competitiveness of the industry of the country (Usha and Annadurai, 2010). The coming phase of technologies specifically in the biotechnology sector related to IPR are emerging with their own complications (Archana, 2013). The competent management of IPRs is thus vital in giving the right inducements for long-lasting technological innovations. IPRs are thus valuable for new business opportunities and for value in the knowledge-based industry. It is high time that India swiftly adapts to the continuously developing technological environment of the world (Commission on Intellectual Property Rights, 2002).

Biotechnology has played a significant role by showing massive commercial significance to turn out a whole new industry and patent protection. Patents are elite rights granted to inventions that satisfy the decisive factors of patentability in the form of novelty, non-obviousness and potential for industrial utility. However, patent law and practice have met serious difficulties in keeping up with the rapid scientific progress in this field and issues such as innovative steps, sufficiency of disclosure and permissible breadth of claims have proved troublesome. There have been many litigations where courts have found it difficult to determine what the general knowledge of a skilled person was at the time invention was made. A process to explore and clone a specific gene of a suitable target organism/plant was a breakthrough at the time when it was first achieved, and not many years later it is now merely a routine procedure. Many special interest groups also create the problem of disagreement against anything involved with genetic engineering and, in particular, against the continuation of patents in this arena (Arnold and Eve, 2002).

Article 27.3(b) of TRIPs covers the patenting of living processes and life-forms. This article is designed to allow the broadest categories of patents, including genetic engineering and other new biotechnologies. The following grounds of this scientific briefing give details why such patents should be banned and revoked.

1. All engross biological processes which are not under direct control of the scientist. They cannot be considered as inventions, but expropriations from life.
2. The hit or miss technologies do not meet the criteria as 'inventions', and are intrinsically hazardous to health and biodiversity.
3. There is no scientific foundation to sustain the patenting of genes, genomes, cells and microorganisms.
4. Many patents are unethical; they contravene basic human rights, destroy livelihoods, create unnecessary suffering in animals or are otherwise contrary to public morality and order.
5. Many patented 'inventions' have been utilized by local communities for millennia and therefore involve acts of plagiarism of indigenous knowledge and bio-piracy of plants/animals bred (Cooper, 1985).
6. Only the procedure itself is patentable not any substance that is made or produced by a chemical process which involves several biotechnological, biochemical and microbiological methods/techniques.

11.6 Patentable Subject Matter in Biotechnology in India

Patent protection is accessible for nearly all biotechnological innovations and thus acts as an incentive for further growth and technical innovations. By tolerating the conventional approach, the new principle of patent law states that *natural life is the creation of God* and could not be patented, but human intellect creating non-natural life by involving the application of human intelligence to natural things could be patentable.

The underlying principle behind patent protection for biotechnology inventions is applicable to the fact that:

- Patents give the leeway to the inventor to recover the development expenses of a new biotech product; and
- Patents have need of the revelation of the new technology for the profit of society. Putting the latest technology into the public domain motivates further innovation and expansion by competing companies, much like scientific publications spawn new-fangled discovery efforts and new lines of enquiry. Thus patents are a very affluent source of technical information.

As with patents, biotechnology patents also include:

- An explanation of the invention itself with precise particulars and the advantages that this invention fetches in contrast with the identified state of the art, and examples;

- A set of claims, which defines the issue for which protection is required. In biotechnology, claims are mainly concerned with:
  - Product claims;
  - Utility claims; and
  - Method of production claims.

As the WTO agreements are mandatory for all the member countries, India is also obliged to accept them. The guidelines produced by the Patent Facilitating Centre (2015) should prove very useful in enhancing the enthusiasm of scientists, universities and research institutions in a variety of R&D projects sponsored by various departments of the Ministry of Science and Technology. The outstanding features of these various guidelines are listed below.

11.6.1 Ownership of intellectual property

With regard to the consequences of R&D, in institutions where the research is approved through funding by central or state governments it is compulsory to search for protection of IPR rights to keep hold on the ownership of such IPRs.

11.6.2 Transfer of technology

Whether it is on an exclusive or non-exclusive basis, the institutions should take the necessary steps to commercially exploit patents.

11.6.3 Royalty to inventors

Owner institutions are allowed to retain the benefits and earnings generated out of the IPR after determining the share of the inventors and other associated persons, which is limited to one-third of such earnings.

11.6.4 Norms for the private industry

IPR initiated through joint research by industrial and institutions on an exclusive or non-exclusive basis may be transferred to a third party for commercialization.

11.6.5 Patent facilitating fund

To set up a patent facilitating fund, the owner institution shall set apart not less than 25% of the revenues produced from IPR. The fund shall be utilized to refurbish the inventions, filing applications for new patents and protecting the IPR against infringement.

11.6.6 Information

The details of the patents acquired the earnings and benefits arising out of the IPR and the turnover of the products shall be submitted by the institution periodically to the department/ministry that had issued the funds (Tripathi, 2002).

11.7 Patenting of Microorganisms and Cells

Patenting of biotechnological practices is not new in the USA; a classical example of this is the patent awarded to Louis Pasteur for a procedure of fermenting beer. Acetic acid fermentation was patented in the early 1800s and consequently in 1895 therapeutic patents in biotechnology were given. The progress of recombinant DNA technology (rDNA), i.e. the proscribed joining of DNA from distinct organisms, has intensely increased the understanding of the genetic and molecular beginnings of life.

The lawsuit judgment of the US Supreme Court in 1980 in *Diamond* v. *Chakrabarty*[1] unlocked the gates for discoveries in the field of biotechnology, predominantly in microorganisms. Ananda Chakrabarty evolved a naturally occurring bacterium into a modified microorganism that has the capability of catabolizing crude oil. The Patent Commissioner in the USA maintained that the topic of the invention was a living organism and was for this reason not patentable, but the US Supreme Court judge affirmed that anything beneath the sun is patentable and determined in favour of the patentee. This landmark verdict paved the way for the grant of numerous biotechnology-related patents. Chakrabarty asserted that the bacteria were rejected on two grounds:

1. Microorganisms are '*products of nature*'; and
2. As living things, microorganisms are not patentable.

The Chakrabarty pronouncement provided a great economic motivation to the patenting of cells and microorganisms, which in turn gave stimulus to the development of industrial biotechnology in the 1980s (Tripathi, 2002).

Globally, most of the patent offices agreed that novel microorganisms isolated for the first time from the natural environment can merely be patented if they are distinct in characteristics from the earlier identified microorganisms and discover a new or better use or function.

11.8 Patentability of Animals and Plants

The majority of the international agreements concentrate on various aspects of agricultural IPRs that are managed by distinct sets of guidelines, rules and policies (Adenle *et al.*, 2012). For example, the International Union for the Protection of New Varieties of Plants (UPOV) put their attention on protecting the right of seed producers (e.g. commercial plant breeders). The Convention on Biological Diversity (CBD) concentrates on protecting the rights of growers using landraces. The Trade-Related Aspects of Intellectual Property Rights (TRIPS Agreement) of the World Trade Organization (WTO) works exclusively on intensification of the rights of inventors (Moschini, 2012).

The Indian Patents (Amendment) Act, 2002, has a section that furnishes that animals and plants and any part of a plant or animal, not including microorganisms but including seeds, are not patentable. Similarly, species, varieties and essentially biological processes used for propagation or production of plants and animals are unpatentable. Last, although the act does not permit patents for a procedure of horticulture, as noted above, the act does clearly specify that any method linking to the treatment of plants is patentable. Thus, the act permits an invention to be patented that is purely a procedure of making plants free of disease, as is a method for enhancing the plant's value or increasing the value of the plant's products. In India a system called the *sui generis* system was evolved for the protection of plant varieties, incorporating the rights of farmers, breeders and village communities. *Sui generis* is a system of rights designed to act as an exclusive alternative to standard patent protection. In India, the PPVFR Act was enacted in 2001 for IPR protection of new plant varieties, and formed auspicious legal conditions for international partnerships in biotechnology R&D (Dewan, 2011).

Some people sense that patenting living things modifies the relationship between humanity and the rest of nature. Animals and plants owned by the farmers who produce them and use them as agricultural commodities, whether of patented or unpatented organisms, are bound to respect animal welfare legislation (Archana, 2013). A suitable internationally accepted legal system with respect to utilizing livestock biodiversity and patenting of novel genes may lead to economic benefit sharing. Synchronization of IP regimes paves the way for the assessment of technologies needed for biotechnological R&D through uncomplicated licensing and/or patent exemptions, particularly for public benefit-oriented research, and also for creating private investments, subsequently encouraging private–public participation (Ramesha, 2011).

Genetically modified organisms together with agriculturally important animals and plants are subject to the Cartagena Protocol on Biosafety, which was taken on in 2000 by parties to the Convention on Biodiversity, and to other domestic regulations in developed as well as in developing countries. In developing countries the authoritarian structure is characteristically less established, and the regulations in force might themselves be viewed, in particular cases, as institutional experiments developing towards feasible implementation (Wright and Pardey, 2006). The exploitation of IPRs in crop research is swiftly evolving, with an rising number of intergovernmental organizations and countries becoming members of and thereby passing the framework furnished by UPOV (Jefferson *et al.*, 2014). The objective of integrating the protection of ancestral farming methods with a desire to encourage agricultural innovation could be more willingly achieved if the provisions of UPOV were joined with other international treaty frameworks, such as that provided by the CBD (Ganguli, 2000). The Indian Patent Law Sections 3i and 3j states that all animals and surgical processes are not patentable; hence, animal models are not patentable in India. It would unlock novel vistas in the research field if the appropriate amendments were made, then animal models could also be patentable in India (Bagle *et al.*, 2012).

11.9 An Overview on Biotechnology Patenting Activity in India

Both the number and scope of biotechnology patentable subject matter were enhanced significantly after the international biotechnology industry's expansion between the 1980s and the 1990s. During this decade, gene patents were extensively accessible in numerous countries (Schilling, 2011).

International policies concerning the patentability of human gene sequences are not standardized, which is one of the currently faced legal challenges to the authority of gene patents (Jamison, 2015).

India also follows the jurisdictions of other countries in that the issue asserted as the invention should be novel, non-obvious, industrially useful and needs sufficient disclosure. One cannot attain a patent on the basis that a new element or compound in nature is discovered, because the shape as it survives in nature is not new. A patent can only be obtained by changing the exposed component through human involvement to make it into a new unique form in which it does not evolve in nature, but has economic significance.

11.10 Insufficiency of Regulations

Access to genetic resources and protection of Plant Breeders' Rights (PBRs) reflects an insufficient level of regulation that is perceived by the stakeholders as a problem due to intrinsic background limitations. Regarding the access to genetic resources, the incipient national experience in applying regulations and designing a system, which is, on the one hand efficient to protect the resources but on the other hand not too bureaucratic, is the key issue. The same refers to mechanisms that ensure unbiased benefit sharing for the exploitation of linked traditional knowledge and genetic resources (Fig. 11.1).

The 1978 Act of the UPOV Convention from which the regulations were derived gives some idea about discrepancies about the deliberation of the concept of 'Essentially Derived Varieties (EDV)', and the scope and the extent of protection of the 'Farmers' Privilege'. EDV balances the rights of biotech companies and plant breeders to modify the previously existing and protected varieties of plant which were continuously exported and unlawfully used for reproduction purposes. Keeping the above mentioned points in mind, the 1991 Act extends the protection to the harvested produce that is utilized illicitly for propagation, and also furnished that the plant breeder does not have any prospects to implement his rights related to this (Table 11.2).

To be entitled for Plant Variety Rights (PVR), the variety must be novel, unique from existing varieties and stable and uniform in its crucial characteristics, and is restricted to commercializing vegetatively or reproductive propagated material derived from a novel variety (Blakeney *et al.*, 1999). It was specified or implied that assured exemptions were authorized to farmers and researchers (breeders); such exemptions under the PVR are termed breeders' privilege and farmers' privilege, respectively.

11.11 Lack of Awareness

The foremost factors affecting the development of regulations are the lack of consciousness concerning the utility to protect the biodiversity and the associated traditional knowledge/knowhow, as well as the esteem for IPR, which prove very beneficial to conserve our resources. For awareness regarding this context two things should be kept in mind: (i) that the biotechnological process and seed sector using traditional knowledge and biogenetic resources to build up novel products should be aware of and

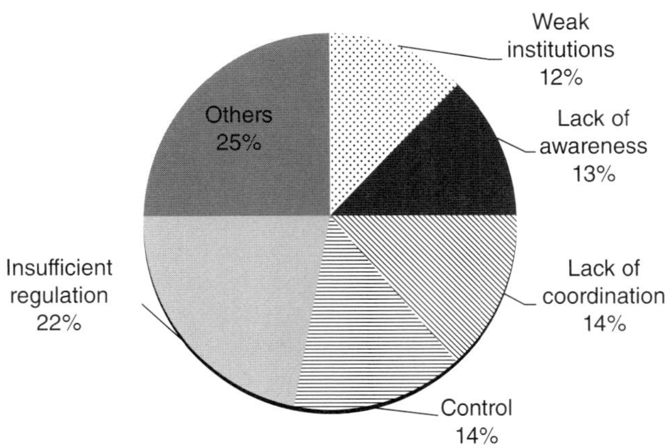

Fig. 11.1. Problems with regulating the access to genetic resources.

Table 11.2. The main provisions of Plant Variety Rights and Patent Law (revised from Van Wijk *et al.*, 1993).

Provisions	Plant Variety Rights (under UPOV, 1991)	Patent Law
Protection coverage	Plant varieties of all genera and species	Inventions
Requirements	Novelty, distinctness, uniformity, stability	Novelty, inventiveness, non-obviousness
Protection term	Minimum 20 years	17–20 years (OECD)
Protection scope	Commercial use of all material of the variety	Commercial use of protected matter
Breeders' exemption (research only)	Yes	Variable
Breeders' exemption (commercial use)	Not for essentially derived varieties	No
Farmers' privilege	Up to national laws	Limited form in prospect in EU countries

UPOV, International Union for the Protection of New Varieties of Plants.

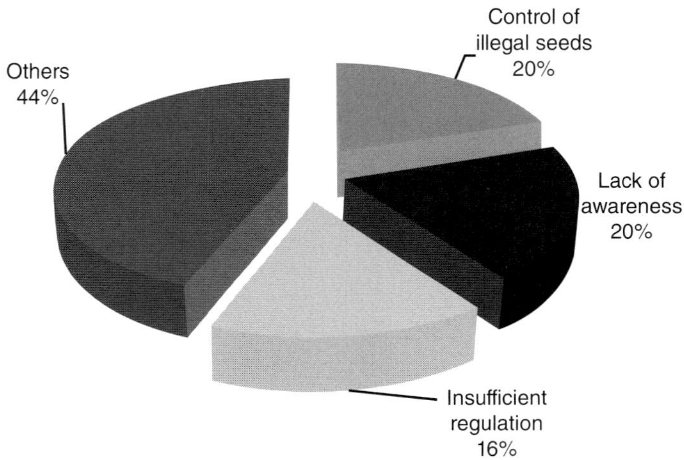

Fig. 11.2. Problems with regulating PBRs.

respect the rights of the traditional consumers of these resources ensuing from their contribution in terms of development and conservation; and (ii) that the producers should also observe the rights of the plant breeders.

11.12 Application and Control

Due to the partial capacities and experiences of the representatives in the application of the new regulations the process concludes in an inadequate execution of the existing regulations. It is very difficult to prove that the utilization of a particular genetic resource is illegal or it is bio-pirated, because the traditional knowledge and knowhow related to the use of that genetic resource entails some sort of technical limitation, a lack of financial and human resources and also a lack of experience to impose this control. This same problem is also revealed by many plant breeders who encounter major difficulties pertaining to the protection of PBRs and also in managing the illegitimate commercialization of seeds (Fig. 11.2).

There are numerous practical examples which highlight these results. In a GTZ project, a novel variety of *Desmanthus virgatus* was developed in Paraguay on which PBR applied but they did not act in accordance with the national legislation accessible to genetic resources. As it was the first instance in Paraguay, the lack of experience of the Office of Registration of New Varieties collaborating with the National Seed Department of the Ministry of Agriculture was possibly unable to point out the right/legal process to arrange commercialization and protection of the new variety. This matter exemplifies the problems present in most countries and its upshots. The problems recognized in this case were that the availability of a new, superior variety for the farmers was delayed due to the lack of knowledge about existing regulations and the lack of experience of the national authorities, which resulted in a lengthy procedure in obtaining PBR (Wendt and Izquierdo, 2000).

Finally, it was concluded that any enhancement of the act of right to use genetic resources and protection of IPRs is strongly linked to the capacity building and development of coordination flanked by the stakeholders on both the national and international level (Byerlee and Fischer, 2001; Salazar, 2001). These capacity-building agenda should be intended for all stakeholders and should emphasize:

- The requirement of conservation of biodiversity;
- The safety of IPRs;
- Accessing bioresources and technology; and
- Teaching to human resources.

11.13 IPRs and Agricultural Biotechnology: Canvas of Interplay

The interactions between IPRs and agricultural biotechnology are multitudinous and intricate and include: biotechnology development facilitated by IPRs; biotechnological growth and development driven by IPRs; and biotechnology that influences IPRs.

11.13.1 Biotechnology development facilitated by intellectual property rights

The beginning of protection will steer two discrete investment pathways in R&D: (i) the present R&D efforts that boost the technology expansion processes without drastic changes in the research model; and (ii) the areas in which private investment will be managed to concentrate on applied aspects, because the public sector will focus on basic research. One example is high-yielding varieties (HYV) of seeds, where private investment is ready to take off, chiefly in pulses. Apart from investment in the evolution of technology, promises to enhance current agricultural practices through improvement in the management and conservation of inputs like vaccines, feed, and pesticides, will encourage increased activity (Thornton, 2010).

11.13.2 Technology development driven by intellectual property rights

In the public or private sector, the degree and nature of protection *inter alia* will affect investment behaviour. The following example elucidates changes in investment resolutions in the public sector. A new regulation necessitates varietal recognition by using a mechanism for ordinance, monitoring and dispute settlement.

A good example is the evolving awareness and amplified investment in technologies such as DNA fingerprinting. Presently, an insignificant 0.5% of the net profits are invested back in the private sector, which is likely to be much smaller in agricultural R&D. The new IPR regime will affect investment resolutions in two ways: (i) private firms will be forced to increase expenses for innovative research; and (ii) technology transfer through joint ventures and/or mergers will need to be increased between domestic and foreign firms.

11.13.3 Biotechnologies that influence intellectual property rights

To overcome operational difficulties in seeking protection, a new set of technologies are being evolved that are staunchly influencing and making the framework of IPRs redundant, which is an utmost but realistic situation. Among all the agricultural technologies, varietal development will receive the maximum enhancing attention (Samuel *et al.*, 2014). The most controversial difficulty faced by biotechnological research is the opposition to genetic engineering biotechnologies, because they are recognized as a tool to force farmers to rely on multinational corporations' seed monopolies. To overcome this controversial problem genetic use restriction technologies (GURTs) evolved and made assured return on investments through the protection of plant varieties (Lombardo, 2014).

Commodities of GURTs are crop varieties with traits whose expression is under outward chemical control. The considered traits may be viability of seeds, germination, flowering, nutritional and savour qualities, sterility and fertility restoration (for hybrid production), resistance to diseases, pests and herbicides, etc. The primary desire of a private innovator is to compel farmers to purchase their seeds for subsequent use.

11.14 Conclusion

India has begun to become aware of the need for greater IP protection, however, these outcomes are only the first steps on the path of India's knowledge-based economy because the patenting of biological material is comparatively new to India. The various biotechnological innovations that have been developed anywhere near to the point of commercialization by public institutions is limited. Further, had more biotechnology innovations been put forward, political and social contra-distinctions to transgenic products in key export markets would probably have blocked commercial adoption in many cases. Thus an explicit statistical analysis of effects of IPR on agricultural biotechnological innovation through to acceptance on farmers' fields is impossible at this time.

Additionally, being one of the biodiversity-rich countries, India should be cautious to protect biotechnological inventions as that would facilitate biotechnology research to compete with global players. Thus, biotechnology must be viewed as a value added tool in a precise system of agriculture rather than a replacement of conventional approaches of its overall improvement strategies. Therefore, an optimum balance between biotechnology and effective conventional efforts is necessary.

India has sailed through the journey from a state of a total lack of IP awareness to the present state of proactive pursuit of IP in frontier areas of technology. Having unleashed India's IT potential in the recent past, the time has now come to harness the tremendous strengths and energies of the country in the biotechnology sector. Biotechnology-enabling environments must be put in place that comprise policy formulation and implementation for creating, supporting and providing the ability building and resources required to evolve open source to sustain agriculture in developing countries. Furthermore, the attention and focus of open source development should not be restricted to one area but it should address other requirements such as health and the environment. The opportunities recommended by open source biotechnology can make a notable impact on sustainable agricultural growth through unobstructed access to modern biotechnology techniques. Open source biotechnology is making slow but steady development in agriculture, but many issues will have to be addressed to enjoy the profits of this new innovation in developing countries.

Note

1 US Supreme Court, *Diamond* v. *Chakrabarty*, 447 U.S. 303 (1980).

References

Adenle, A.A., Sowe, S.K., Parayil, G. and Aginam, O. (2012) Analysis of open source biotechnology in developing countries: An emerging framework for sustainable agriculture. *Technology in Society*, 34, 256–269.

Agnihotri, A. (2008) IPR issues in Agricultural Biotechnology. In: *Proceedings of Workshop on IPR Management in Indian Agriculture under changing R&D Scenario-Implications of WTO, PVP&FR Act and Biodiversity Acts*, 17 October 2008. Organized by Amity Institute of Microbial Technology.

Archana, K. (2013) Do we need patent protection to biotechnology inventions? *International Journal of Scientific and Research Publications* 3(4), 1–5.

Arnold, B.E. and Eve, O.Z. (2002) Patenting genes and genetic research tools: Good or bad for innovation. *Annual Review of Genomics and Human Genetics* 3, 415–432.

Bagle, T.R., Kunkulol, R.R., Baig, M.S. and More, S.Y. (2012) Transgenic animals and their application in medicine. *International Journal of Medical Research & Health Sciences* 2, 107–116.

Blakeney, M. (2011) Recent development in intellectual property and power in the private sector related to food and agriculture. *Food Policy* 36(Suppl. 1), S109–113.

Blakeney, M., Cohen, J.I. and Crespi, S. (1999) Intellectual property rights and agricultural biotechnology. In: Cohen, J.I. (ed.) *Managing Agricultural Biotechnology Addressing Research Program Needs and Policy Implications.* CAB International, Wallingford, UK, pp. 209–227.

Byerlee, D. and Fischer, K. (2001) Accessing Modern Science: Policy and Institutional Options for Agricultural Biotechnology in Developing Countries. IP Strategy Today [online]. n. 1. Available at: http://www.biodevelopments.org/ip/ipst1.pdf (accessed 15 December 2015).

Commission on Intellectual Property Rights (2002) *Integrating IPR and Development Policy – Report of the Commission on Intellectual Property Rights*. CIPR, London.

Cooper, I.P. (1985) *Biotechnology and the Law*. Clark Boardman, New York.

Dewan, M. (2011) IPR Protection in Agriculture: An Overview. *Journal of Intellectual Property Rights* 16, 131–138.

Fernandez-Cornejo, J. (2004) The seed industry in U.S. agriculture: An exploration of data and information on crop seed markets, regulation, industry structure, and research and development. *Agriculture Information Bulletin* AIB 786. US Department of Agriculture Economic Research Service, Washington, DC.

Fuglie, K., Ballenger, N., Day, K., Klotz, C., Ollinger, M., Reilly, J., Vasavada, U. and Yee, J. (1996) Agricultural research and development: Public and private investments under alternative markets and institutions. *Agricultural Economics Report* 735. US Department of Agriculture Economic Research Service, Washington, DC.

Ganguli, P. (2000) Intellectual property rights: Mothering innovations to markets. *World Patent Information, pp.* 22–43.

Jamison, M. (2015) Patent Harmonization in Biotechnology: Towards International Reconciliation of the Gene Patent Debate. *Chicago Journal of International Law* 15(2), 690–720.

Jefferson, D.J., Camacho, A.B. and Chi-Ham, C.L. (2014) Towards a balanced regime of intellectual property rights for agricultural innovations. *Journal of Intellectual Property Rights* 19, 395–403.

Kumar, A. and Das, G. (2010) *Biodiversity, Biotechnology and Traditional Knowledge: Understanding Intellectual Property Rights.* Narosa, New Delhi.

Lombardo, L. (2014) Genetic use restriction technologies: a review. *Plant Biotechnology Journal* 12(8), 995–1005.

Moschini, G.C. (2012) Intellectual property rights and the World Trade Organization: retrospect and prospects. Available at: http://www.card.iastate.edu/faculty/profiles/giancarlo_moschini/moschini-trips-preprintoct-04.pdf (accessed 28 May 2012).

Patent Facilitating Centre (2015) Instructions for Technology Transfer and IPR. Technology Information, Forecasting and Assessment Council. Available at: http://www.pfc.org.in/info/tt_ipr.htm (accessed December 2015).

Raju, C.B. (2006) *Intellectual Property Rights*, 1st edn. Serials Publications, New Delhi.

Ramesha, K.P. (2011) Intellectual Property Rights Regime for Livestock Agriculture in India – Present Status and Future Prospects. *Journal of Intellectual Property Rights* 16(2), 154–162.

Safrin, S. (2002) Treaties in collision? The biosafety protocol and the World Trade Organization agreements. *The American Journal of International Law* 96, 22.

Salazar, S. (2001) Experiencias de ISNAR en la Construcción de Capacidades en la Gestión de los Derechos de Propiedad Intelectual en Biotechnología. *Proceedings of the IV Latin-American Meeting on Plant Biotechnology*. REDBIO/FAO, Goiania, Brazil, 4–8 June.

Samuel, M.P., Sastry, R.K. and Venkattakumar, R. (2014) Status and Prospects of IP Regime in India: Implications of Agricultural Education. *Journal of Intellectual Property Rights* 19, 189–201.

Schilling, S.H. (2011) DNA as Patentable Subject Matter and a Narrow Framework for Addressing the Perceived Problems Caused by Gene Patents. *Duke Law Journal* 61(3), 731–773.

Science and Technology Policy (2001) Implementation Strategy (Draft Document: Action Plan: version 3). Available at: http://www.clionline.org/busserv/biotechnology/policy.htm (accessed 29 October 2001).

Science and Technology Policy (2003) Available at: http://www.dst.gov.in/doc/STP2003.doc (accessed 9 December 2015).

Sharma, M. (2001) India: Biotechnology research & development. In: Persley, G.J. and Lantin, M.M. (eds) *Agricultural Biotechnology and the Poor: An International Conference on Biotechnology*, Washington, D.C., USA, pp. 51–57.

Shiva, V. (2001) *Patents: Myths and Reality.* Penguin Publication, New Delhi.

Thornton, P.K. (2010) Livestock production: recent trends, future prospects. *Philosophical Transactions of the Royal Society, B* 365(1554), 2853–2867.

Tripathi, K.K. (2002) Biotechnology: Government of India Initiatives. *SaketInd Digest, pp.* 49–53.

Usha, S. and Annadurai, C. (2010) The Intellectual Property Rights and Pharmaceutical Biotechnology in India. *Journal of Pharmacy Research* 3(10), 2466–2469.

Van Wijk, J., Cohen, J.I. and Komen, J. (1993) Intellectual Property Rights for Agricultural Biotechnology: Options and Implications for Developing Countries. *ISNAR Research Report* No. 3. International Service for National Agricultural Research, The Hague.

Wendt, J. and Izquierdo, J. (2000) La Práctica del Acceso a los Recursos Genéticos y de los Derechos de Obtenciones Vegetales en América Latina. FAO, Santiago, Chile. Available at: http://www.rlc.fao.org/prior/recnat/pdf/biogen.pdf (accessed 9 December 2015).

Wright, B.D. and Pardey, P.G. (2006) Changing intellectual property regimes: implications for developing country agriculture. *International Journal of Technology and Globalisation* 2(1/2), 93–114.

Wright, B.D., Pardey, P.G., Nottenburg, C. and Koo, B. (2007) Agricultural innovation: investments and incentives. In: Evenson, R. and Pingali, P. (eds) *Handbook of Agricultural Economics* vol. 3. Elsevier, Amsterdam, the Netherlands, pp. 2533–2603.

12 DNA Patenting

MICHAEL BLAKENEY*

Faculty of Law, University of Western Australia, Crawley, Western Australia

12.1 Introduction

Historically, patent law distinguished between patentable inventions and non-patentable discoveries. The evolution of recombinant DNA technology in the 1970s made it possible for scientists to identify useful DNA. Developments in patent law have made it possible to commodify and appropriate commercially useful DNA. The privatization of DNA has important implications for food security, when private corporations can secure the ownership of DNA which is critical for agricultural innovations at a time of climate change. The patenting of DNA also has equally important implications for modern medicine. This chapter examines the development of patent law beyond the protection of mechanical, electronic and pharmaceutical inventions to embrace the patenting of DNA.

12.2 International Intellectual Property Infrastructure

The details of the rules in relation to the patenting of DNA differ between countries, but the underlying patent principles between countries are fairly harmonized. This is because most countries are members of the World Trade Organization (WTO) and are obliged to introduce intellectual property (IP) laws which implement the norms prescribed by the WTO Agreement on Trade-Related Aspects of Intellectual Property Rights ('TRIPS Agreement').[1] Key provisions in the present context are Article 27.1, which provides that 'patents shall be available for any inventions, whether products or processes, in all fields of technology, provided that they are new, involve an inventive step and are capable of industrial application.' This provision requires also that 'patents shall be available and patent rights enjoyable without discrimination as to the ... field of technology'. The effect of this provision is to establish a patenting regime that extends to all WTO members. Additionally, Article 27.3(b) of the TRIPS Agreement requires that WTO members 'shall provide for the protection of plant varieties either by patents or by an effective *sui generis* system or by any combination thereof.'

12.3 Patenting of DNA

In relation to agriculture, the development of recombinant DNA technology has enabled the engineering of desirable genetic traits from useful plant species. These include: (i) pest control traits such as insect, virus and nematode resistance as well as herbicide tolerance; post-harvest traits such as delayed ripening of spoilage-prone fruits; (ii) agronomic traits such as nitrogen fixation and utilization, restricted branching, environmental stress tolerance, male and/or seed sterility for hybrid systems; and (iii) output traits such as plant colour and vitamin enrichment. The production of transgenic plants has become possible through the development of a number of enabling and transformation technologies.

A key issue around the patenting of genetic resources has been whether a DNA sequence could be characterized as an 'invention'. It will be recalled that Article 27.1 of the TRIPS Agreement requires that patents are made available for 'inventions'. In the early history of patent law an invention was thought to involve some kind of technical innovation and a distinction was drawn between patentable inventions and non-patentable discoveries. The US Supreme Court in its 1980 determination *Diamond* v. *Chakrabarty*[2] held in a

*Corresponding author e-mail: michael.blakeney@uwa.edu.au

4:3 majority decision that a bacterium genetically engineered to degrade crude oil was an invention. This decision provided the legal underpinning for the US biotechnology industry. The European Parliament's belated response in 1998 was its Biotechnology Directive, which provided in Article 3.2 that 'biological material which is isolated from its natural environment or produced by means of a technical process is deemed to be an invention even if this material previously occurred in nature.'

The patentability of genetic materials and gene fragments, such as expressed sequence tags (ESTs) and single nucleotide polymorphisms (SNPs), as well as enabling gene-based technologies led to what has been described as a 'genomic gold rush' in the 1990s as vast numbers of gene-based patent applications were filed, particularly in the USA (Joly, 2003). Significant misgivings were expressed by numerous commentators. Probably the most influential among these were Heller and Eisenberg (1998), who suggested that genetic research tool patents could create a 'tragedy of the anticommons' in which multiple patent owners would tie-up genetic materials in a thicket of IP patent rights. This was perceived by Correa (2009) to be a particular problem for the genetic improvement of crops since this is an incremental process and each new patent would constrain the 'freedom to operate', particularly of public agricultural research institutes.

Arguably, this gold rush has been brought to an end, at least in the USA, by 2005 in the *In* re *Fisher* decision of the US Court of Appeals for the Federal Circuit,[3] which upheld a ruling by the US Board of Patent Appeals and Interferences refusing to allow patent applications made on behalf of Monsanto Co. on five ESTs encoding protein and protein fragments in maize plants grown by the Asgrow Seed Company of Des Moines, Iowa. Joly (2006) suggested that the *Fisher* case was used by Monsanto Co., a significant downstream user of research tools, to urge upon the court a higher patentability standard in order to eliminate the thousands of research tool patents which were cluttering research efforts. Mainly for this reason, the case attracted *amicus* briefs filed by academic institutions as well as major biotechnology and pharmaceutical companies. The Board of Patent Appeals and Interferences was unable to identify any 'substantial utility' or usefulness in the application for patentability of the ESTs. The Appeal Court agreed with this approach, stating that that claimed inventions 'ought to have a specific and substantial utility' to satisfy the requirements of the US patent statute. The court observed that the application comprised asserted uses based upon 'merely hypothetical possibilities' which had not yet been achieved in the real world. As the applicant did not identify the function for the underlying protein-encoding genes, the court held that 'the claimed ESTs have not been researched and understood to the point of providing an immediate, well-defined, real world benefit to the public meriting the grant of a patent'.

Although this decision imposed a higher patent standard, which might result in the invalidation of previously granted patents over research tools, this was not specifically addressed by the court. However, Joly (2006) optimistically suggested that 'academic researchers as well as a considerable portion of the biotechnology and pharmaceutical industry will be satisfied by this judgment as it should reduce the number of parasite patents on gene sequences, in the United States.'

12.4 Myriad Genetics in the USA

The patentability of DNA was recently addressed in the USA in litigation concerning the patenting of isolated DNA for use in medical diagnostic kits. In *Association for Molecular Pathology* v. *USPTO*,[4] a judge of the US District Court for the Southern District of New York delivered a summary judgment which invalidated patents related to the *BRCA 1* and 2 breast and ovarian cancer susceptibility genes, which had been obtained by the company Myriad Genetics. The trial judge ruled that the claims to DNA sequences in isolation were held to be insufficiently distinct from naturally occurring genes in the body and were thus products of nature rather than inventions. He observed that DNA represented the 'physical embodiment of biological information, distinct in its essential characteristics from any other chemical found in nature and that DNA in an 'isolated' form alters neither this fundamental quality as it exists in the body nor the information it encodes'.

This decision was successfully appealed to the US Court of Appeals for the Federal Circuit (CAFC).[5] The Appeal Court considered that the District Court judge had fallen into error in considering not whether the isolated DNAs were markedly different from naturally occurring DNAs, but rather whether they had the same informational content as native DNA sequences. Nevertheless, the CAFC considered that the District Court was correct in holding that Myriad's claims directed to comparing and analysing gene sequences were not patentable, as these claims contained no transformative steps and covered only patent ineligible abstract steps.

The Supreme Court reversed the decision of the CAFC ruling that the DNA in question was not 'patent eligible'.[6] The court held that a naturally occurring DNA segment is a product of nature and not patent eligible merely because it has been isolated. The court noted that it was undisputed that Myriad did not create or alter any of the genetic information encoded in the *BRCA1* and *BRCA2* genes. The location and order of the nucleotides existed in nature before Myriad found them. Nor did Myriad create or alter the genetic structure of DNA. Instead, Myriad's principal contribution was uncovering the precise location and genetic sequence of the *BRCA1* and *BRCA2* genes within chromosomes 17 and 13.

However, the court observed that 'cDNA does not present the same obstacles to patentability as naturally occurring, isolated DNA segments' and that had Myriad created an innovative method of manipulating genes while searching for the *BRCA1* and *BRCA2* genes, it could possibly have sought a method patent. But at the time of Myriad's patents the processes used by Myriad to isolate DNA 'were well understood, widely used, and fairly uniform insofar as any scientist engaged in the search for a gene would likely have utilized a similar approach,' and were not at issue in this case.

12.5 Myriad Genetics in Australia

The same issue was considered by the Australian Federal Court. In *CancerVoices Australia* v. *Myriad Genetics Inc*[7] the trial judge found that isolated human DNA was patentable because the breaking of covalent bonds linking the gene to the rest of the chromosome made it sufficiently different from the gene as found in the human body. The question in Australia is whether a claimed invention can be described as a 'manner of new manufacture' within s18(1)(a) of the Patents Act 1990 (Cth). The Australian High Court had previously held in *National Research Development Corporation* v. *Commissioner of Patents*[8] that the subject matter of a patent claim had to consist of an 'artificially created state of affairs', which provided a new and useful effect of economic significance. The trial judge applied this decision in observing that isolated DNA was the product of human intervention and involved processes of extraction and purification, and immense research and intellectual effort.

The Full Federal Court followed the same approach as the trial court in upholding the patent (*D'Arcy* v. *Myriad Genetics Inc*).[9] By the time that the case came before the Full Federal Court, the US Supreme Court had ruled that a naturally occurring DNA segment was a product of nature and not patent eligible merely because it has been isolated. However, the court noted that in Australia, there was no statutory or jurisprudential limitation of patentability to exclude 'products of nature' and that in the *NRDC* case the High Court had specifically rejected such an approach (paras 204–206). The Full Federal Court also noted that Myriad's patent claim was 'not, as the US Supreme Court considered, concerned "*primarily with the information contained in the genetic sequence* [rather than] *with the specific chemical composition of a particular molecule*"' (para. 2016). It observed that following the *Chakrabarty* and *NRDC* cases, the isolated nucleic acid in the patent under consideration had 'markedly different characteristics from any found in nature'; Myriad did not merely 'separate that gene from its surrounding genetic material' (para. 2016).

Both Australian courts acknowledged the report of the Australian Law Reform Commission (ALRC) into gene patenting (ALRC, 2004), which had concluded that a new approach to the patentability of genetic materials was not then warranted (para. 6.53). The ALRC had taken the view that this would represent a significant and undesirable departure from accepted international practice with respect to genetic inventions, and may adversely affect investment in the Australian biotechnology industry.

The case was appealed to the High Court of Australia, which handed down its decision on 7 October 2015.[10] The High Court reversed the court below, the majority observing that despite the formulation of the claimed invention as a class of product, its substance was 'information embodied in arrangements of nucleotides [which was] not "made" by human action' (p. 4). The majority observed that: where an affirmative application of the concept is likely to result in the creation of important rights as against the world, to involve far-reaching questions of public policy and to affect the balance of important conflicting interests, the question must be asked whether that application is best left for legislative determination. The patentability of nucleotide sequences derived from human DNA is in that category. The inherent patentability of the invention as claimed would powerfully imply patentability of any claim for an isolated nucleic acid coding for a specified polypeptide.

As a matter of policy the majority observed that there was a real risk that the chilling effect of the claims, on the use of any isolation process in relation to the *BRCA1* gene, would lead to the creation of an exorbitant and unwarranted *de facto* monopoly on all methods of isolating nucleic acids containing the sequences coding for the BRCA1 protein.

As was the situation in the US decision, the Australian High Court's decision leaves open the possibility of patenting isolated DNA where the claimant has modified its chemical structure.

12.6 DNA Patenting and Agriculture

The cultivation by farmers of genetically modified (GM) crops has on occasion led to IP liability, where GM seed is patented and the cultivation of that seed by the patentee is unauthorized. The cases are divided between those where farmers knowingly cultivate patented GM seed and those where the cultivation of patented seed is apparently inadvertent, for example, where crops are apparently pollinated by wind or insect-borne pollen.

An example of the first category of case is *Monsanto Co.* v. *Scruggs*,[11] which concerned Monsanto's patented Roundup Ready ('RuR') glyphosate-tolerant seeds. This was licensed to seed companies, who were obliged to sell the seed to growers who signed technology licence agreements acknowledging Monsanto's patent and on condition that they could only be used by growers for a single commercial crop, i.e. growers could not save seed produced from a harvested crop for replanting during the following growing season. Scruggs, who had not signed a technology licensing agreement, purchased a small quantity of RuR soybeans and cotton seeds which were cultivated and from which he saved seed for further plantings. The court decided that Monsanto's patent had been infringed by Scruggs, rejecting his defence that neither Monsanto's biotechnology nor the plants in their fields were covered by the patent and that the first sale of the seed embodying the invention exhausted the patent rights of Monsanto. The court noted that Monsanto never made an unrestricted sale of its seed technology, as it licensed its technology to seed companies with a proviso: subsequent sales of seed containing its transgenic trait must be limited to growers who obtained a licence from Monsanto and for only a single growing season.

A recent variant of these facts occurred in *Monsanto Co.* v. *Bowman*,[12] where a farmer, Bowman, purchased commodity seeds from a local grain elevator which were not subject to a technology agreement. Following the application of glyphosate to the crops grown from these seeds, Bowman identified those which were glyphosate resistant and these were saved and replanted in subsequent years which enabled Bowman to use glyphosate-based herbicide. Monsanto filed a patent infringement claim against Bowman and in September 2009, the district court in Indiana granted summary judgment on patent infringement for Monsanto. Bowman appealed to the Court of Appeals for the Federal Circuit. Bowman argued that Monsanto's patent rights were exhausted under the first sale doctrine in relation to all second-generation RuR soybean seeds that were present in the grain elevators. He cited the 2008 Supreme Court case of *Quanta Computer, Inc.* v. *LG Electronics, Inc.*[13] In this case the Supreme Court held that sales of products that 'substantially embody' the disputed patents will also be considered sales that exhaust the patent right. Bowman argued that the court should hold that subsequent generations of the seeds are 'substantial embodiments' of the first-generation seeds, and thus the sales of these seeds would be exhausting sales. The appeal court held that even if Monsanto's patent rights in the commodity seeds were exhausted, such a conclusion would be of no consequence because once a grower, like Bowman, planted the commodity seeds containing Monsanto's RuR technology and the next generation of seed developed, the grower had created a newly infringing article. It observed that 'The fact that a patented technology can replicate itself does not give a purchaser the right to use replicated copies of the technology. Applying the first sale doctrine to subsequent generations of self-replicating technology would eviscerate the rights of the patent holder.'

A case of apparently inadvertent infringement is illustrated by the Canadian litigation between Monsanto Canada, Inc and a farmer, Percy Schmeiser. Schmeiser grew canola commercially in Saskatchewan. He had never purchased Monsanto's patented RuR canola nor did he obtain a licence to plant it. Yet, in 1998, tests revealed that 95–98% of his 1000 acres of canola crop was made up of RuR plants. The origin of the plants is unclear. They may have been derived from RuR seed that blew on to or near Schmeiser's land. Monsanto brought an action for patent infringement. In finding patent infringement the trial judge ruled that the growth of the seed, reproducing the patented gene and cell, and sale of the harvested crop constituted taking the essence of Monsanto's invention, using it, without permission and in so doing infringed the patent. By a majority of 5:4 the Federal Court of Appeal ruled that Schmeiser's saving and planting seed, then harvesting and selling plants that contained the patented cells and genes appeared to the court, on a common sense view, to constitute 'utilization' of the patented material for production and advantage, within the meaning of s. 42 of the Canadian Patent Act (*Monsanto Canada, Inc.* v. *Schmeiser*).[14] The argument that the infringing seed had merely grown, as the result of wind pollination or through the pollinating activities of birds and bees was rejected by the majority of judges as denying 'the realities of modern agriculture.' What was at stake in this case was sowing and cultivation, 'which necessarily involves deliberate and careful activity on the part of the farmer'. They noted that he had actively cultivated RuR canola as part of his business operations, thus in light of all of the relevant considerations, Schmeiser had used the patented genes and cells, and infringement was established.

The Schmeiser litigation and some 147 other suits against farmers filed by Monsanto (Monsanto, 2015) induced a coalition of 38 farmers, seed sellers and agricultural organizations led by the Organic Seed Growers and Trade Association, to seek declaratory judgments of non-infringement and invalidity in the District Court for the Southern District of New York (*Organic Seed Growers & Trade Ass'n* v. *Monsanto Co*),[15] with respect to 23 patents owned by Monsanto Co. and Monsanto Technology, LLC (collectively, 'Monsanto').

The applicants described themselves as growers, seed-selling businesses and agricultural organizations which grow, use or sell conventional seeds, and many of whom have organic certification and who did not want to use or sell transgenic seed incorporating Monsanto's technologies. Their principal concern was that, given Monsanto's patent enforcement policy, if their crops became contaminated by transgenic seed they could perversely be accused of patent infringement by the company responsible for the transgenic seed that contaminated them.

The plaintiffs' application for a declaration was refused by the District Court on the ground that there was no justiciable dispute between the parties. It noted that there was no evidence that Monsanto had commenced litigation against inadvertent users of patented seed and there was no evidence that any of the plaintiffs had experienced contamination from Monsanto's seed, or had ever been threatened by Monsanto for patent infringement. Shortly after they initiated the lawsuit, the plaintiffs had asked Monsanto for an express undertaking not to sue. While refusing to enter into such an undertaking Monsanto referred the plaintiffs to its website, which contained the statement that 'It has never been, nor will it be Monsanto policy to exercise its patent rights where trace amounts of our patented seeds or traits are present in farmer's fields as a result of inadvertent means.' Monsanto's attorneys by letter further expanded on the company's absence of any intent to sue persons in the position of the plaintiffs, declaring that:

> Monsanto is unaware of any circumstances that would give rise to any claim for patent infringement or any lawsuit against your clients. Monsanto therefore does not assert and has no intention of asserting patent-infringement claims against your clients. You represent that 'none of your clients intend to possess, use or sell any transgenic seed, including any transgenic seed potentially covered by Monsanto's patents.' Taking your representation as true, any fear of suit or other action is unreasonable, and any decision not to grow certain crops unjustified.

These representations were also taken into account by the District Court in ruling that there was no imminent dispute between the parties.

These factors were equally influential in the determination of the Court of Appeals that there was no justiciable controversy between the parties.[16] The US Supreme Court refused the grant of *certiorari* to allow an appeal to it.[17]

Monsanto's undertaking not to bring patent infringement actions in cases of inadvertent contamination disposed of the plaintiffs' declaratory action, but the District Court noted the inevitability that conventional crops would be contaminated by trace amounts of windblown pollen or seeds from genetically modified crops or other sources.[18] The Court of Appeals noted Monsanto's acknowledgement that conventional crops could be exposed to 'cross-pollination from nearby fields where biotech crops are grown' and that they 'might inadvertently contain traces of Monsanto biotech genes (because, for example, some transgenic seed or pollen blew onto the grower's land)'.[19]

The Court of Appeals referred to a study finding that, despite stringent precautionary measures meant to prevent any commingling of modified and conventional seed crops, a large majority of conventional seed samples had become contaminated by Monsanto's Roundup resistance trait. The District Court [20] found that due to contamination:

> ...some unlicensed – and unintended – use of transgenic seeds is inevitable. Like any other seeds, transgenic seeds may contaminate non-transgenic crops through a variety of means, including seed drift or scatter, crosspollination, and commingling via tainted equipment during harvest or postharvest activities, processing, transportation, and storage.

The Court of Appeals observed that GM seeds cannot easily be separated from conventional seeds; thus, a grower who harvests and uses or sells contaminated crops risks incurring infringement liability. The Court of Appeals observed that 'both parties seem to concede that at a minimum, using or selling patented seeds without a license is potentially infringing activity'.[21] Thus for the purposes of the appeal before it the court assumed '(without deciding) that using or selling windblown seeds would infringe any patents covering those seeds, regardless of whether the alleged infringer intended to benefit from the patented technologies.'

The first consideration by the US Supreme Court of inadvertent contamination occurred in *Monsanto Co* v. *Geertson Seed Farms*.[22] This case arose out of a decision by the Animal and Plant Health Inspection Service (APHIS) to deregulate a variety of genetically engineered lucerne. The US Supreme Court in holding that conventional farmers had standing to challenge the administrative deregulation of RuR lucerne recognized that there is a risk of 'gene flow' from GM crops into conventional crops.

12.7 Patenting of Stress-Tolerant Genes

Somvanshi in a 2009 study identified 30 patents relating to drought-tolerant genes (Somvanshi, 2009). These included: (i) patents related to proline biosynthesis; (ii) patented dehydration responsive element binding factors (DREB) and C-repeat sequences binding factors (CBF); (iii) patents related to protein kinases; (iv) various patents awarded for transcription factors involved in improving drought-stress tolerance in plants; and (v) patents related to miscellaneous drought tolerance genes. A 2008 study by the ETC Group identified 55 patent 'families' or related patent applications and/or issued patents published in more than one country or patent office (ETC, 2008). A total of 532 patent documents were identified, which represented applications to patent offices by a group of biotechnology companies on so-called 'climate-ready' genes around the world.

The ETC 2010 update of this study 'examined patents containing claims concerned with abiotic stress tolerance (i.e. traits related to environmental stress, such as drought, salinity, heat, cold, chilling, freezing, nutrient levels, high light intensity, ozone and anaerobic stresses' (ETC, 2010). It noted 'a dramatic upsurge in the number of patents published (both applications and issued patents) related to "climate-ready" genetically engineered crops from June 30, 2008 to June 30, 2010, identifying 262 patent families and 1663 patent documents' (ETC, 2010, Appendix A).

The 2010 report of the ETC contrasts the ownership of 9% patent families by public sector institutions (9% of the total) with the private sector, which holds 91% of the total. As is the case with biotechnological patenting generally, proprietary biotechnologies are concentrated in the same few corporations (Lesser, 1998). The 2010 report points out that 'just three companies – DuPont, BASF, Monsanto – account for two-thirds (173 or 66%) of the total.' This level of market concentration gives cause for concern for those who espouse the positive role of competition.

In addition to the possible adverse impacts this market concentration might have upon the vigour of competition, the market dominance of these private corporations also has an important influence upon the sort of biotechnological research which is undertaken. For example, to what extent will the dominance of private corporations in biomedical and agricultural research direct that research towards Northern concerns away from Southern food priorities (Alston *et al.*, 1998)? It has been estimated that only 1% of research and development budgets of multinational corporations is spent on crops of interest that could be useful in the developing world (Pingali and Traxler, 2002). Almost entirely neglected by these corporations are the five most important crops of the poorest, arid countries: sorghum, millet, pigeon pea, chickpea and groundnut (Human Rights Council, 2008).

12.8 Patenting of Plant Varieties

The development of new plant varieties is protectable in most countries as a species of intellectual property right (IPR) derived from the International Convention for the Protection of New Varieties of Plants (UPOV). Countries which are members of the World Trade Organization (WTO) are obliged by Article 27.3(b) of the WTO Agreement on Trade Related Aspects of Intellectual Property Rights (TRIPS) to 'provide for the protection of plant varieties either by patents or by an effective *sui generis* system or by any combination thereof.' The TRIPS Agreement does not specify which '*sui generis* system' will meet its requirements, but most of the members of the WTO have promulgated domestic legislation based upon the 1991 version of UPOV.

UPOV allows the protection of new varieties of plants that are distinct, uniform and stable. A variety is considered to be new if it has not been commercialized for more than 1 year in the country of protection. A variety is distinct if it differs from all other known varieties by one or more important botanical characteristics. A variety is uniform if the plant characteristics are consistent from plant to plant within the variety. A variety is stable if the plant characteristics are genetically fixed and therefore remain the same from generation to generation, or after a cycle of reproduction in the case of hybrid varieties. The 1991 version of UPOV recognizes the right of breeders to use protected varieties to create new varieties. However, this exception is itself restricted to such new varieties as are not 'essentially derived' from protected varieties. The drafters added this restriction to prevent second-generation breeders from making merely cosmetic changes to existing varieties in order to claim protection for a new variety. From the perspective of farmers, probably the most contentious aspect of the 1991 Act is the limitation of the farmers' privilege to save seed for propagating the product of the harvest they obtained by planting a protected variety 'on their own holdings' 'within reasonable limits and subject to the safeguarding of the legitimate interests of the breeder'. Earlier versions of UPOV permitted farmers to sell or exchange seeds with other farmers for propagating purposes.

The seed-saving privilege and the permitted development of non-essentially derived new varieties from protected material were compromises built into the legislation to take account of public policy concerns. It was appreciated that permitting individuals to privatize food varieties might compromise food security if breeding material was locked up and if farmers were prevented from saving seed for further harvests. However, from the perspective of plant breeders any derivation of new varieties from their protected varieties, whether essential or non-essential, was inconvenient for them and any seed saving by farmers deprived them of new sales. Consequently, they looked to patents law, which does not contain these exceptions, to protect their new varieties.

Plant varieties can be protected in the USA under a system of plant patents, or under a system of utility patents or under the Plant Variety Protection Act (PVPA). The Plant Patent Act makes available patent protection to new varieties of asexually reproduced plants. Under this scheme a plant variety must be novel and distinct and the invention, discovery or reproduction of the plant variety must not be obvious. One of the disadvantages of the scheme is that only one claim, covering the plant variety, is permitted in each application. The Federal Circuit Court of Appeals resolved any potential conflict between patent protection and protection under the Plant Variety Protection Act (PVPA) in its decision in *Pioneer Hi-Bred International Inc.* v. *J.E.M. Ag Supply Inc.*[23]

Pioneer's patents covered the manufacture, use, sale and offer for sale of the company's inbred and hybrid maize seed products as well as certificates of protection under the PVPA for the same seed-produced varieties of maize. The defendants argued that the enactment of the PVPA had removed seed-produced plants from the realm of patentable subject matter in the Patents Act. The Federal Circuit rejected this argument noting that the Supreme Court held that 'when two statutes are capable of co-existence, it is the duty of the courts ... to regard each as effective'.

This was illustrated by *Monsanto Co.* v. *McFarling*,[24] which concerned Monsanto's patent for glyphosate-tolerant plants, the GM seeds for such plants, the specific modified genes, and the method of producing the GM plants. Monsanto required that sellers of the patented seeds obtained from purchasers a 'Technology Agreement', in which they agreed that the seeds were to be used 'for planting a commercial crop only in a single season', that the purchaser would not 'save any crop produced from this seed for replanting, or supply saved seeds to anyone for replanting.'

Mr McFarling, a farmer in Mississippi, purchased RuR soybean seed in 1997 and again in 1998; he signed the Technology Agreement. He saved 1500 bushels of the patented soybeans from his harvest during one season, and instead of selling these soybeans as crop he planted them as seed in the next season. He repeated this activity in the following growing season. This saved seed retained the genetic modifications of the RuR seed. Mr McFarling did not dispute that he violated the terms of the Technology Agreement but claimed that the contractual prohibition against using the patented seed to produce new seed for planting, when he produced only enough new seed for his own use the following season, violated the seed-saving provision of the PVPA. The court declined to limit the patent law by reference to the PVPA and Mr McFarling was found to have infringed Monsanto's patent.

12.9 Patenting of Plant-Breeding Methods

In addition to the patenting of the products of plant breeding, some patent laws allow for the patenting of plant breeding methods. For example, in the USA a patent has been obtained for the 'method for selective increase of the anticarcinogenic glucosinolates in brassica species'[25] and an application published concerning a 'method for breeding tomatoes having reduced water content'.[26] This raises the possibility that methods of crop breeding to withstand climate stress can be privatized in the USA, which permits so-called methods patents.

On the other hand in Europe the exclusion in its patent legislation of 'essentially biological processes for the production of plants or animals' defined in Article 2.2 of the EU Biotechnology Directive as consisting 'entirely of natural phenomena such as crossing or selection', resulted in the denial of patent protection for the same methods for breeding brassica and tomatoes (Blakeney, 2012). The Board of Appeals of the European Patent Organization (EBA) observed that with the creation of new plant varieties, for which a special property right was going to be introduced under the subsequent UPOV Convention in 1960, the legislative architects of the European Patent Convention were concerned with excluding from patentability the kind of plant breeding processes which were the conventional methods for the breeding of plant varieties of that time. These conventional methods included in particular those based on the sexual crossing of plants deemed suitable for the purpose pursued and on the subsequent selection of the plants having the desired trait(s). These processes were characterized by the fact that the traits of the plants resulting from the crossing were determined by the underlying natural phenomenon of meiosis. On the other hand processes for changing the genome of plants by technical means such as irradiation were cited by the EBA as examples of patentable technical processes. On the other hand it pointed out that the provision of a technical step, be it explicit or implicit, in a process which is based on the sexual crossing of plants and on subsequent selection does not cause the claimed invention to escape the exclusion from patentability if that technical step only serves to perform the process steps of the breeding process.

Thus if a process of sexual crossing and selection includes within it an additional step of a technical nature, which step by itself introduces a trait into the genome or modifies a trait in the genome of the plant produced, so that the introduction or modification of that trait is not the result of the mixing of the genes of the plants chosen for sexual crossing, then that process leaves the realm of the plant breeding and, consequently, is not excluded from patentability in Europe. This principle applies only where the additional step is performed within the steps of sexual crossing and selection, independently from the number of repetitions, otherwise the exclusion of sexual crossing and selection processes from patentability could be circumvented simply by adding steps which do not properly pertain to the crossing and selection process, being either upstream steps dealing with the preparation of the plant(s) to be crossed or downstream steps dealing with the further treatment of the plant resulting from the crossing and selection process. The EBA noted that for the previous or subsequent steps *per se* patent protection was available. This will be the case for genetic engineering techniques applied to plants which differ from conventional breeding techniques, as they work primarily through the deliberate insertion and/or modification of one or more genes in a plant.

12.10 Conclusion

The application of IPRs to genetic resources has become a pronounced feature of agricultural innovation in the past decade. The FAO Panel of Eminent Experts on Ethics in Food and Agriculture has observed that 'while most innovation for food and agriculture does not depend on IPRs, the acquisition and exercise of IPRs in this field raise a variety of ethical concerns' (FAO, 2005). These include the fact that 'IPRs protection may just mean the lack of access to innovations for the poor' and the concerns raised by the 'patenting of merely isolated genes, the basic building blocks of life', which 'are not invented, but are part of nature.' More practically, the ability of individuals and corporations to obtain proprietary rights over agricultural innovations has important implications for food security (Blakeney, 2009), particularly as the expense and general transactional costs has tended to concentrate such IPRs in a few hands. In particular, IPRs on genetic resources (GR) may impede their use by third parties for further research and breeding during the term of protection, and thereby inhibit the development of new products and the capacity to address emerging problems, such as agricultural stresses caused by climate change.

The IPRs landscape which confronts countries is of course dominated by the TRIPS Agreement. However, that agreement contains a number of flexibilities. First, it allows the exclusion from patent protection of plants and animals (whether genetically modified or not). Second, the criteria under which patents are granted, i.e. novelty, and inventive step and industrial applicability may exclude materials identified through the application of traditional knowledge (TK) or GRs that exist in nature (even if isolated) as well as microorganisms. Article 30 of the TRIPS Agreement provides that 'Members may provide limited exceptions to the exclusive rights conferred by a patent, provided that such exceptions do not unreasonably conflict with a normal exploitation of the patent and do not unreasonably prejudice the legitimate interests of the patent owner, taking account of the legitimate interests of third parties.' Thus patent laws may allow third parties to undertake research and breeding during the patent term and farmers may be granted the right to save and re-use seeds where plant varieties, or certain components thereof, are subject to patent protection, in a way similar to the 'farmers' privilege' under PVP.

Over the last few years there has been a significant amount of patenting in relation to genetic material which might be useful in permitting organisms to resist the stresses of climate change. This patenting mirrors the high market concentration levels which have already been observed in the seed industry and the control of patent thickets by a small number of companies. It should be noted in this regard, both in relation to patent rights and PVP, that national laws may provide for compulsory licences in situations of national emergency. There is also the possibility for the intervention of the competition authorities to remedy abuses in the exercise of patent rights.

The practical effects of the application of IPRs to genetic resources are reflected in the actions which are brought for infringements of IPRs. To date, these actions have mainly been brought against farmers who have cultivated patented GM crops without the permission of the relevant rights holder, as well as actions against importers of products containing patented GM ingredients. Potential IPR liability lies against governments, research institutes (international and national) and seed breeders who supply or utilize proprietary technologies. The TRIPS Agreement establishes machinery to deal with the international trade in infringing goods and this machinery is currently being supplemented by the proposed Anti-Counterfeiting Trade Agreement (ACTA), of which the final text was settled on 3 December 2010.

Notes

1 *Agreement on Trade-Related Aspects of Intellectual Property Rights.* The TRIPS Agreement is Annex 1C of the Marrakesh Agreement Establishing the World Trade Organization, signed in Marrakesh, Morocco on 15 April 1994. Text available at: https://www.wto.org/english/tratop_e/trips_e/t_agm0_e.htm (accessed December 2015).

2 US Supreme Court, *Diamond* v. *Chakrabarty*, 447 U.S. 303 (1980), No. 79-136, argued 17 March 1980, decided 16 June 1980.

3 US Court of Appeals for the Federal Circuit, *In* re *Fisher*, 421 F.3d 1365 (Fed. Cir. 2005).

4 United States District Court, S.D. New York, *Association For Molecular Pathology, et al.,* v. *United States Patent and Trademark Office, et al.*, 702 F. Supp.2d 181, 2010 WL 1233416, 94 U.S.P.Q.2d 1683 (S.D.N.Y. 2010), 29 March 2010, as amended 5 April 2010.

5 United States Court of Appeals, Federal Circuit, *Association For Molecular Pathology, et al.,* v. *United States Patent and Trademark Office, et al.*, 689 F.3d 1303 (Fed. Cir. 2012).

6 US Supreme Court, *Association For Molecular Pathology, et al.,* v. *Myriad Genetics, Inc, et al.*, 569 U.S. (2013), 13 June 2013.

7 Federal Court of Australia, *Cancer Voices Australia* v. *Myriad Genetics Inc*, [2013] FCA 65, 15 February 2013.

8 *National Research Development Corporation* v. *Commissioner of Patents* ('NRDC case'), [1959] HCA 67; (1959) 102 CLR 252 (16 December 1959).

9 Federal Court of Australia, *D'Arcy* v. *Myriad Genetics Inc*, [2014] FCAFC 115, 5 September 2014.
10 High Court of Australia, *D'Arcy* v. *Myriad Genetics Inc*, [2015] HCA 35, 7 October 2015, Case Number: S28/2015.
11 US District Court for the Northern District of Mississippi, *Monsanto Co.* v. *Scruggs*, 342 F. Supp. 2d 584 (N.D. Miss. 2004), 6 July 2004.
12 US Court of Appeals for the Federal Circuit, *Monsanto Co.* v. *Bowman*, No. 10-1068 (Fed. Cir. 2011) 21 September 2011.
13 US Supreme Court, *Quanta Computer, Inc.* v. *LG Electronics, Inc.*, 553 U.S. 617 (2008), 9 June 2008.
14 Supreme Court of Canada, *Monsanto Canada Inc.* v. *Schmeiser*, [2004] 1 S.C.R. 902, 2004 SCC 34, 21 May 2004.
15 United States District Court, *Organic Seed Growers & Trade Ass'n* v. *Monsanto Co*, 851 F.Supp.2d 544 (S.D.N.Y. 2012) (No. 11-CV-2163).
16 US Court of Appeals for the Federal Circuit, *Organic Seed Growers & Trade Ass'n* v. *Monsanto Co.*, 718 F.3d 1350 (2013).
17 *Organic Seed Growers & Trade Ass'n* v. *Monsanto Co.*, 718 F.3d 1350, 1358-1359 (Fed. Cir. 2013), *cert. denied*, 134 S. Ct. 901 (2014).
18 United States District Court, *Organic Seed Growers & Trade Ass'n* v. *Monsanto Co.*, 851 F.Supp.2d 544 (S.D.N.Y. 2012).
19 *Organic Seed Growers & Trade Ass'n* v. *Monsanto Co.*, 718 F.3d 1350 at 1354 (2013).
20 United States District Court, *Organic Seed Growers & Trade Ass'n* v. *Monsanto Co.*, 851 F.Supp.2d 548 (S.D.N.Y. 2012).
21 *Organic Seed Growers & Trade Ass'n* v. *Monsanto Co.*, 718 F.3d 1350 at 1358 (2013).
22 US Supreme Court, *Monsanto Co.* v. *Geertson Seed Farms*, 130 S.Ct. 2743 (2010).
23 *Pioneer Hi-Bred International Inc.* v. *J.E.M. Ag Supply Inc.*, 200 F.3d 1374 (Fed. Cir. 2000), cert. granted, 148 L. Ed. 2d 954 (2001).
24 United States Court of Appeals, Federal Circuit, *Monsanto Co.* v. *McFarling*, 302 F.3d 1291 (Fed. Cir. 2002).
25 Mithen, R., Faulkner, K. and Williamson, G., 22 January 2002. *Method for selective increase of the anticarcinogenic glucosinolates in Brassica sp.* US6340784 B1.
26 Schaffer, A.A., 15 April, 2010, *Method for breeding tomatoes having reduced water content and product of the method*. US20100095393 A1.

References

ALRC (Australian Law Reform Commission) (2004) *Genes and Ingenuity: Gene Patenting and Human Health* (ALRC Report 99). ALRC, Canberra.

Alston, J., Pardey, G. and Rosenboom, J. (1998) Financing agricultural research: international investment patterns and policy perspectives. *World Development* 26(6) 1057–1071.

Blakeney, M. (2009) *Intellectual Property Rights and Food Security*. CAB International, Wallingford, UK.

Blakeney, M. (2012) Patenting of plant varieties and plant breeding methods. *Journal of Experimental Botany* 63(3), 1069–1074.

Correa, C.M. (2009) *Trends in Intellectual Property Rights Relating to Genetic Resources for Food and Agriculture*. Background Study Paper 49, Commission on Genetic Resources for Food and Agriculture, Rome.

ETC (2008) Patenting the 'climate genes'...and capturing the climate agenda. *Communiqué*, no.99, May/June 2008, Available at: http://www.etcgroup.org/content/patenting-climate-genes-and-capturing-climate-agenda (accessed 13 August 2015).

ETC (2010) Gene giants stockpile patents on 'Climate-ready' crops in bid to become 'Biomassters'. *Patent Grab Threatens Biodiversity, Food Sovereignty 106*, October 2010. Available at: http://www.etcgroup.org/upload/publication/pdf_file/FINAL_climate-ready-Comm_106_2010.pdf (accessed 13 August 2015).

FAO (2005) *Panel of Eminent Experts on Ethics in Food and Agriculture, 3rd Report.* Available at: http://www.fao.org/docrep/010/a0697e/a0697e00.htm (accessed 10 December 2015).

Heller, M.A. and Eisenberg, R.S. (1998) Can patents deter innovation? The anticommons in biomedical research. *Science* 280, 698–701.

Human Rights Council (2008) *Report of the Special Rapporteur on the Right to Food, Jean Ziegler*, A/HRC/7/5.

Joly, Y. (2003) Accès aux mèdicaments: le système international des brevets empêchera-t'il les pays du tiers monde de bénéficier des avantages de la pharmacogénomique? *Les cahiers de propriété intellectuelle* 16, 131–185.

Joly, Y. (2006) Wind of change: in re Fisher and the evolution of the American biotechnology patent law. *Law in Context* 24(1), 67–84.

Lesser, W. (1998) Intellectual property rights and concentration in agricultural biotechnology. *AgBioForum* 1(2), 56–62.

Monsanto (2015) *Saved Seed and Farmer Lawsuits*. Available at: http://www.monsanto.com/newsviews/pages/saved-seed-farmer-lawsuits.aspx (accessed 13 August 2015).

Pingali, P.L. and Traxler, G. (2002) Changing focus of agricultural research: will the poor benefit from biotechnology and privatization trends? *Food Policy* 27, 223–238.

Somvanshi, V.S. (2009) Patenting Drought Tolerance in Organisms 3. *Recent Patents on DNA & Gene Sequences* 3(1), 16–25.

13 The Development of Patentability of Genetic Patent in Mainland China and Taiwan

Jui-Chu Lin,[1] Tzu-Hsun Hung[2]* and Chien-Te Fan[3]

[1]*College of Intellectual Property Studies, National Taiwan University of Science & Technology;* [2]*Attorney-at-Law, Taiwan;* [3]*Institute of Law for Science & Technology, National Tsing Hua University, Taiwan*

13.1 Introduction

'Biotechnology' literally means technologies related to biology, including agriculture, food and medical sciences. It was defined as 'Biotechnology means any technological application that uses biological systems, living organisms, or derivatives thereof, to make or modify products or processes for specific use' (Wang, 2007). At the UN Convention on Biological Diversity held in Brazil in 1992, Taiwan's Ministry of Economic Affairs, in accordance with the international trend and opinions from all quarters, defines it as 'the sciences and technologies based on life sciences, such as molecular biology, cell biology, immunology, genomics, and proteomics, and technologies such as genetic engineering, protein engineering, cell engineering, and tissue engineering to research and develop, manufacture, or improve product quality in order to improve the quality of human life' (Taiwan Biotech Industry White Paper, 2014). To sum up, major modern biotechnology includes (among others) recombinant DNA technology, gene therapy, gene replication, gene mapping, DNA sequencing and protein engineering. In this chapter, biological science refers to 'biotechnology related to genes or DNA sequences and the methods for their isolation'.

However, the question of whether the development of biotechnology should be complemented by the protection of a patent system has been controversial. The main controversy lies in whether a gene itself is patentable, a question that has long been an important concern in biotechnology. Under public scrutiny, the US Supreme Court handed down its verdict on the *AMP* v. *Myriad* case,[1] which focused on the patentability of genes, on 13 June 2013. The nine justices of the US Supreme Court decided unanimously that any chromosomal DNA that has the same expression as the natural state of existing human DNA cannot be patented, which also applies to DNA isolated from human bodies by scientific methods (Lee, 2014a). However, the Supreme Court's decision did not negate the patentability of manufactured genes, and indicated that isolated gene sequences that are different from naturally existing ones should be regarded as patentable human inventions. One example of manufactured genes is complementary DNA (cDNA), which is produced using a reverse transcriptase and takes RNA (which is free of non-coding intron sequences) as a template (Lee, 2014b).

The USA has made a decision after the verdict on whether a gene is patent eligible; this decision also reflects the values and choice of the majority of patent practice in the USA. From observation of the current development of the cross-Taiwan-Strait biotechnology industry, it is evident that research and development of biotechnology and new capital

*Corresponding author, e-mail: starfever1211@gmail.com

investment are necessary, and the question of whether the output of science and technology should be protected by patent system should focus our attention. Currently, the patent practices on both sides of the strait (Mainland China and Taiwan) for gene patentability focus on doctrine and have not been put into substantive practice. This chapter is a study of the current cross-strait patent laws and benchmarking reviews, and explores whether gene sequencing and its related technologies are patentable – and if so, whether there are further restrictions.

13.2 The Importance of Biotechnology in Cross-Strait Industries

The biotechnology industry has developed rapidly in recent years in both Taiwan and mainland China. In Taiwan, for example, the promotion of the biotechnology industry began in the 1980s in order to promote drugs, medical management services and medical equipment, and in 2013 Taiwan approved the 'Taiwan biotech industry taking-off action program' to implement and improve industrial development. According to a 2014 Taiwan Biotech Industry White Paper, the total turnover of Taiwan's biotech industry amounted to NT$276.9 billion in 2013, an increase of 5.29% compared to 2012 and nearly twice of that of 2004 (NT$144.9 billion). In total turnover, the turnover of applications for biotechnology (namely genetic engineering products) reached NT$78.2 billion in 2013, an increase of 128% as compared to the NT$34.3 billion it generated in 2004.

In accordance with the 'Act for the Development of Biotech and New Pharmaceuticals Industry' promulgated in 2007, Taiwan's biotech industry authorities provide preferential measures on technical, personnel and funds to encourage manufacturers to invest in biotech industry development. Consequently, compared to 2012, investment in biotech drug companies increased 65% in 2013, showing that the implementation of the act significantly increased investment in the biotech industry in Taiwan.

On the other side of the strait, the development of the biotech industry in mainland China is also booming. In 2012, encouraged by policies included in the 12th Five-Year Plan for the pharmaceutical industry, people returning from overseas and the Thousand Talents Program, the revenue generated by the biotech industry on the mainland has reached RMB2 trillion, becoming the world's second-largest pharmaceutical market, with an annual growth of 15% (BIOONNEWS, 2012). This is an indication that the development of a cross-strait biotechnology industry is inevitable, and the promotion of investment in the emerging biotech market is the developing direction of cross-strait policy.

A patent protection system is quite important for the development of most industries. The system aims to encourage, protect and apply inventions in order to promote the development of the industry (Republic of China, 2014). Genes are integral to the application of pharmaceutical sciences, clinical diagnoses and treatment evaluation. According to the US researcher Mark Schankerman, in most cases, profits generated from patent rights account for 15–25% of research funds. Thus, there is no doubt that introducing a patent protection system will promote the development of scientific and technological research (Schankerman, 1998).

In addition, the biopharmaceutical industry combines the invention of individual genes with pharmacology, through comparing individual differences in genes so as to provide suitable drugs and treatment methods for different individuals, and even prevent diseases. This is called 'personalized medicine'. Therefore, whether genes and related technologies should be granted patent protection deserves legislators' consideration. It is worthwhile to explore whether the attitudes towards this kind of science and technology embodied in cross-strait law systems meet the needs of the industry, and the concerns of the companies and investors who want to invest in the cross-strait biotech industry.

13.3 The Development of Gene-Related Patents

13.3.1 The international development of patentability of genes

As stated above, the patentability of genes is promising as a means of promoting the development of the industry – so why do doubts still exist among legislators on whether it should be open or not? The discussion on gene patentability can be traced back to the *Diamond* v. *Chakrabarty* case in 1980.[2] In its decision on whether a living organism should be patentable, the US Supreme Court held that if an artificially cultivated fungus has distinctively different characteristics as compared to natural ones, it is a patentable invention (Lee, 2014b). The USA later issued the first gene patent in 1982 and the first 'expressed sequence tags' (ESTs) in 1998 (Robertson, 1999). However, for fear that gene patent applications may extend the patent eligibility, the United States Patent and Trademark Office (USPTO) stipulates that a patent will be granted only when the functions of the gene are clearly explained (USPTO, 2001).

In 2010, the US Association of Molecular Pathology applied to the Federal District Court to withdraw the *BRCA1/BRCA2* gene patent held by Myriad. The main dispute is whether the gene patent should be classified as not patentable due to natural law. The Federal District Court decided that Myriad's patent is not patentable because it is only a discovery of a natural phenomenon. The function of an isolated gene is not markedly different from native genes, and the gene is acquired by personnel versed with technology. Meanwhile, the patented methods held by Myriad, including the application of 'identifying potential cancer treatments by observing cell growth rate', are 'merely data collection' and 'basic scientific principles', and are therefore not eligible for patent.[3]

The verdict of the District Court was overturned when the case was appealed in the Court of Appeals for the Federal Circuit (CAFC). The CAFC decided that isolated genes are patentable because, although their functions are only slightly different from natural genes, they have unique chemical structures, and are therefore markedly different from natural ones and thus patent eligible. After appealing the case to the Supreme Court, the verdict of the CAFC was withdrawn. Since CAFC still upheld the verdict, the Supreme Court made its final decision on 13 June 2013, ruling that DNA sequences isolated from human bodies are not patent eligible if they are naturally existing, but they are patent eligible if they are markedly different from natural sequences due to the use of certain experimental methods. For example, complementary DNA sequences are accordingly patent eligible.

In Europe, taking ethical and socio-economic disputes into consideration, the provisions of Article 53 of the European Patent Convention (EPC) of 1973 stipulate that methods including violating public order or morality and violating major ways of producing plants and animals biologically are unpatentable, but the methods of producing microorganisms are excluded. The stipulations on unpatentability are retained even in the amendments made to the EPC in 2000.

Implementing Regulations to the Convention on the Grant of European Patents (IRCGEP) Rule 23d explains the provisions regarding the uncertainty over unpatentability and stipulates four cases that are not biotechnology patents: (i) human cloning; (ii) modifying the germ line genetic identity of human beings; (iii) using human embryos for industrial or commercial purposes; and (iv) modifying animal genetic characterization and harming the animal without benefiting the animal or humans. IRCGEP Rule 23c stipulates that substances isolated from the natural environment and plants or animals (not confined to a particular plant or animal variety) are patent eligible if they are not simply discovery of one of its elements, including the sequence or partial sequence of a gene, and IRCGEP Rule 23e(2) stipulates that substances, including gene sequences and parts of sequences, isolated from humans in scientific and technological ways are to be patented, even if the substances are naturally occurring. In addition, European Council Directive No. 98/44 provides that gene-related inventions should be protected by patent laws in member states. These regulations demonstrate that genes are not prohibited from being patented in Europe, even if the sequences of the gene are the same as natural ones.

13.3.2 The related controversies over patentability of genes

Following the global development of personalized medicine, the patentability of genes will greatly benefit the development of the emerging biotechnology industry. Research and development in biomedical technology industry is expensive, risky and time-consuming. Therefore, it is difficult to attract new investors and develop the industry if there is not enough incentive to ensure the interests of investors and researchers. There are also many disputes over gene patentability, and questions on whether industrial development can be well balanced if genes are patentable.

Some scholars hold that genes and their related technologies, materials and methods are characterized by 'low substitutability but high general application'. Patenting genes may hamper the follow-up research and development of related technologies, especially when patent holders are reluctant to provide information or demand excessively high prices for patent licences. Thus, approving gene patentability easily is contrary to the purpose of developing the industrial patent system (Chen *et al.*, 2010). Some cite Directive No. 98/44, holding that genes and their related inventions are patentable (Xie *et al.*, 2005), while others insist that whether genes are patentable lies in their non-obviousness or industrial use rather than patent eligibility – that is, on a policy level, gene patentability should be included with 'cooperative values' to balance the interests of the public and patent holders (Shen, 2012).

Aside from the considerations at the industry and policy levels, the invention and application of biotechnology involving life forms such as gene sequences, transfer reproductive technology and so on, are all beyond the expectation of the original patent mechanism and the question of whether and how they may be suited to the current patent system should be closely and thoroughly examined, especially as they may lead to ethical issues and violate public orders. As to ethical issues, Taiwan's patent law has special provisions (Article 24) to examine whether inventions meet ethical standards and public orders. If not, the invention is not granted a patent, even if it is progressive, innovative and practical. The Patent Law of the People's Republic of China also lists relevant provisions on inventions that may not be granted patents in Article 5(1)): (i) any invention-creation that is contrary to the laws or social morality, or that is detrimental to public interest; and (ii) any invention or creation where the acquisition or use of genetic resources (on which the development of the invention or creation relies) violates the provisions of the laws or administrative regulations.

In recent years, the global discussion on the limitation of gene patents has been constant. More and more scholars, arguing from a public health perspective, suggest that the medical field should limit the application of gene patents and narrow patent-right holders' exclusive rights through experimental implementation and a compulsory authorization system (Li, 2010). Certain incentives should be provided to gene patent holders to avoid obstacles in future research and development related to science, and to realize the goal of promoting the development of the industry. We should also pay attention to these considerations. The following section will study the legal system for the cross-strait development of gene patentability, and then further reflect on future improvement.

13.4 The Development of a Cross-Strait Patent Legal System on Gene Patentability

13.4.1 The development of Taiwan's patent legal system on gene patentability

Can genetic patents be regarded as inventions?

Since its enactment in 1944, the Taiwan Patent Act (TPA) has undergone 12 amendments in the following 60 years (as of 2014). The last amendment was confirmed on 22 January 2014, to alter provisions in Article 143 regarding the responsibility of patent authorities to maintain patent application files, and revised and enlarged provisions 97-1 to 97-4 on customs detainment about patent infringement. The patent act in Taiwan before 1986 stipulated that chemicals, pharmaceuticals and their reconciliations are unpatentable (Republic of China, 1979). This provision was deleted in 1986 to reflect that patent law should protect and support public health and the public interest. The patent act was revised and updated to state that animals, plants and microorganisms could not be patented as inventions, except for the cultivating methods of new plants and microorganisms. Restrictions on the provisions for microorganisms have been deleted since Taiwan joined the General Agreement on Tariffs and Trade (GATT, the predecessor of the WTO) and signed the Agreement on Trade Related Aspects of Intellectual Property Rights (TRIPS).

In accordance with the provision in Section 27.3(b) of TRIPS, EU Directive 98/44 and Article 53(b) of the EPC, the TPA amended and explicitly provided in 2003 that 'plants and animals' and 'essentially biological production methods' are excluded from patent protection. As to the questions about whether partial gene sequences may be patented, we should first explore whether genes and their related technologies are the creation of technical ideas involving natural laws, namely inventions. If they are, we should judge whether they fall into the category of Article 24(1) of the patent law, and thus cannot be patented. The key of the former is to clarify whether isolated gene fragments and the methods used to isolate them are 'discoveries' or 'inventions': the key of the latter lies in clarifying the ambiguous legal concepts of 'essentially biological production methods' and 'plants and animals'. If they are unfixed legal concepts, then they should be further clarified.

In accordance with the manual of patent examination of Taiwan, regarding a substantive review standard for invention patents, the discovery of objects existing in natural patterns does not belong in the realm of the creation of technical ideas involving natural laws, even if the objects are previously unknown. Isolated objects and the methods used to isolate them may be patented if the objects are isolated for the first time, their structures, patterns, or other physical and chemical properties are different from any known object, and they can be clearly identified. The review standard uses an existing natural gene or microorganism as an example to show that if that gene or microorganism has been acquired by a special separation process it is defined as an invention.

Are genes subject to patent protection?

Two factors need to be taken into account in determining whether genes may be patented. The first is the question of whether gene patents are restricted by Article 24(1) of TPA where the key issue is what the essentially biological approach about animals and plants as well as the production of animals and plants is. The manual of patent examination explains that 'animals and plants' includes transgenic animals and plants. If a biological approach relating to the production of animals and plants involves sexual hybridization of a whole genome, the animals and plants selected by this approach will be deemed as by essentially biological approaches. Even if an extra technical character is included, it is still an essentially biological approach if the genetic marker is used to select parental and progeny generation. However, if the whole genome is not involved but a single gene or trait is introduced through genetic engineering to improve the animal or plant's characteristics and production, or to promote or repress growth, the approach is primarily a non-essentially biological one.

As for the process of isolating human genes, gene sequences with specific functions are selected from the 3 billion base pairs and isolated after artificial sequencing, purification and classification. This procedure is complicated and involves human intervention. However, can the explained procedures that failed to clarify themselves in the manual of patent examination – the structure, form and other physical and chemical properties that are different from the known ones – be explained clearly? If the isolated gene sequences are the same as natural ones, can they be patented? Moreover, if the isolated gene sequences differ from naturally occurring ones but have the same functions, can they be granted patent protection? This has not yet been addressed in detail.

Taking the controversial patent application in the Myriad case as an example, if it happened in Taiwan, what would be the result under Article 24(1) of the TPA? It is known that the legislative intent of Article 24 in TPA follows the international development trend: its main references are TRIPs and EU directives and conventions. If the intentional explanation is followed and the non-discriminatory principle of Article 27(1) in TRIPs is referred to, inventions from different technical fields, domestic or foreign inventions, and articles and approaches should all be subject to the same treatment in the patent system. Furthermore, according to the European Parliament and Council Directive 98/44/EC on the legal protection of biotechnological inventions, EU member states should implement their domestic patent laws to protect gene technology. Therefore, the legislative intent and direction of Taiwan's Patent Law should not restrict genes and their related technologies from being patented. In other words, the patent of the controversial example of the Myriad case ought to be passed via the application of Article 24(1) in Taiwan.

In addition to the stipulation of Article 24(1) of the TPA, the question of whether genetic information such as the sequencing of gene fragments, specific gene fragments and so on, with their special features, should be added to the property right system is quite controversial. The particularity of genetic information is that its content is specific and replicable (Yan, 2001). Technology now allows us to use it to reveal personal physiological information. Furthermore, it may be used to clone a human being. Therefore, some think genes are one of the 'carriers' of human dignity. If genes are treated as patented objects, it is probable that human dignity will be violated (Cai, 2001) and thus lead to ethical controversy. If this is taken into consideration, genes and their related technology should fall under Article 24(3) of the TPA and not be treated as patentable.

Article 24(3) of the TPA stipulates that anything that impairs the public order and good social customs will not be granted a patent. According to the legislative intent of this article, which refers to the European Patent Convention, if the result of a gene invention is involved in each of the following objects: (1) cloning a human embryo; (2) division technology; or (3) changing the gene characteristics of a human reproductive system, such as using gene therapy to change germ cells (such as embryos) with genetic disease factors, which contributed to the removal of hereditary gene defects in the progeny of sick patients, it will be deemed as being contrary to public order and good morality. Therefore, the manual of patent examination in Taiwan indicates that cloning humans, cloning methods and changing hereditary human traits such as the reproductive system are categorized as inventions that impair the public order and good social customs, and will not be granted patents under Article 24(3) of the TPA. However, when the genetic technology in question does not involve changes to reproductive characteristics, whether it should be restricted by this article on public order and good morality is currently the subject of dispute in Taiwan's judicial practices, according to the manual of patent examination. One reason for this is that there are few relevant cases brought before Taiwanese courts. Besides, the public order and good social customs themselves reflect the values and standards of virtue of the contemporary society.

According to Article 24(2) of the TPA, gene therapy is categorized as medical treatment involving living human beings, animals and plants, so it cannot be granted an invention patent. However, the manual of patent examination includes modifying genes outside the living body, detecting and analysing biological materials outside of the living body, and genes, carriers for gene therapy, and recombinant genes as objects of invention patents, and so they should not be excluded.

In conclusion, genes and related technology are probably admitted as inventions under the current TPA. According to the existing manual of patent examination, they are not excluded by Article 24(1) of the TPA. If a human embryo and changes of hereditary characteristics such as the reproductive system are not involved, and Article 24(3) on public order and good social morals is followed, genetic technology can be patented. Finally, analysis of biological material outside the living body is not restricted by Article 24(2), and so patent applications may also be filed on those grounds.

13.4.2 The development of patentability of genes in the patent legal system of mainland China

The research and development of biotechnology and its related industries has been in existence for over 30 years. In recent years, the proportion of these industries is increasing daily in the global medical market. China has surpassed its neighbour Japan and is now the second largest biotechnology market in the world (BIOONNEWS, 2012), indicating that the development of China's future biotechnology industry is promising. The government also supports the development of the biotechnology industry. They promoted the goal of increasing scientific research growth in the 12th Five-Year Plan, issued the rollout of Talents' Acting Guiding Principle and initiated the Thousand Talents Program to recruit foreign experts and bring in foreign genetic and medical technology (*United Daily News*, 2013).

Another question that deserves exploration is, when China fully develops its biotechnology industry, will the patent legal system play an accordingly auxiliary role, or will it prove to be a stumbling block by setting too many restrictions on biotechnology? The Patent Law of the People's Republic of China (PLPRC) was passed in 1984 and formally implemented in 1985. Three amendments were later issued, in 1992, 2001 and 2008. The 1992 amendment included chemicals in the PLPRC's protection range. The State Intellectual Property Office of the PRC included genes in the range of patent protection for the first time in the 2001 Manual of Patent Examination (Li and Cai, 2014).

The present Article 25 of the PLPRC stipulates that no patent right shall be granted to any of the following: scientific discoveries, rules and methods for cognitive activities, methods for the diagnosis or treatment of diseases, animal and plant varieties, substances obtained by means of nuclear transformation, and designs, patterns, or colours serving mainly as indicators on two-dimensional printed goods. Regarding the patent-related content in this stipulation, it should first be explored whether the identification of a gene falls under the category of scientific discoveries. According to the instructions in the manual of patent examination, if the applicant wishes to patent a gene or DNA fragment existing in a natural state in the natural world, it represents a scientific discovery under Article 25 of the PLPRC and thus may not be patented (Cao, 2015).

The special instruction in the manual of patent examination states that if a gene or DNA fragment is separated or extracted from its natural conditions for the first time which has an exact function, value and use in the biotechnology industry, that gene or DNA fragment, as well as the means used to isolate it, are patentable. However, this type of gene patent has special restrictions. According to the provisions of Articles 20(1)(2) and 26(4) of the PLPRC, its implementing rules need to be in line with the following format: (i) the base sequence should be directly restricted; (ii) for a structural gene, the amino acid sequence of the polypeptide or protein coded by the mentioned gene should be restricted; (iii) when the base sequence of the gene or the amino acid sequence of the polypeptide or protein coded by the gene is recorded in the sequence table or the figure attached to the instruction, it must be possible to refer to the sequence table or the attached figure directly; and (iv) the specific functions of the gene should be restricted (Cao, 2015).

Secondly, it must be determined whether genetic technology is involved in diagnosing and/or treating a disease. The State Intellectual Property Office of the PRC pointed out in the 2010 Manual of Patent Examination that if any living human or animal body is treated as an object in an attempt to achieve a medical result or a physical condition, the method used shall not be granted a patent. Although a sample *in vitro* may be treated as an object, if the intervention is designed to obtain a diagnosis or physical result it shall not be granted a patent. This makes it clear that gene detection and screening for diagnoses cannot be patented.

Given the rapid development of the economy and society, in order to improve and drive the new systems of property rights and intellectual property rights (IPR) protections, and to promote the commercialization of research findings, the standpoint with its relatively strict restrictions was amended in the draft amendment of the patent law (consultative draft) during the fourth overall amendment's study and preparation work, focused on an 11-aspect monographic study starting in the second half of 2014. In light of the rapid development of the aquaculture and the livestock and poultry breeding industries, and the fact that these two industries' demand for patents on methods of diagnosing and treating disease in farm animals increases accordingly, it is suggested that the present patent law's prohibition on patenting methods of diagnosing and treating disease in humans and animals be reversed and the original strictly limited regulations be restored in the amendment, particularly considering the norms and attitudes toward the matter in the USA, Australia, Japan, South Korea and Europe (Draft Amendment of the Patent Law of People's Public of China, 2015).

According to the Manual of Patent Examination of State Intellectual Property Office of PRC, production methods used with animals and plants are patentable under Article 25 (2) of the PLPRC if it is not the essential method for biotechnology. Furthermore, transgenic animals or plants produced through genetic engineering using recombined DNA technology still belong to the range of animals and plants covered in Clause 4 of Article 25(1) and should not be granted patents. In addition to the stipulation of Article 25, as biotechnology often involves highly sensitive ethical or legal issues, related patent applications are also restricted under Article 5(1) of the PLPRC. The manual of patent examination stipulates that biotechnologies, including human stem cells obtained from embryos, methods of changing the identity of a human being's reproductive heredity or body, cloned humans and methods of cloning humans or human embryos for industrial and business purposes, are all regarded as conflicting with social morality and therefore may not be patented.

In conclusion, it is known that the PLPRC does not limit patents on genes or DNA fragments and their segregation.[4] However, if such material is obtained as a pure natural scientific finding or is used to diagnose or treat diseased animals or plants, it is subject to restrictions and may not be patented. Furthermore, its required instruction of patent right should conform to a specified format and reveal its appropriate function before it can be granted a patent. Besides, if the related technology is suspected of being against public morality or public interest, it shall not be considered for a patent.

13.5 Conclusion

On the issue of whether gene or DNA fragments and related technology should be granted patents, the Taiwan Patent Act and the Patent Law of the People's Republic of China have something in common – they both have an open attitude in principle and set limitations under certain circumstances. In the general direction of actively improving biotechnology between Taiwan and mainland China, such a legislative direction and policy attitude seems beneficial to push forward investment in new biotechnology and the industry and promote related research and development.

On the issue of granting patent protection to a specific industry, whether the protection can benefit the development of the industry should be taken into account; the objective of the development of the industry is to improve human welfare and development of technology. To decide whether a patent on genes is granted is actually a value selection. The government first considers the industry's situation and the people's welfare to make the decision on whether the patent should be granted. The open attitude of the legal systems in Taiwan and mainland China is in line with the international trend.

It is worth mentioning that in both Taiwan and mainland China, the patent law and related examination regulations on biotechnology contain a number of uncertain legal concepts. To distinguish the details requires clarification by practice.

The disputable patent content in the Myriad case, for instance, includes the *BRCA1/BRCA2* sequence, the comparison and analysis in the method of separating DNA sequences, as well as the method of finding potential cancer treatment medications by observing the cell's growth rate. As for the gene sequence and its analysis methods for the isolated gene sequence, as far as the legal systems for patents between Taiwan and mainland China are concerned, they take a more open point of view than that of the Supreme Court of the USA. The point of view is that even though the gene sequence in its natural state has the same structure, it should be granted a patent if the isolation process involves artificial intervention and the specified function can be proven. It seems, however, that the issue of whether it is proper to grant individuals monopoly over the carrier of humanity deserves detailed consideration.

To observe a cell's growth rate to find a method to treat cancer involves gene detection. It is required to distinguish on the patent application between gene sequencing as a method to assist in seeking medication for treatment and gene sequences as a treatment method. In Taiwan, based on the manual of patent examination, detections *in vitro* will be granted a patent, regardless of which – the former or the latter – is employed. Under the legal system of mainland China, even *in vitro* detection, if the purpose is to diagnose a disease and its physical condition, will not be granted a patent. Therefore, although both do the same in practical operation, there would be a dilemma of opposite results in the two different ways to apply for patents.

In addition to the more detailed performance considering the specific character of genes, more care should be taken when a patent is granted to specific people and exclusive rights are given; and this practice may essentially hinder competition, progress and development in future science and technology. Legislators should carefully consider whether corresponding steps such as test exemption, compulsory licensing etc., should be taken when opening gene patenting as an incentive for investment and research and development to maintain the impetus of an industry's development and consider the rights and interests of patent holders as well as public interest.

Notes

1 US Supreme Court, *Association For Molecular Pathology, et al.*, v. *Myriad Genetics, Inc, et al.*, 569 U.S. (2013), 13 June 2013.

2 US Supreme Court, *Diamond* v. *Chakrabarty*, 447 U.S. 303 (1980), No. 79-136, argued 17 March 1980, decided 16 June 1980.

3 United States District Court, S.D. New York, *Association For Molecular Pathology, et al.*, v. *United States Patent and Trademark Office, et al.*, 702 F. Supp. 2d 181, 2010.

4 It also requires other patent-related features such as novelty, advancement and practicability. As the paper mainly focuses on the topic of patentability, these aspects are not further discussed here.

References

BIOONNEWS (2012) IMS review of medical market of China in 2012. Available at: http://www.bioon.com/industry/market/564017.shtml (accessed 12 June 2015).

Cai, W.Y. (2001) A proposal of the concept 'quasi-subject of rights'. *Cheng Kung Law Review* 2, 41–74.

Cao, L.R. (2015) On China's definition of gene patent protection range. *Journal of Nanjing University of Science and Technology* 28(2), 39–45.

Chen, J.-H., Chong, C.-H., Chang, N.-W., Lin, F.-Y. and Cheng, Y-C. (2010) Research on patents grant of gene related research tools: focus on review procedure and criteria of gene patents. *NTU Law Forum* 39(1), 403–447.

Lee, C.H. (2014a) Rediscussion on gene patents: American experience reflections. *The Taiwan Law Review* 224, 304–307.

Lee, J.A. (2014b) Gene patents in the aftermath of AMP vs. Myriad Genetics, *Inc. NCCU Intellectual Property Review* 12(2), 1–46.

Li, S.H. (2010) Development trend of genes and gene medicine legal system. In: Xie (ed.) *Genome Medicine Research and Innovation in Intellectual Property.* Yuan Zhao, Taipei City, Taiwan, pp. 159–182.

Li, W. and Cai, L.S. (2014) The scope of patent protection for gene technology in China. *Nature Biotechnology* 32, 1001–1003.

Republic of China (1979) Taiwan Patent Act. Available at: http://law.moj.gov.tw/Eng/LawClass/LawHistory.aspx?PCode=J0070007 (accessed 16 March 2016).

Republic of China (2014) Taiwan Patent Act. Available at: http://law.moj.gov.tw/Eng/LawClass/LawAll.aspx?PCode=J0070007 (accessed 16 March 2016).

Robertson, D. (1999) EST patent granted for human kinase homologs. *Biotechnology* 17, 125.

Schankerman, M. (1998) How valuable is patent protection? Estimates by technology field. *Rand Journal of Economics* 29, 1.

Shen, Z.Y. (2012) Patent eligibility of genes – starting from myraid case. *The Law of Nations* 181, 51–60.

Taiwan Biotech Industry White Paper (2014) Available at: https://www.moeaidb.gov.tw/external/ctlr?PRO=publication.PublicationView&id=2419 (accessed 16 March 2016).

United Daily News (2013) Thousand talents program implemented by mainland to attract foreign top talents. Available at: http://vision.udn.com/storypage.php?ART_ID=794 (accessed 14 May 2015).

USPTO (2001) *Utility Examination Guidelines.* Available at: http://www.uspto.gov/web/offices/com/sol/og/2001/week05/patutil.htm (accessed 16 March 2016).

Wang, X.-.G. (2007) *Introduction to the Biotech Industry* [in Chinese]. New Wen Jing, New Taipei City, Taiwan.

Xie, M.Y., Li, S.H. and Song, H.Z. (2005) Views on patentability of stem cell-related inventions from the Europeans. *The Taiwan Law Review* 118, 63–89.

Yan, J.A. (2001) Property, personality, or information? On the legal status of human genome. *National Taiwan University Law Journals* 31(1), 1–44.

14 Bioprospecting Microbial Diversity: Intellectual Property Rights Issues

Om Prakash,* Aabheejeet Pansare and Sunil K. Dhar

Microbial Culture Collection, National Centre for Cell Science, Pune, India

14.1 Introduction

According to the definition of the Organization for Economic Cooperation and Development (OECD), 'Biotechnology is defined as the application of science and technology to living organisms as well as parts, products and models thereof, to alter living or non-living materials for the production of knowledge, goods and services'. Biotechnology is continuously healing, fuelling, feeding and enriching human society and civilization. Based on area of specialization and demarcation, biotechnology is divided into five basic groups: red (related to medical and health), green (agricultural biotechnology), white (industrial biotechnology), blue (biotechnology of freshwater and marine systems) and grey biotechnology or environmental biotechnology. The progress of biotechnology is based on the exploration of biodiversity using the principle of bioprospection. Bioprospecting is the process of the discovery and use of new biomolecules from biological resources for its commercialization and use or in other words exploitation of biological diversity of genetically and biochemically valuable biodiversity for commercial exploitation (Ferrer *et al.*, 2007; Li *et al.*, 2008; Jeon *et al.*, 2009; Joint *et al.*, 2010; Lee *et al.*, 2010; Satpute *et al.*, 2010; Singh, 2010; Zhang and Kim, 2010). Thus, bioprospection is a part of biotechnology and the growth of biotechnology and efficiency of bioprospection depends upon the diversity (richness) and quality of genetic resources (biodiversity) of a country. In general, bioprospecting includes four different phases: (i) collection of sample; (ii) isolation and characterization of organisms of interest; (iii) screening for biomolecules; and (iv) product development and commercialization including the patent, trials, sales and marketing. Although the contribution of biological systems like animals and plants in the growth of biotechnology cannot be neglected, microorganisms, due to their small size, single cellular nature, rapid generation time and huge diversity, are considered the pillars of biotechnology and have immensely contributed in all the areas of biotechnology. Due to decades of adaptive evolution and various kinds of selective pressures the environment of the earth has inexhaustible reserves of microbial diversity (Rappé and Giovannoni, 2003; Riesenfeld *et al.*, 2004; Lewis *et al.*, 2010). Discovery of modern methods of cultivation and the culture-independent metagenomic approach has revealed the immense but not yet cultured diversity of microorganisms and successfully brought some of them for research, bioprospection and commercial exploration (Zengler *et al.*, 2002; Lee *et al.*, 2006; Nichols *et al.*, 2010; Prakash *et al.*, 2013b). Collaboration of public and private partnerships for the exploration of biological diversity is an attractive and efficient option, but due to blurred laws on intellectual property rights (IPR) issues it is underexploited. There is a strong correlation between the preservation of biological resources and bioprospection, because both are interrelated.

*Corresponding author e-mail: omprakash@nccs.res.in

 Intellectual Property Issues in Biotechnology (Eds H.B. Singh, A. Jha and C. Keswani)

14.2 Importance of Microbes

Due to metabolic versatility and genome plasticity microbes constitute the richest diversity among all the living organisms on earth. They constitute up to 60% of earth's total biomass, supply nearly 50% of earth's oxygen and are a key component of biogeochemical cycling of materials (Prakash *et al.*, 2013b, 2015). They regulate the composition of atmospheric gases and play an important role in global climate change. It is believed that cultivation of the untapped hidden diversity of microbes contains numerous hidden biotechnological potentials and major resources for bioprospection (Lee *et al.*, 2006; Singh, 2010; Tichet *et al.*, 2010). Microbes produce fuels like biodiesel, bio-methane and bioethanol from renewable energy sources like agro-wastes and participate in production and mitigation of greenhouse gases (Rawat *et al.*, 2008; Christensen *et al.*, 2009). It is estimated that, up to 2020, 20% of global chemicals and 60% of fine chemicals supply of the world will come from bioprospection of microbial diversity. Microbes are the source of new biomolecules, next-generation antibiotics, anti-tumour, immunosuppressant, antidiabetic, anticancer and antilipidaemic compounds (Lee *et al.*, 2006; Liu *et al.*, 2010). They are also used as human and animal vaccine, a source of vitamins, foods, feeds, neutraceuticals and pro- and prebiotics. More than 50% of the world's antibiotics come from *Actinobacteria* (Singh, 2010). Agricultural sustainability and the production of a global supply of nutritious food are the current challenge for agricultural microbiologists. Published data indicated that plant-associated microbes play an important role in stress tolerance, disease resistance, increase nutrient uptake efficiency of plants and also promote plant growth (Prakash *et al.*, 2015). Degradation of agricultural chemicals and biological control of plant disease is also governed by the beneficial activity of the microbes. Plant–microbe interactions are understudied, but are a very hot area of interest among the agricultural microbiologists (Berg, 2009). Several bioinoculants based on the bacteria, cyanobacteria and fungi are available in the market, which control plant disease and promote crop health and productivity (Priya *et al.*, 2015; Tripathi *et al.*, 2015). The progress of environmental biotechnology or grey biotechnology is also based on the activities of microbes. Microbes play important roles in management of environmental resources such as water, soil and sewage, and are prime constituents of solid waste management and promote or mitigate the generation of greenhouse gases (Weiland, 2010). The use of microbes in the development of clean water technology and bioremediation of pollutants such as degradation of oil spills and pesticides has been used for decades (Kostka *et al.*, 2011). Microbial based waste in energy programmes with generation of green energy and compost from municipal solid wastes is the current hot topic (Weiland, 2010).

Approximately 70% of the earth is covered by ocean, which is a source of huge microbial diversity due to the great variety of micro-habitats in the ocean. It contains from three to six living phyla belonging to domains Bacteria, Archaea and Eukarya (Lee *et al.*, 2010; Tichet *et al.*, 2010; Zhang and Kim, 2010). The oceans contain a large amount of untapped and unexplored microbial diversity. The temperature of ocean varies from 35°C to −35°C. It is believed that the oceanic microbial diversity is most important from the bioprospection point of view because up to the present, bioprospecting has only focused on terrestrial diversity and the chance of getting novel biomolecules are 500 times more in ocean than in its terrestrial counterpart (Dionisi *et al.*, 2012). The oceans contain a variety of habitats, such as hydrothermal vents, deep-sea sediment, sinking particles, animal gut, marine invertebrates and sponge-associated diversity. The diversity of the intertidal marine ecosystem is also important from a bioprospection point of view because it faces more physiological and physiochemical turbulences.

14.3 Culturomics and Metagenomics are the Basis of Bioprospection

Culture-independent metagenomic studies from different environmental niches have shown the presence of immense but untapped microbial diversity (Van Lanen and Shen, 2006). Study of databases of cultured organisms to date indicated that only 1–10% of microbial diversity has been cultured and the other 90–99% are not yet cultured (Zengler *et al.*, 2002; Prakash *et al.*, 2015). Thus the biotechnological and bioprospection potentials of the major part of microbial biodiversity remain unexplored (Joint *et al.*, 2010). Traditional microbiological cultivation methods generally employed only limited numbers of organic-rich culture media and a short incubation time and the least variation in culture conditions (Prakash *et al.*, 2013b). The above mentioned strategy only resulted in the growth of the fast-growing r-strategist weed types of environmental organism, but the slow growing, environmentally more relevant k-strategist types of organism remain uncultured. Considering the importance of microbial diversity, the concept of microbial cultivation has been revisited by modifying the cultivation conditions and approaches (Rappé *et al.*, 2002; Rappé and Giovannoni, 2003; Giovannoni and Stingl, 2007; Prakash *et al.*, 2013b). In recent decades several novel methods of cultivation have been developed, consequently several groups of previously uncultured organism have been successfully cultivated and studied under laboratory conditions (Zengler *et al.*, 2002; Nichols *et al.*, 2010; Prakash *et al.*, 2013b). Along with metagenomics, proteomics and transcriptomics, the concept of culturomics is also emerging in the jargon of microbiology. Even researchers started thinking that culturomics is superior to metagenomics, because metagenomics only tells what is there but is unable to say what they are doing. In contrast, culturomics gives a pure culture of the organism for study of functionality, physiology, genomics and application or bioprospection purposes.

Construction of a metagenomic library in a heterologous host like *E. coli* and screening the library for valuable product and processes is another aspect of bioprospection (Warren *et al.*, 2008). Different kinds of cloning vectors, such as plasmid, cosmid, pasmid and bacterial artificial chromosome (BAC), are used for construction of a metagenomic library depending on the researcher's strategy (Dionisi *et al.*, 2012; Thomas *et al.*, 2012). For example, a plasmid vector can accommodate a smaller gene fragment but gives better expression, while BAC can accommodate an entire gene or metabolic pathway but rate of expression will be low. A clone library can be screened using function as well as sequence-based attributes (Warnecke and Hess, 2009; Dionisi *et al.*, 2012). In functional attributes clones are screened for desired activity such as production of industrial enzymes like protease or lipase while in sequence-based approaches, metagenomic libraries are screened using PCR primers and probes designed from a conserved region of a desired gene sequence (Tang and Zhao, 2009; Uchiyama and Miyazaki, 2009). Expression of foreign protein in a heterologous host often creates problems due to the recognition of foreign genes in the *E. coli* host. According to bioinformatics analysis, <50% of the activities of foreign proteins have been recovered in a heterologous host such as *E. coli*. To solve the problem of expression of foreign protein in a heterologous host, different host cells such as *Streptomyces*, *Pseudomonas* and *Rhizobium* etc. have been tested by different groups. In addition, expression of psychrophilic enzymes in a mesophyllic host at normal temperature creates the problem of protein miss-folding and loss of function and activities (Feller *et al.*, 1998). Although the cultivation of the host at lower temperature regains the folding and activities of the enzymes, simultaneously it slows down the growth of organisms and reduces the rate of enzyme production. The above problem has been addressed using the *E. coli* strain with chaperone gene from *Oleispira antarcatica*. Currently, next generation sequencing is also being used to obtain the novel genes for biotechnological products and pathways (Metzker, 2010; Miller *et al.*, 2010).

14.4 Role of Microbial Resource Centres in Bioprospection Research

It has been discussed in the previous sections that microbial diversity is the raw material for bioprospection research in all areas of biotechnology, including the red, green, blue, white and grey (Singh, 2010; Prakash *et al.*, 2013b, 2015). Therefore, along with cultivation of the novel microorganism from untapped microbial diversity, its scientific preservation for future biotechnological exploitation is equally important (Prakash *et al.*, 2013a). Microbial Resource Centres (MRCs), also known as Microbial Depository or Microbial Culture Collections, play an important role in cultivation, identification and preservation of microbial diversity and supply them for academics, industry and research institutions (Cypess, 2003; Heylen *et al.*, 2012). A microbial repository also works as a microbial seed bank or microbial library, which maintains the different diversity of the microorganisms, copies them (revive) and supplies them to researchers on demand (Lennon and Jones, 2011; Prakash *et al.*, 2013b). A well-equipped MRC contains diverse expertise and facilities to handle a diverse range of microbial diversity including proteobacteria, actinobacteria, fungi, archaea, anaerobes, photoautotrophs and cyanobacteria. Some of the culture collections also maintain the genomic DNA, clones and phages for different activities. It is desirable that MRCs should not only focus on activities like culture preservation and supply, but also strengthen their research related to cultivation of novel but not yet cultured organisms, optimize the preservation strategy for problematic groups and design and initiate the research related to genetic changes in microorganisms during the time of preservation (Griffith, 2012; Prakash *et al.*, 2013a). Extreme environments include the Arctic and Antarctic polar regions, high seas, deep seas, hydrothermal vents, acid mine drainage, hot-water spring submarine trenches, cold seeps etc. (Dionisi *et al.*, 2012). Extremophiles are the unique group of microorganisms adopted for extreme conditions of temperature, pressure, salinity and pH, where normal organisms could not survive. Extremophiles are the hope for novel enzymes and biomolecules. Cultivation and preservation of extremophile organisms, such as acidophiles, alkaliphiles, halophiles, psychrophiles, barophiles and thermophiles from ocean, acid mine drainage, permanently cold habitats such as polar habitats, hot spring and deep-sea hydrothermal vents, have immense biotechnological and bioprospecting potentials and culture collections should try to develop expertise in that area (Bodelier, 2011; Dionisi *et al.*, 2012). In addition, culture collection should also promote networking and collaboration with national laboratories to promote the conservation of biodiversity of the nation and protect the valuable cultures from loss and extinction. Most of the small laboratories lack proper preservation methods and facilities and ultimately lose the valuable cultures due to lack of proper attention, expertise and care.

14.5 Biopiracy

Biopiracy is a recent term in the jargon of biological science which describes 'The illegal and unfair appropriation (patent claim) and exploitation of genetic and biological resources of specific regions by corporations' (Gallant, 2009). The term biopiracy seems recent, but the practice of biopiracy is very old in human civilization and has occasionally been used by several countries and invaders from thousands of years back (Gallant, 2009).

For example, *Medicago sativa* (lucerne) was acquired by Greeks from Persia and commercialization of Peru's quinquina plant (*Cinchona officinalis*) by Spanish Jesuits are good examples of biopiracy. Sometimes multinational companies from developed countries clam the ownership on certain varieties of biodiversity, specific genes and their by-product or extract that are not native to their own geographical location or even not modified enough using the modern approach of genetic engineering (genetically modified varieties). These companies pirate the knowledge from the traditional communities, patent them and make huge benefits, without sharing them with the traditional community or country from which information has been gathered. In addition, due to holding of patents, traditional communities and countries are deprived from their benefit as well as from free use (Gallant, 2009; Cressey, 2014). Due to being biodiversity-rich nations, India, the Philippines, Brazil, Gabon and Malaya have faced the problem of biopiracy in the past. With the growing problems of biopiracy, governments from the above-mentioned countries and from other organizations such as the France-based Action Group for an Alternative to Biopiracy, the Europe-based Genetic Resource International and the Asia-based Third World Network started fighting against the problem of biopiracy and successfully managed to cancel several patents filed by the multinational companies. There are several examples of biopiracy that have occurred in recent times, and patents that have been obtained by foreign companies have been cancelled due to the voice raised against them, for example the cancellation of a US patent on the wound-healing property of turmeric (*Curcuma longa*) filed by the University of Mississippi in 1998. The success story of the Indian network 'Navdanya'[1] against Basmati rice and Nap Hal variety of rice in 2001 and 2004, respectively, patented by Rice Tech Inc (America) and Monsanto is a good example of biopiracy. Similarly, the fungicidal properties of neem (*Azadirachta indica*) patented by the American company W.R. Grace were cancelled in 2005 after the intervention of The Research Foundation for Science, Technology and Ecology (RFSTE) and the International Federation of Organic Agriculture Movements (IFOAM; Gallant, 2009).

14.6 Intellectual Property Rights

In brief, property means something valuable which is owned or possessed by someone in a tangible or intangible form. When something is created or invented using human intellect or brain then it comes under intellectual property (IP). It is an intangible form of property and includes scientific inventions, literary and artistic works, designs, symbols and logos etc. In order to provide the benefit and encourage the inventor or creator the IP should be protected from infringement and other violations of right. The concept of IPR arose in the 18th century when traders refused to gather and to share their ideas, goods and preparation methods and started avoiding participation in invention-related international seminars due to the fear of someone else stealing their ideas. Due to above aforementioned issues, strictly enforcing IPR among the signatory countries could offer significant advantages in improving their economic status. Conventions are formal meetings or assemblies among the delegates/members of different countries to discuss matters of common interest or concern. In other words it is an agreement to seek legislative protection in different signatory countries for their creation. The Paris and Bern conventions are well-known for IP-related issues. Various international organizations such as the World Intellectual Property Organization (WIPO) and the World Trade Organization (WTO) come forward in order to improve trade globally in the correct way and different treaties, such as the Budapest Treaty on the International Recognition of the Deposit of Microorganisms for the Purposes of Patent Procedure, have been signed. In 1893, an international organization called the United International Bureaux for the Protection of Intellectual Property (BIRPI) was set up to deal with IP-related issues. BIRPI is the predecessor of today's WIPO. WIPO handles the administrative work on information related to IP matters. Currently, 21 international treaties governed by WIPO are all related to IP. WIPO is also engaged in setting up rules that may be enforced in all countries. The Agreement on Trade-Related Aspects of Intellectual Property Rights (TRIPS) is an international agreement to set the minimum standard for IP-related issues and is jointly structured by WTO and WIPO (WIPO, 2004, 2015).

14.7 Patent and Microbial Patenting

Invention is the essence of science, society, civilization and industry. Without invention and creativity the wheel of economic and industrial development will stop and the pleasure of living will diminish. Patent is a form of IPR provided by countries in order to exploit the benefit of their own invention(s) for 20 years. The person holding a patent is known as the patentee and the patentee invests his time, money and intellect to find out something completely new or better than currently in existence. The patent right prevents others from making, using, selling and licensing, without the permission of the patentee. Once the patent term has expired, the invention automatically comes into the public domain and anybody can utilize the invention. A patent can be acquired for any invention but the invention must stick to laws and rules and regulation of state statute. All over the world, the rules and laws for granting the patent are almost the same with minor differences, but it should have novelty (prior art), efficiency, non-obviousness, industrial applicability and sufficient disclosure (Roy, 2009; Prakash *et al.*, 2013b).

In addition, patents cannot be granted to all inventions, particularly when it comes under the category of patentable threat such as hazardous to society, natural law and the environment. Inventions which are not for direct public use and related to atomic energy, mathematical or business method and computer programs are also not patentable. IP law also does not allow granting the patent on any whole animal or plant or any part thereof other than microorganisms, but it includes seed varieties and species. It also does not allow patenting of any biological process which leads to production and propagation of animals. Before granting the patent it is essential to know whether the product is of natural origin or produced due to intervention of human activities, because natural products are non-patentable.

The rules of microbial patenting are more or less similar throughout the world with a few exceptions. In 1873 Louis Pasteur obtained a patent for a process that involved microbes. Initially only the product and processes related to microorganisms were patentable, but after the *Diamond* v. *Chakrabarty* case[2] the US Supreme Court passed the law that live microorganisms are also patentable and Dr Ananda M. Chakrabarty received the first patent on *Pseudomonas putida*, a genetically modified organism which is capable of degrading oil spillage. After that, various other inventors came forward and claimed patents on pure cultures of microorganisms that do not exist in nature in pure form. Microbes from Danubian mud[3] and a microorganism isolated from Indian soil[4] (US Pat No 4232 316 (1982)) are good examples. Novel species and existing microbial products and processes are not patentable, but microorganisms isolated using a special procedure and not discovered before are patentable. Along with providing the detailed description of invention, microorganisms used for patenting must be deposited in an International Depository recognized under the Budapest Treaty (Chopra and Taneja, 2013; Prakash *et al.*, 2013b). Description of the International Depositary Authority (IDA) under the Budapest Treaty and essential features of microbial patenting have been discussed elsewhere (Prakash *et al.*, 2013b). Microbial patenting is an interesting subject matter, and any microbiologist working in an area related to microbial bioprospection should have basic knowledge of the subject in order to protect their IP.

14.8 Convention on Biological Diversity

The concept of the Convention on Biological Diversity (CBD) originated due to the commitment of the scientific community to make an international law for the conservation and sustainable use of biodiversity for a sustainable earth and environment. The first meeting was held on 29 January 2000 and the chairman was Ambassador Yang of Cameroon. The CBD started working within a year of its formulation. In 1991, this expert group was named the Intergovernmental Negotiating Committee with the main objectives: (i) to conserve biological diversity; (ii) sustainable use of its product or components; and (iii) to ensure the benefits of the product or component are shared fairly (CBD, 2015). There are some general measures that need to be considered for conservation and sustainable use of biodiversity. For example, all the contracting agencies should follow the strategies that have been formulated by the CBD, or, they can make their own strategies for the conservation and sustainable use of biodiversity. In addition to the above, agencies may collaborate within relevant sectors or possibly some other sectors for the conservation and sustainable use of biodiversity (CBD, 2014a). The CBD has considered the current need of developing countries to promote the concept of biodiversity conservation and its sustainable use. According to the CBD, all the contracting agencies must promote the technical and scientific education related to biodiversity conservation and also boost the research and use of scientific tools related to the area of biodiversity conservation and its sustainable use.

It has been decided under various articles of the CBD that all the contracting agencies should share and transfer the scientific technologies and information related to biodiversity conservation and its sustainable use (Dutfield, 2002; Oliva, 2011). In particular, this should be transferred to developing countries from developed countries, and technologies registered under patent or IPR can be used under mutually agreed terms. The exchange of information from all sources should be done by placing it in the public domain. In addition, any benefits arising from such experiments should be shared on mutually agreed terms. Financial assistance is the other factor that comes into the picture for such kinds of programmes. It is recommended by the CBD that developed countries must support developing countries for scientific experiments and all the expenses that are utilized to achieve the objectives of the CBD. There should be a mechanism, which will be under competent authority, through which these developing countries will receive financial aid, and that competent authority will be accountable to the conference of the parties for purposes of this convention.

The CBD, once it came into force on 29 December 1993, played a very important role among culture collections in the world. Before this period, culture collections used to work in their own way. CBD considers the sovereign rights of the state on their own biodiversity. According to CBD regulations, at the time of sampling one needs to take the permission from the competent authority of the state, i.e. they have to get the prior informed consent (PIC). For the transfer of genetic resources there is the provision of the Material Transfer Agreement (MTA); recently, however, an efficient tool known as MOSAICC (Microorganisms Sustainable Use and Access Regulation International Code of Conduct) has been implemented by the CBD, which is providing guidelines for PIC and maintains a monitoring system on the MTA. Through MOSAICC, the scientific community has received much-awaited clarity on these matters (Janssens *et al.*, 2010).

14.9 Cartagena and Nagoya Protocol Related to Biodiversity Conservation

During the whole process of the formation of the CBD, the leaders of the convention did not made any protocols for the implementation of the CBD's objectives. As a result, it lacked that much needed force to control the loss or disruption of biodiversity (Kursar, 2011). In order to implement the first and third objectives of the CBD, the Cartagena and Nagoya Protocols were established, respectively.

The Cartagena Protocol was adopted in 2000 by the CBD, where 170 contracting agencies were present and an Advanced Informed Agreement (AIA) was formed. This protocol was first made to protect biodiversity from the threats posed by living genetically modified organisms (GMO) and later on it also included food safety (Lusser *et al.*, 2013). The implementation of the protocol was as follows: if a country does not have enough scientific proof that the GMO is safe for biodiversity, then, under this protocol, the committee will not allow the import of that particular product to another country. Therefore risk assessment is mandatory for all the GMOs that need to be sent to another country. There are two simple points which need to be considered for the import of GMOs: (i) the party should be informed about the timings of the import of the GMOs and it should be processed at the same time; and (ii) the import of the GMOs will not be under advanced informed agreement procedure. While importing the GMOs, contracting agencies may need to consider socio-economic impacts of the GMO (if any) on the local communities as per the Cartagena Protocol. Financial assistance will be as per the CBD policies. This protocol came into force on 11 September 2003 (CBD, 2003).

To fulfil the third objective of the CBD, the Nagoya Protocol was adopted on 29 October 2010 in Nagoya, Japan, where 59 contracting agencies were present. This protocol provides a legal basis and transparent conditions between the genetic resource provider and the user for their working (Kursar, 2011). This protocol came into force on 12 October 2014 (CBD, 2014b). Under Article 15 of the CBD, the Nagoya Protocol shall apply only for genetic resources and the benefits such as knowledge arising from the genetic resources. While accessing the natural resources of any country, the law of the land needs to be followed. The contracting agency should take note of whether prior informed consent has been obtained from the competent authority. There are also some special considerations, which are mentioned in the Nagoya Protocol: (i) all the contracting agencies need to promote the conditions for the research with regards to conservation and sustainable development of biodiversity; and (ii) contracting agencies need to give attention towards the critical issues such as health or agriculture issues.

The contacting parties must motivate the providers and users to take the benefits which are arising when there is utilization of genetic resources and must run some awareness programmes about the importance of genetic resources. Article 14 of the Nagoya Protocol states that an access and benefit sharing clearing house has to be established, which will serve as a platform for the sharing of the information and knowledge arising from the benefit sharing of the genetic resources. Contracting agencies should request non-contracting agencies to follow the protocol and share the relevant information with the clearing house (CBD, 2014c).

To implement any protocol in general and the Nagoya Protocol in particular, all the contracting agencies need to come together and cooperate with each other on various aspects of the convention. Developed countries need to provide the resources, which may be logistical or intellectual, to underdeveloped countries for the successful convention. In the same regard, developed countries may utilize the help of local communities for better process. Article 35 clearly states that if any contracting agency needs to withdraw from the protocol they may do so, by giving written notification to the depository. This may be done any time after 2 years from the date this protocol came into force (CBD, 2014d).

14.10 Conclusion

Due to their small size and rapid generation, microbes are the key components of bioprospection. Available data indicate that their huge microbial diversity with an immense biotechnological potential are not yet cultured. In order to harness their biotechnological potential, it is essential to bring them into culture using modern approaches of culturomics. Clubbing the genomics, culturomics and metagenomics together can bring more information and bring hidden microbes in culture collections for use and exploitation. In addition to culturomics and screening the cultured organisms for compounds of biotechnological importance, authentic preservation of the organisms is equally important. Therefore countries with rich biodiversity such as India should promote the establishment of high quality microbial culture collections with all the basic facilities for handling and preservation of their biological resources. Biopiracy is an emerging issue in the area of bioprospection. Therefore researchers and culture collections dealing with bioprospection and diversity-related research should have a basic knowledge of microbial patenting and laws pertaining to the use of biodiversity.

Notes

1 http://www.navdanya.org
2 US Supreme Court, *Diamond* v. *Chakrabarty*, 447 U.S. 303 (1980), No. 79-136, argued 17 March 1980, decided 16 June 1980.
3 Jarai, M., 1 February 1983. *Micromonospora culture*. US4371622 A.
4 Cavalleri, B., Pagani, H. and Volpe, G., 4 May 1982. *Biologically pure culture of actinoplanes capable of producing antibiotic A/16686*. US4328316 A.

Acknowledgement

This work was supported by the Department of Biotechnology (DBT; Grant no. BT/PR/0054/NDB/52/94/2007), Government of India, under the project 'Establishment of microbial culture collection'.

References

Berg, G. (2009) Plant-microbe interactions promoting plant growth and health: perspectives for controlled use of microorganisms in agriculture. *Applied Microbiology and Biotechnology* 84, 11–18.

Bodelier, P.L. (2011) Toward understanding, managing and protecting microbial ecosystems. *Frontiers in Microbiology* 2, 1–8.

CBD (2003) About the Protocol. Available at: http://bch.cbd.int/protocol/background (accessed December 2015).

CBD (2014a) Article 6. General Measures for Conservation and Sustainable Use. Available at: https://www.cbd.int/convention/articles/default.shtml?a=cbd-06 (accessed December 2015).

CBD (2014b) About the Nagoya protocol. Available at: http://www.cbd.int/abs/about (accessed December 2015).

CBD (2014c) Text of the Nagoya Protocol. Article 14. The Access and Benefit-sharing Clearing-House and Information Sharing. Available at: https://www.cbd.int/abs/text/articles/default.shtml?sec=abs-14 (accessed December 2015).

CBD (2014d) Text of the Nagoya Protocol. Article 35. Withdrawal. Available at: https://www.cbd.int/abs/text/articles/default.shtml?sec=abs-35 (accessed December 2015).

CBD (2015) History of the Convention. Available at: http://www.cbd.int/history/default.shtml (accessed December 2015).

Chopra, S. and Taneja, A. (2013) *Patent Agent Examination – A Book for Students and Industry Professionals*. Lexis-Nexis, India.

Christensen, T., Simion, F., Tonini, D. and Moller, J. (2009) Global warming factors modeled for 40 generic municipal waste management scenarios. *Waste Management Research* 27, 871–884.

Cressey, D. (2014) Biopiracy ban stirs red-tape fears. *Nature* 2, 514, 14–15.

Cypess, R. (2003) *Biological Resource Centers: From Concept to Reality.* American Type Culture Collection, Manassas, Virginia.

Dionisi, H.M., Lozada, M. and Olivera, N.L. (2012) Bioprospection of marine microorganisms: biotechnological applications and methods. *Revista Argentina de Microbiología* 44, 49–60.

Dutfield, G. (2002) Sharing the benefits of biodiversity. *The Journal of World Intellectual Property* 5(6), 899–931.

Feller, G., Le Bussy, O. and Gerday, C. (1998) Expression of psychrophilic genes in mesophilic hosts: assessment of the folding state of a recombinant alpha-amylase. *Applied and Environmental Microbiology* 64, 1163–1165.

Ferrer, M., Golyshina, O., Beloqui, A. and Golyshin, P.N. (2007) Mining enzymes from extreme environments. *Current Opinion in Microbiology* 10, 207–214.

Gallant, D. (2009) Biopiracy and bioprospection: a new terminology for an old problem. Rede Verde. Available at: http://www.rede-verde.org/index.php?option=com_content&view=article&id=50 (accessed December 2015).

Giovannoni, S. and Stingl, U. (2007) The importance of culturing bacterioplankton in the 'omics' age. *Nature Reviews* 5, 820–826.

Griffith, G.W. (2012) Do we need a global strategy for microbial conservation? *Trends in Ecology & Evolution* 27, 1–2.

Heylen, K., Hoefman, S., Vekeman, B., Peiren, J. and De Vos, P. (2012) Safeguarding bacterial resources promotes biotechnological innovation. *Applied Microbiology and Biotechnology* 94, 565–574.

Janssens, D., Arahal, D.R., Bizet, C. and Garay, E. (2010) The role of public biological resource centers in providing a basic infrastructure for microbial research. *Research in Microbiology* 16, 422–429.

Jeon, J.H., Kim, J.T., Kim, Y.J., Kim, H.K., Lee, H.S., Kang, S.G., Kim, S.J. and Lee, J.H. (2009) Cloning and characterization of a new cold – active lipase from a deep-sea sediment metagenome. *Applied Microbiology and Biotechnology* 81, 865–874.

Joint, I., Muhling, M. and Querellou, J. (2010) Culturing marine bacteria – an essential prerequisite for biodiscovery. *Microbial Biotechnology* 3, 564–575.

Kostka, J.E., Prakash, O., Overholt, W.A., Green, S.J., Freyer, G., Canion, A. and Huettel, M. (2011) Hydrocarbon-degrading bacteria and the bacterial community response in Gulf of Mexico beach sands impacted by the Deepwater Horizon oil spill. *Applied and Environmental Microbiology* 77, 7962–7974.

Kursar, T.A. (2011) What are the implications of the Nagoya Protocol for research on biodiversity? *BioScience* 61(4), 256–257.

Lee, H.S., Kwon, K.K., Kang, S.G., Cha, S.S., Kim, S.J. and Lee, J.H. (2010) Approaches for novel enzyme discovery from marine environments. *Current Opinion in Biotechnology* 21, 353–357.

Lee, M.-H., Lee, C.-H., Oh, T.-K., Song, J.K. and Yoon, J.-H. (2006) Isolation and characterization of a novel lipase from a metagenomic library of tidal flat sediments: evidence for a new family of bacterial lipases. *Applied and Environmental Microbiology* 72, 7406–7409.

Lennon, J.T. and Jones, S.E. (2011) Microbial seed banks: the ecological and evolutionary implications of dormancy. *Nature Reviews Microbiology* 9, 119–130.

Lewis, K., Epstein, S., D'Onofrio, A. and Ling, L.L. (2010) Uncultured microorganisms as a source of secondary metabolites. *Journal of Antibiotics (Tokyo)* 63(8), 468–476.

Li, Q., Du, W. and Liu, D. (2008) Perspectives of microbial oils for biodiesel production. *Applied Microbiology and Biotechnology* 80, 749–756.

Liu, X., Ashforth, E., Ren, B., Song, F., Dai, H., Liu, M., Wang, J., Xie, Q. and Zhang, L. (2010) Bioprospecting microbial natural product libraries from the marine environment for drug discovery. *Journal of Antibiotics (Tokyo)* 63, 415–422.

Lusser, M., Raney, T., Tillie, P., Dillen, K. and Cerezo, E.R. (2013) *International workshop on socio-economic impacts of genetically modified crops co-organised by JRC-IPTS and FAO* (No. JRC69363). Institute for Prospective and Technological Studies, Joint Research Centre.

Metzker, M.L. (2010) Sequencing technologies – the next generation. *Nature Reviews Genetics* 11, 31–46.

Miller, J.R., Koren, S. and Sutton, G. (2010) Assembly algorithms for next-generation sequencing data. *Genomics* 95, 315–327.

Nichols, D., Cahoon, N., Trakhtenberg, E.M., Pham, L., Mehta, A., Belanger, A., Kanigan, T., Lewis, K. and Epstein, S.S. (2010) Use of I-chip for high-throughput in situ cultivation of 'uncultivable' microbial species. *Applied Environmental Microbiology* 76, 2445–2450.

Oliva, M.J. (2011) Sharing the benefits of biodiversity: a new international protocol and its implications for research and development. *Planta Medica* 77, 1221–1227.

Prakash, O., Nimonkar, Y. and Shouche, Y.S. (2013a) Practice and prospects of microbial preservation. *FEMS Microbiology Letters* 339(1), 1–9.

Prakash, O., Shouche, Y.S. and Jangid, K. (2013b) Microbial cultivation and the role of microbial resource centers in the omics era. *Applied Microbiology and Biotechnology* 53, 247–252.

Prakash, O., Sharma, R., Rahi, P. and Karthikeyan, N. (2015) Role of microorganisms in plant nutrition and health. In: *Nutrient Use Efficiency: from Basics to Advances*. Springer, India, pp. 125–161.

Priya, H., Prasanna, R., Ramakrishnan, B., Bidyarani, N., Babu, S., Thapa, S. and Renuka, N. (2015) Influence of cyanobacterial inoculation on the culturable microbiome and growth of rice. *Microbiological Research*. 171, 78–89.

Rappé, M.S. and Giovannoni, S.J. (2003) The uncultured microbial majority. *Annual Review of Microbiology* 57, 369–394, doi:10.1146/annurev.micro.57.030502.090759.

Rappé, M.S., Connon, S.A., Vergin, K.L. and Giovannoni, S.J. (2002) Cultivation of the ubiquitous SAR11 marine bacterioplankton clade. *Nature* 418, 630–633.

Rawat, M., Singh, U.K., Mishra, A.K. and Subramanian, V. (2008) Methane emission from landfill areas of India. *Environmental Monitoring and Assessment* 137, 67–74.

Riesenfeld, C.S., Goodman, R.M. and Handelsman, J. (2004) Uncultured soil bacteria are a reservoir of new antibiotic resistance genes. *Environmental Microbiology* 6, 981–989.

Roy, S.C. (2009) *Lecture on IP Law*. *Eastern Book Company, India, pp.* 13–28, 139–168.

Satpute, S.K., Banat, I.M., Dhakephalkar, P.K., Banpurkar, A.G. and Chopade, B.A. (2010) Biosurfactants, bioemulsifiers and exopolysaccharides from marine microorganisms. *Biotechnology Advances* 28, 436–450.

Singh, B.K. (2010) Exploring microbial diversity for biotechnology: the way forward. *Trends in Biotechnology* 28, 111–116.

Tang, W.L. and Zhao, H. (2009) Industrial biotechnology: tools and applications. *Journal of Biotechnology* 4, 1725–1739.

Thomas, T., Gilbert, J. and Meyer, F. (2012) Metagenomics-a guide from sampling to data analysis. *Microbial Informatics and Experimentation* 2(3), 1–12.

Tichet, C., Nguyen, H.K., Yaakoubi, S.E. and Bloch, J.F. (2010) Opinion: commercial product exploitation from marine microbial biodiversity: some legal and IP issues. *Microbiology and Biotechnology* 3, 507–513.

Tripathi, S., Das, A., Chandra, A. and Varma, A. (2015) Development of carrier-based formulation of root endophyte *Piriformospora indica* and its evaluation on *Phaseolus vulgaris* L. *World Journal of Microbiology and Biotechnology* 31, 337–344.

Uchiyama, T. and Miyazaki, K. (2009) Functional metagenomics for enzyme discovery: challenges to efficient screening. *Current Opinion in Biotechnology* 20, 616–622.

Van Lanen, S.G. and Shen, B. (2006) Microbial genomics for the improvement of natural product discovery. *Current Opinion in Microbiology* 9, 252–260.

Warnecke, F. and Hess, M. (2009) A perspective: metatranscriptomics as a tool for the discovery of novel biocatalysts. *Journal of Biotechnology* 142, 91–95.

Warren, R.L., Freeman, J.D., Levesque, R.C., Smailus, D.E., Flibotte, S. and Holt, R.A. (2008) Transcription of foreign DNA in Escherichia coli. *Genome Research* 18, 1798–1805.

Weiland, P. (2010) Biogas production: current state and perspectives. *Applied Microbiology and Biotechnology* 85, 849–860.

WIPO (2004) WIPO-WTO Joint Initiative. WIPO Intellectual Property Handbook: Policy, Law and Use. Available at: http://www.wipo.int/about-ip/en/iprm/pdf/ch5.pdf

WIPO (2015) Depositary Institutions Having Acquired the Status of International Depositary Authority under the Budapest Treaty. Available at: http://www.wipo.int/treaties/en/registration/budapest/pdf/ida.pdf.

Zengler, K., Toledo, G., Rappe, M., Elkins, J., Mathur, E.J., Short, J.M. and Keller, M. (2002) Cultivating the uncultured. *The Proceedings of the National Academy of Sciences USA* 99, 15681–15686.

Zhang, C. and Kim, S.K. (2010) Research and application of marine microbial enzymes: status and prospects. *Marine Drugs* 8, 1920–1934.

15 CRISPR/Cas9 System, A Revolutionary Technology for Genome Editing: Applications and Intellectual Property Disputes

Kartikay Bisen,[1] Chetan Keswani,[1]* Akanksha Singh,[2] Rakesh Pandey,[2] Sandhya Mishra,[1] Birichi K. Sarma[1] and Harikesh B. Singh[1]

[1]*Department of Mycology and Plant Pathology, Institute of Agricultural Sciences, Banaras Hindu University, Varanasi;* [2]*Department of Microbial Technology and Nematology, Central Institute of Medicinal and Aromatic Plants, Lucknow, India*

15.1 Introduction

Genetic manipulation through recombinant DNA technology laid the foundation for the manipulation of genetic code and harnessing new knowledge to develop novel medicines and crop verities. Alteration in the targeted genome sequence of living cells and organisms is a powerful tool with potential application in therapeutics and the agriculture sector. Specific gene insertions and deletion can induce desired changes and incorporation of such useful traits in agricultural crops and livestock development. With recent advancement in genome engineering and editing technologies researchers can now directly edit or alter the function of DNA sequences in any organism of choice. However, the technology inducing precise modification to the genome was limited to certain organisms. A recent development of RNA-guided genome editing technology, CRISPR/Cas9, has paved the way for efficient and easy genomic modifications. This technology has potential application across many fields, including biotechnology, biological science and medicine.

In their attempt to study the protein-encoding gene in *E. coli* scientists in 1987 first discovered CRISPR, pattern DNA sequences separated by non-repeating short 'spacer' DNA sequences. The presences of these repeats were reported over a few years in many other bacteria. The term CRISPR was coined in 2012 and stands for 'clustered regularly interspaced short palindromic repeats'. Later researchers found out that the repeating DNA patterns, along with Cas proteins and RNA molecules, play a significant role in prokaryotes' immune systems and considered the entire complex as the CRISPR/Cas system. In 2012, a new gene-editing tool was claimed by a team of scientists led by Jennifer Doudna and Emmanuelle Charpentier using CRISPR/Cas immune system from bacteria. CRISPR/Cas9 gene editing system involved CRISPR, hybrid RNA and Cas9 protein that can be used to identify and precisely excise any gene sequence.

Recently, genetic modification of human spleen using gene editing technology CRISPR and CRISPR-associated protein 9 (Cas9) has surprised the global science communities. Chinese scientists Junjiu Huang and his colleagues from Sun Yat-sen University reported the first ever genetic modification of human embryo. They published their research

*Corresponding author e-mail: chetankeswani@rediffmail.com

findings in *Protein and Cell Journal* (Liang *et al.*, 2014) as previously they had been rejected by the journals *Nature* and *Science* over ethical concerns. However, CRISPR technology had already been used in adult human cells and animal embryos, but it was the first time that this technology had been used in *in vivo* experiments (Lanphier *et al.*, 2015). Recently, the genome of various living organisms including humans, animals and plants were edited by CRISPR/Cas9. Forming a breakthrough in genome engineering technology based on the class of RNA-guided endonucleases, CRISPR/Cas9 has revolutionized biological and medical research. Easy application and target specificity of the CRISPR/Cas9 system has led to its widespread applications in recombinant DNA technology such as clinical application and treatment of infectious diseases, correction of genetic disorders, plant genome editing, generating germline animal models, somatic genome engineering and functional genome screening (Fig. 15.1).

15.2 Application of CRISPR/Cas9

15.2.1 Treatment of genetic disorders

Application of the CRISPR/Cas9 technology in the cure of genetic disorders is one of the most appealing fields. The CRISPR/Cas9 system has been successfully used in a cataracts mouse model for correction of a dominant *Crygc* gene mutation by injecting Cas9 mRNA and sgRNA (Wu *et al.*, 2015). The CRISPR/Cas9 system has been successfully used in modification of the *Crygc* gene in spermatogonial stem cells (SSCs) and modified SSCs can undergo spermatogenesis and generate phenotypically corrected offspring (Wu *et al.*, 2015). Correction of the dystrophin gene mutation, which is responsible for muscular dystrophy in mutant mice, was successfully performed by injecting Cas9 sgRNA into zygotes (Schwank *et al.*, 2013). Furthermore, CRISPR/Cas9 has been successfully used in correction of adult stem cells from patients diagnosed with single-gene hereditary disorder and thereby suggesting its promising role in human gene therapy.

15.2.2 Treatment of infectious disease

Deletion of the pathogen genome from an infected organism by CRISPR/Cas9 may serve as an adaptive immune system and it is useful in the treatment of infectious diseases. In recent studies scientists reported the elimination of the HIV-1 genome and prevention of its further infection by the CRISPR/Cas system (Ebina *et al.*, 2013; Hu *et al.*, 2014). sgRNA expression vector when transfected into a HIV-infected human cell efficiently cleaved the long terminal repeats (LTR) of HIV-1 and suppressed viral gene expression. Furthermore, the CRISPR/Cas system has also been demonstrated to delete viral genes from the host cell chromosome (Ebina

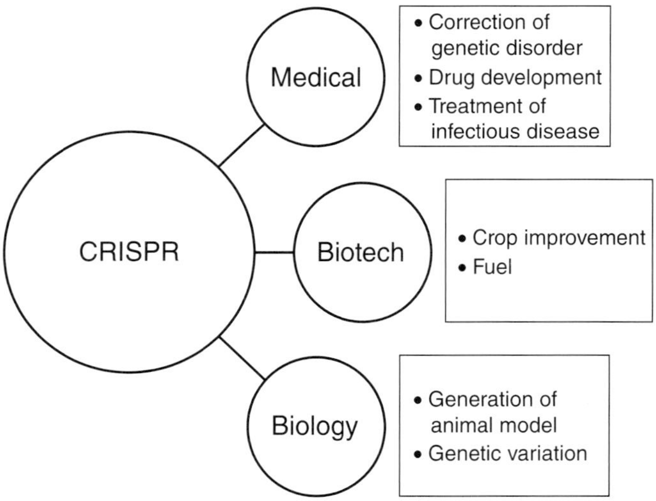

Fig. 15.1. Application of CRISPR/Cas9.

et al., 2013). Target specificity of Cas9 has been recently demonstrated in HIV-1 genome editing (Hu *et al.*, 2014). Cas9 sgRNAs inactivated the replication and gene expression of HIV in infected T cells, microglial and pro-monocytic cells. Recently Cas9-mediated genome editing has been reported to immunize cells against HIV-1 infection (Hu *et al.*, 2014). All these experiments indicate the significance of CRISPR/Cas9 technology in clinical applications to cure infectious diseases.

15.2.3 Genetic engineering in plants

The CRISPR/Cas9 gene editing technology has been applied successfully in various plants, including *Nicotiana benthamiana* (Belhaj *et al.*, 2013; Jiang *et al.*, 2013; Li *et al.*, 2013; Nekrasov *et al.*, 2013), *N. tabacum* (Gao *et al.*, 2015), *Arabidopsis* (Feng *et al.*, 2013; Jiang *et al.*, 2013, 2014; Li *et al.*, 2013; Fauser *et al.*, 2014; Feng *et al.*, 2014), wheat (Shan *et al.*, 2013; Upadhyay *et al.*, 2013; Wang *et al.*, 2014), maize (Liang *et al.*, 2014), rice (Feng *et al.*, 2013; Miao *et al.*, 2013; Shan *et al.*, 2013; Xie and Yang, 2013; Xu *et al.*, 2014; Zhang *et al.*, 2014; Zhou *et al.*, 2014), sorghum (Jiang *et al.*, 2014), tomato (Brooks *et al.*, 2014; Ron *et al.*, 2014) and sweet orange (Jia and Wang, 2014). Bi-allelic and homozygous mutations in first-generation transgenics induced by CRISPR/Cas9 have been reported in *Arabidopsis* (Feng *et al.*, 2013), rice (Xu *et al.*, 2014; Zhou *et al.*, 2014) and tomato (Brooks *et al.*, 2014).

15.2.4 Generation of animal models for research and laboratory purposes

Recent studies have established CRISPR/Cas9 technology as a tool for rapid generation of a modified genome of a wide range of organisms (Bassett *et al.*, 2013; Chang *et al.*, 2013; Cheng *et al.*, 2013; Gratz *et al.*, 2013; Hwang *et al.*, 2013; Wang *et al.*, 2013; Yang *et al.*, 2013; Platt *et al.*, 2014). Microinjecting mRNA-encoded Cas9 and sgRNA into zebrafish one-cell stage embryo resulted in efficient and rapid modification of the target genes *in vivo* (Chang *et al.*, 2013; Hwang *et al.*, 2013). Moreover, transgenic mice have been produced by using sgRNAs that act on two separate sites in the same gene (Wang *et al.*, 2013). Prior experiments have confirmed that CRISPR/Cas9 technology can be efficiently used for the generation of mutant mice carrying gene deletions, insertions and conditional alleles at different loci. The CRISPR/Cas9 system has been successfully used for genome editing in many other organisms such as *Drosophila* (Bassett *et al.*, 2013; Gratz *et al.*, 2013), *Caenorhabditis elegans* (Friedland *et al.*, 2013), axolotl (Flowers *et al.*, 2014), *Xenopus tropicalis* (Blitz *et al.*, 2013; Nakayama *et al.*, 2013), rat (Hu *et al.*, 2013), pig (Whitworth *et al.*, 2014) and monkey (Niu *et al.*, 2014).

15.2.5 Somatic genome editing

CRISPR/Cas9 system-based genome engineering allows functional characterization of assumed disease genes in various mice models through somatic genome engineering (Maddalo *et al.*, 2014; Sánchez-Rivera *et al.*, 2014; Xue *et al.*, 2014). A CRISPR plasmid (expressing Cas9) and sgRNAs are injected into the liver for mediating *Pten* mutation with or without *p53* mutation; thereby exhibiting the effects of *Pten* and *p53* gene knockout (Xue *et al.*, 2014). Recent studies have demonstrated the transfer of a mutant β-catenin gene into hepatocytes by injecting Cas9 plasmids expressing sgRNAs that target the β-catenin gene and a point mutation activated by oligonucleotide donor carrying β-catenin (Xue *et al.*, 2014).

15.3 Global Market of CRISPR/Cas9

The market for CRISPR/Cas9-based genome engineering is experiencing significant and fast growth in the medical and agricultural sectors. These markets include genetically engineered crops, gene therapy, cell therapy, immune therapy, generation of animal models for medical research and drugs etc. Recently, various private companies have been involved in financing Cas9-based genetic engineering research. The current market for CRISPR-based therapeutics is estimated to be approximately US$25 million, of which the sales of Caribou Biosciences are estimated at US$2.9 million, Novartis US$2 million, Recombinetics Inc. US$5 million and Intellia Therapeutics US$15 million. The market size is soon expected to exceed US$600 million (van Erp *et al.*, 2015).

The pace of growth is notable. The first patent for CRISPR technology in eukaryotic cells was awarded on 15 April 2014 (Tables 15.1, 15.2). The research tools market sector is now also taking up CRISPR-based technology. Sigma-Aldrich has licensed CRISPR technology for generation of modified plant and animal models, cell line and

Table 15.1. Recent patents employing the CRISPR/Cas9 system for genome editing.

Current areas of focus		Intellectual property	Company
Research and development	Research tools	US14/216 655	System Biosciences
		Broad License	GE Healthcare
		Cellectis Sublicense	Thermo Fisher
	Expression system	PCT/US2013/073307	Sigma-Aldrich
	Animals model	Broad License	Taconic
		Caribou, Broad License	Sage
Pharmaceuticals	Therapeutics	PCT/US2013/032589	Crispr Therapeutics
		Caribou License Broad, Duke, MGH Licenses	Intellia
			Editas
	Pharmaceuticals	Caribou License	Novartis
	In vitro applications only	Boston Children's Hosp., Institut Pasteur License	Cellectis
	Target validation	Open Innovation Model	AstraZeneca
	Monogenic diseases	PCT/US2013/032381 PCT/US2013/039979 PCT/US2013/028348	Sangamo Biosciences
Agriculture and dairy	Crops	Boston Children's Hosp.,	Cellectis Plant Sciences
		Institut Pasteur License	Dow Agrosciences
		PCT/US2013/039979 co-owned with Sangamo Biosciences	
	Livestock	PCT/US2014/0201857	Recombinetics
	Yogurt, cheese	7 919 277; 8 361 725; 13/722 539; 11/990 885	Danisco (DuPont)

Table 15.2. Recent patents employing the CRISP/Cas9 system (prepared 10 December 2015 using Google Patent search: https://patents.google.com/?q=crisp+cas9).

Inventor/original assignee	Publication no.	Application	Publication date
Feng Zhang The Broad Institute, Inc., Massachusetts Institute of Technology	US8697359 B1	CRISPR/Cas systems and methods for altering expression of gene products	15 April 2014
Feng Zhang The Broad Institute, Inc., Massachusetts Institute of Technology	WO2014204726 A9	Delivery and use of the CRISPR/Cas systems, vectors and compositions for hepatic targeting and therapy	5 March 2015
Daniel F. Voytas, Paul Atkins, Nicholas J. Baltes Regents of the University of Minnesota	US20140273235 A1 WO2014144155 A1	Engineering Plant Genomes Using CRISPR/Cas Systems	18 September 2014
Blake Wiedenheft Montana State University	WO2014093479 A1	CRISPR (clustered regularly interspaced short palindromic repeats) RNA-guided control of gene regulation	19 June 2014
Feng Zhang, Neville Espi Sanjana, Ophir Shalem	WO2014093701 A1 US20140357530 A1	Functional genomics using CRISPR/Cas systems, compositions, methods, knock out libraries and applications thereof	19 June 2014
Fuqiang Chen, Gregory Davis Sigma-Aldrich Co., Llc	US20140273233 A1	CRISPR-based genome modification and regulation	18 September 2014
Fangting Wu System Biosciences, Llc	US20140273226 A1	CRISPR/Cas systems for genomic modification and gene modulation	18 September 2014

genetic screening. Financial activity in these sectors is possibly sharpened by early-stage market development and, given the possibility of current applications of CRISPR across multiple industries, there is limitless scope for research and financial promises in this space. Pharmaceuticals giants are also licensing CRISPR/Cas9 intellectual property (IP) for commercial interest. Novartis has united with Atlas Ventures to assign $15 million to initiate Intellia Therapeutics, and Pfizer and Cellectis SA will produce T-cells with chimeric antigen receptors using CRISPR/Cas9. In January 2015, AstraZeneca declared partnerships with academia for use of Cas9 to authorize new drug targets.

15.4 Raising Concern Over CRISPR-based Genome Editing

An effective gene editing technology is the greatest distinct advantage to hit science since PCR. Be that as it may, with its immense potential come pressing concerns. CRISPR is bringing about a noteworthy change in biomedical and agriculture research. Unlike other gene editing technologies, CRISPR is fast, cheap and simple to utilize, and it has been cleared through labs around the globe as an outcome. Scientists now tend to utilize CRISPR technology in various fields such as human gene modification in response to elimination of genetic disorders and production of superior plant varieties. CRISPR is now regarded as the greatest discovery in molecular biology since PCR. Like PCR, gene amplification technology which has hugely influenced the biological science after its development in 1985, CRISPR is influencing the life sciences in such a large number of ways.

Yet, in spite of the fact that CRISPR has much to offer, the scientific community is concerned over the quick expansion of this technology leaving little time to deal with ethical and biosafety parameters. This issue was pushed into the spotlight in April 2015 when news broke that researchers have employed CRISPR to edit human embryo *in vivo*. The used embryo was not able to result into a viable birth, yet the report has produced heated verbal confrontation about whether and how CRISPR ought to be used to roll out heritable changes in the human genome. There are other concerns about the possibly hazardous genome edits and altered life forms which could potentially disturb the entire ecosystem.

Certain biomedical challenges need to be focused on before using this technology in human embryos because insertion and deletion of specific genes in the genome of a developing embryo is a complicated process. A given gene may influence different characteristics and may lead to unexpected consequences (MacKellar and Bechtel, 2014). Changes would happen in all succeeding generations resulting from the procedure, thus representing a germline modification.

Another ethical intricacy to the gene editing of a human embryo is that it is dependent on the stage at which change occurs. If a change happens in gametes prior to the fertilization as in the formation of pronuclear zygotes, a new individual is brought into life that would not have otherwise existed. Similarly, if change occurs after fertilization, new concerns may arise. For instance, if a particular gene from a two-cell embryo was edited, then the issue would be that the genetic changes would bring about a new individual while the original embryo would no longer exist, or whether it would be possible to consider that the original embryo continues to exist and is simply altered (Chapman and Frankel, 2003).

Target specificity of CRISPR/Cas9 nuclease-based genetic editing is a major concern. Studies have demonstrated the potential off-site targeting through mismatch between guide RNA and complementary target sequence. Chromatin immunoprecipitation sequencing analysis has been applied in evaluation of Cas9 cutting specificity and even more impartial and unbiased experiments are urgently needed to explain the Cas9 binding and cleavage specificity.

15.5 Patent Dispute Over Rights

The application of CRISPR-Cas9 in various fields has led to two challenging patent claims on this novel gene-editing technology. Patents for CRISPR/Cas9 are the matter of a disagreement between Broad Institute of the Massachusetts Institute of Technology (MIT) and Harvard, Massachusetts, USA and scientists at the University of California (UC), Berkeley, USA. The first claim came from Jennifer Doudna (UC Berkeley) and Emmanuelle Charpentier, Helmholtz Centre for Infection Research in Germany, for using CRISPR/Cas9 technology for genome editing *in vitro*. They applied for patent on 25 May 2012, which included 155 claims covering various uses of the technology in cell types (Jinek, 2013). The second claim came from Feng Zhang, Broad Institute of MIT and Harvard, Massachusetts,

USA on the method for using CRISPR/Cas9 for genome editing in eukaryotic cells. Zhang applied for patent on 12 December 2012 (Zhang *et al.*, 2014). Several other CRISPR/Cas9 patents cover various applications of this technology including Huntington's disease treatment and yogurt production; the first CRISPR patent of Zhang represents the technology for a wide range of applications. However, the US Patent and Trademark Office (USPTO) verdict on patent claim was not received by UC Berkeley and hence they had asked the USPTO to reconsider the decision. The Doudna/Charpentier claim would only be awarded with major revisions and it would limit its scope, whereas Zhang's patent has already reached the books. On the other hand the Doudna/Charpentier claim which is still under assessment could be awarded and would nullify some of Zhang's claims, upsetting his existing benefits. It could take several years to fully settle the CRISPR IP dispute, however, scientists and companies are utilizing various tactics to make use of this gene-editing technology.

15.6 Conclusion

Under 'first to file' rules, whoever applied first for the patent right gets the right, which means an easy going for Doudna and Charpentier because they applied 7 months before Zhang. However, due to the date of discoveries the case is being accepted under 'first to invent' rules where one has to show that he was first to make an invention work. Potential application across many fields, including biotechnology, biological science, agriculture and medicine has led to the emergence of a billion dollar worldwide industry. However, a potential patent dispute over rights to CRISPR between two parties has left the scientific community pondering its outcome.

Acknowledgements

KB is grateful to BHU for financial support. BKS and CK are thankful to ICAR-AMAAS for providing financial assistance (Project No. P-27/131).

References

Bassett, A.R., Tibbit, C., Ponting, C.P. and Liu, J.L. (2013) Highly efficient targeted mutagenesis of *Drosophila* with the CRISPR/Cas9 system. *Cell Reports* 4, 220–228.

Belhaj, K., Chaparro-Garcia, A., Kamoun, S. and Nekrasov, V. (2013) Plant genome editing made easy: targeted mutagenesis in model and crop plants using the CRISPR/Cas system. *Plant Methods* 9, 39.

Blitz, I.L., Biesinger, J., Xie, X. and Cho, K.W.Y. (2013) Biallelic genome modification in F_0 *Xenopus tropicalis* embryos using the CRISPR/Cas system. *Genesis* 51, 827–834.

Brooks, C., Nekrasov, V., Lippman, Z.B. and Van Eck, J. (2014) Efficient gene editing in tomato in the first generation using the clustered regularly interspaced short palindromic repeats/CRISPR-associated 9 system. *Plant Physiology* 166, 1292–1297.

Chang, N., Sun, C., Gao, L., Zhu, D., Xu, X., Zhu, X., Xiong, J.W. and Jeff, J. (2013) Genome editing with RNA-guided Cas9 nuclease in zebrafish embryos. *Cell Research* 23, 465–472.

Chapman, A.R. and Frankel, M.S. (eds) (2003) *Designing our Descendants: the promises and perils of genetic modifications*. JHU Press, Baltimore, Maryland.

Cheng, A.W., Wang, H., Yang, H., Shi, L., Katz, Y., Theunissen, T.W., Rangarajan, S., Shivalila, C.S., Dadon, D.B. and Jaenisch, R. (2013) Multiplexed activation of endogenous genes by CRISPR-on, an RNA-guided transcriptional activator system. *Cell Research* 23, 1163–1171.

Ebina, H., Misawa, N., Kanemura, Y. and Koyanagi, Y. (2013) Harnessing the CRISPR/Cas9 system to disrupt latent HIV-1 provirus. *Scientific Reports* 3, 1–7.

Fauser, F., Schiml, S. and Puchta, H. (2014) Both CRISPR/Cas-based nucleases and nickases can be used efficiently for genome engineering in *Arabidopsis thaliana*. *The Plant Journal* 79, 348–359.

Feng, Z., Zhang, B., Ding, W., Liu, X., Yang, D. L., Wei, P., Cao, F., Zhu, S., Zhang, F., Mao, Y. and Zhu, J.K. (2013) Efficient genome editing in plants using a CRISPR/Cas system. *Cell Research* 23, 1229.

Feng, Z., Mao, Y., Xu, N., Zhang, B., Wei, P., Yang, D.L., Wang, Z., Zhang, Z., Zheng, R., Yang, L., Zeng, L., Liu, X. and Zhu, J.K. (2014) Multigeneration analysis reveals the inheritance, specificity, and patterns of CRISPR/Cas-induced gene modifications in Arabidopsis. *Proceedings of the National Academy of Sciences USA* 111, 4632–4637.

Flowers, G.P., Timberlake, A.T., McLean, K.C., Monaghan, J.R. and Crews, C.M. (2014) Highly efficient targeted mutagenesis in axolotl using Cas9 RNA-guided nuclease. *Development* 141, 2165–2171.

Friedland, A.E., Tzur, Y.B., Esvelt, K.M., Colaiácovo, M.P., Church, G.M. and Calarco, J.A. (2013) Heritable genome editing in *C. elegans via* a CRISPR-Cas9 system. *Natural Methods* 10, 741–743.

Gao, J., Wang, G., Ma, S., Xie, X., Wu, X., Zhang, X., Wu, Y., Zhao, P. and Xia, Q. (2015) CRISPR/Cas9-mediated targeted mutagenesis in *Nicotiana tabacum*. *Plant Molecular Biology* 87, 99–110.

Gratz, S.J., Cummings, A.M., Nguyen, J.N., Hamm, D.C., Donohue, L.K., Harrison, M.M., Wildonger, J. and O'Connor-Giles, K.M. (2013) Genome engineering of *Drosophila* with the CRISPR RNA-guided Cas9 nuclease. *Genetics* 194, 1029–1035.

Hu, W., Kaminski, R., Yang, F., Zhang, Y., Cosentino, L., Li, F., Luo, B., Alvarez-Carbonell, D., Garcia-Mesa, Y., Karn, J., Mo, X. and Khalili, K. (2014) RNA-directed gene editing specifically eradicates latent and prevents new HIV-1 infection. *Proceedings of National Academy of Sciences USA* 111, 11461–11466.

Hu, X., Chang, N., Wang, X., Zhou, F., Zhou, X., Zhu, X. and Xiong, J.W. (2013) Heritable gene-targeting with gRNA/Cas9 in rats. *Cell Research* 23, 1322–1325.

Hwang, W.Y., Fu, Y., Reyon, D., Maeder, M.L., Tsai, S.Q., Sander, J.D., Peterson, R.T., Yeh, J.R.J. and Joung, J.K. (2013) Efficient genome editing in zebrafish using a CRISPR-Cas system. *Nature Biotechnology* 31, 227–229.

Jia, H. and Wang, N. (2014) Targeted genome editing of sweet orange using Cas9/sgRNA. *PLoS One* 9, e93806.

Jiang, W., Zhou, H., Bi, H., Fromm, M., Yang, B. and Weeks, D.P. (2013) Demonstration of CRISPR/Cas9/sgRNA-mediated targeted gene modification in Arabidopsis, tobacco, sorghum and rice. *Nucleic Acids Research*, doi: 10.1093/nar/gkt780.

Jiang, W., Yang, B. and Weeks, D.P. (2014) Efficient CRISPR/Cas9-mediated gene editing in *Arabidopsis thaliana* and inheritance of modified genes in the T2 and T3 generations. *PLoS One*, e99225.

Jinek, M., East, A., Cheng, A., Lin, S., Ma, E. and Doudna, J. (2013) RNA-programmed genome editing in human cells. *Elife* 2, e00471.

Lanphier, E., Urnov, F., Haecker, S.E., Werner, M. and Smolenski, J. (2015) Don't edit the human germ line. *Nature* 519(7544), 410.

Li, J.F., Norville, J.E., Aach, J., McCormack, M., Zhang, D., Bush, J., Church, G.M. and Sheen, J. (2013) Multiplex and homologous recombination-mediated genome editing in Arabidopsis and *Nicotiana benthamiana* using guide RNA and Cas9. *Nature Biotechnology* 31, 688–691.

Liang, Z., Zhang, K., Chen, K. and Gao, C. (2014) Targeted mutagenesis in *Zea mays* using TALENs and the CRISPR/Cas system. *Journal of Genetics and Genomics* 41, 63–68.

MacKellar, C. and Bechtel, C. (eds) (2014) *The Ethics of the New Eugenics*. Berghahn Books, Oxford, UK.

Maddalo, D., Manchado, E., Concepcion, C.P., Bonetti, C., Vidigal, J.A., Han, Y.C., Ogrodowski, P., Crippa, A., Rekhtman, N., de Stanchina, E., Lowe, S.W. and Ventura, A. (2014) *In vivo* engineering of oncogenic chromosomal rearrangements with the CRISPR/Cas9 system. *Nature* 516, 423–427.

Miao, J., Guo, D., Zhang, J., Huang, Q., Qin, G., Zhang, X., Wan, J., Gu, H. and Qu, L.J. (2013) Targeted mutagenesis in rice using CRISPR-Cas system. *Cell Research* 23, 1233.

Nakayama, T., Fish, M.B., Fisher, M., Oomen-Hajagos, J., Thomsen, G.H. and Grainger, R.M. (2013) Simple and efficient CRISPR/Cas9 mediated targeted mutagenesis in *Xenopus tropicalis*. *Genesis* 51, 835–843.

Nekrasov, V., Staskawicz, B., Weigel, D., Jones, J.D. and Kamoun, S. (2013) Targeted mutagenesis in the model plant *Nicotiana benthamiana* using Cas9 RNA-guided endonuclease. *Nature Biotechnology* 31, 691–693.

Niu, Y., Shen, B., Cui, Y., Chen, Y., Wang, J., Wang, L., Kang, Y., Zhao, X., Si, W., Li, W., Xiang, A.P., Zhou, J., *et al*. (2014) Generation of gene-modified cynomolgus monkey via Cas9/RNA-mediated gene targeting in one-cell embryos. *Cell* 156, 836–843.

Platt, R.J., Chen, S., Zhou, Y., Yim, M.J., Swiech, L., Kempton, H.R., James, E. Dahlman, E.J., Parnas, O., Eisenhaure, T.M., Jovanovic, M., Graham, D.B., *et al*. (2014) CRISPR-Cas9 knockin mice for genome editing and cancer modeling. *Cell* 159, 440–455.

Ron, M., Kajala, K., Pauluzzi, G., Wang, D., Reynoso, M.A., Zumstein, K., Garcha, J., Winte, S., Masson, H., Inagaki, S., Federici, F., Sinha, N., Deal, R.B., Bailey-Serres, J. and Brady, S.M. (2014) Hairy root transformation using *Agrobacterium* rhizogenes as a tool for exploring cell type-specific gene expression and function using tomato as a model. *Plant Physiology* 166, 455–469.

Sánchez-Rivera, F.J., Papagiannakopoulos, T., Romero, R., Tammela, T., Bauer, M.R., Bhutkar, A., Joshi, N.S., Subbaraj, L., Bronson, R.T., Xue, W. and Jacks, T. (2014) Rapid modeling of cooperating genetic events in cancer through somatic genome editing. *Nature* 516, 428–431.

Schwank, G., Koo, B.K., Sasselli, V., Dekkers, J.F., Heo, I., Demircan, T., Sasaki, N., Boymans, S., Edwin Cuppen, E., van der Ent, C.K., Nieuwenhuis, E.E.S., Beekman, J.M. and Clevers, H. (2013) Functional repair of CFTR by CRISPR/Cas9 in intestinal stem cell organoids of cystic fibrosis patients. *Cell Stem Cell* 13, 653–658.

Shan, Q., Wang, Y., Li, J., Zhang, Y., Chen, K., Liang, Z., Zhang, K., Liu, J., Xi, J.J., Qiu, J.L. and Gao, C. (2013) Targeted genome modification of crop plants using a CRISPR-Cas system. *Nature Biotechnology* 31, 686–688.

Upadhyay, S.K., Kumar, J., Alok, A. and Tuli, R. (2013) RNA-guided genome editing for target gene mutations in wheat. *G3: Genes–Genomes–Genetics* 3, 2233–2238.

van Erp, P.B., Bloomer, G., Wilkinson, R. and Wiedenheft, B. (2015) The history and market impact of CRISPR RNA-guided nucleases. *Current Opinion in Virology* 12, 85–90.

Wang, H., Yang, H., Shivalila, C.S., Dawlaty, M.M., Cheng, A.W., Zhang, F. and Jaenisch, R. (2013)

One-step generation of mice carrying mutations in multiple genes by CRISPR/Cas-mediated genome engineering. *Cell* 153, 910–918.

Wang, Y., Cheng, X., Shan, Q., Zhang, Y., Liu, J., Gao, C. and Qiu, J.L. (2014) Simultaneous editing of three homoeoalleles in hexaploid bread wheat confers heritable resistance to powdery mildew. *Nature Biotechnology* 32, 947–951.

Whitworth, K.M., Lee, K., Benne, J.A., Beaton, B.P., Spate, L.D., Murphy, S.L., Samuel, M.S., Mao, J., O'Gorman, C., Walters, E.M., Murphy, C.N., Driver, J.P., Mileham, A., McLaren, D., Wells, K.D. and Prather, R.S. (2014) Use of the CRISPR/Cas9 system to produce genetically engineered pigs from *in vitro*-derived oocytes and embryos. *Biology of Reproduction* 78, 1–13.

Wu, Y., Zhou, H., Fan, X., Zhang, Y., Zhang, M., Wang Y., Xie, Z., Bai, M., Qi, Y., Liang, D., Tang, W., Liao, J., Zhou, C., Liu, W., Zhu, P., Guo, H., Pan, H., Wu, C., Shi, H., Wu, L., Tang, F. and Jinsong, L. (2015) Correction of a genetic disease by CRISPR-Cas9-mediated gene editing in mouse spermatogonial stem cells. *Cell Research* 25, 67–79.

Xie, K. and Yang, Y. (2013) RNA-guided genome editing in plants using a CRISPR–Cas system. *Molecular Plant* 6, 1975–1983.

Xu, R., Li, H., Qin, R., Wang, L., Li, L., Wei, P. and Yang, J. (2014) Gene targeting using the *Agrobacterium tumefaciens*-mediated CRISPR-Cas system in rice. *Rice* 7, 5.

Xue, W., Chen, S., Yin, H., Tammela, T., Papagiannakopoulos, T., Joshi, N.S., Cai, W., Yang, G., Bronson, R., Crowley, D.G., Zhang, F., Anderson, D.G., *et al*. (2014) CRISPR-mediated direct mutation of cancer genes in the mouse liver. *Nature* 514, 380–384.

Yang, H., Wang, H., Shivalila, C.S., Cheng, A.W., Shi, L. and Jaenisch, R. (2013) One-step generation of mice carrying reporter and conditional alleles by CRISPR/Cas-mediated genome engineering. *Cell* 154, 1370–1379.

Zhang, H., Zhang, J., Wei, P., Zhang, B., Gou, F., Feng, Z., Mao, Y., Yang, L., Zhang, H., Xu, N. and Zhu, J.K. (2014) The CRISPR/Cas9 system produces specific and homozygous targeted gene editing in rice in one generation. *Plant Biotechnology Journal* 12, 797–807.

Zhou, H., Liu, B., Weeks, D.P., Spalding, M.H. and Yang, B. (2014) Large chromosomal deletions and heritable small genetic changes induced by CRISPR/Cas9 in rice. *Nucleic Acids Research* doi: 10.1093/nar/gku806.

16 Healthcare Innovation, Personalization and the Patent System: Where is the Public Interest?

Graham Dutfield*

School of Law, University of Leeds, Leeds, UK

16.1 Introduction: Current Trends in Healthcare Innovation, and the Personalization Turn

The relationships between health, innovation and the law are dynamic and arguably changing faster now than ever before. To understand what is going on and why things are happening now, we need to take account of five major drivers of change. These drivers are not all new but their prominence has become more noticeable since the beginning of the modern health biotechnology era in the early 1980s, a time when the pharmaceutical industry was in the midst of a slump with fewer genuinely original drug products entering the market than in the 'Golden Age' of the 1960s and 1970s. These are internationalization, complexification, personalization, digitalization and fragmentation.

By internationalization, I mean a number of things. For one, emerging economies such as China and recently developed ones like Singapore and South Korea are either producing health innovations of the kind they were not doing earlier, or are making large investments in the life sciences which will bear fruit sooner or later. This expansion in the geography of expertise and investment has entailed an increase in international research and commercial collaborations. It has also vastly increased the interest of pharmaceutical companies in the intellectual property (IP) rules of other countries, especially those that are markets, or potential markets, for their products – or which are large producers and exporters of generic drugs like India.

Second, the disease profiles of many developing countries are become more similar to those of the developed countries. This is not just because wealthier people anywhere in the world live longer than the poor, and diseases that tend to affect the elderly most are broadly similar everywhere: non-communicable ones such as cancer, cardio-vascular diseases, Alzheimer's, etc. It is also because there is a certain amount of uniformity in the lifestyles including diets of people having similar levels of income and purchasing power *wherever they happen to live*. This means diseases affecting people are become more and more the same wherever they live. This prospect for industry of the global similarity of disease profiles and life expectancies with their attendant market and economies of scale opportunities appears to be quite new.

There has also been an internationalization of IP rules relating to medicines. Since the mid-1990s, most countries of the world have been held to a somewhat uniform set of IP standards prescribed by the World Trade Organization's TRIPS Agreement (Agreement on Trade-related Aspects of Intellectual Property Rights). Previously, the countries of the world were pretty much free to provide IP protection for medicines as they desired: a lot of protection,

*Corresponding author e-mail: g.m.dutfield@leeds.ac.uk

 Intellectual Property Issues in Biotechnology (Eds H.B. Singh, A. Jha and C. Keswani)

a little, or none. This is no longer the case. This has led to criticism that access to medicines, especially in the developing countries, has been hindered by these uniform standards, which are largely transplanted from the developed countries into the developing ones *albeit typically without the public health safeguards that balance out the otherwise strong protection.*

Complexity explains personalization, digitalization and fragmentation. It also has great implications for how businesses acquire and manage intellectual property rights (IPRs). A trend towards further complexity is evident in a number of ways, beyond the adage that the more one knows about something, the more one realizes how much more there is out there that we do not or cannot know – true as that is. For one thing, understanding at the molecular level of specific malfunctions which lead to disease has increased dramatically, for example in cancer. But this does not lead automatically to cures. Indeed, insights into disease causation are the start and not the end of a journey that may be long indeed. The likelihood is that it will engage the expertise of many individuals from numerous disciplinary fields not all of whom are likely to be working in the same laboratory, company or place of employment. There is still room for individual brilliance but teamwork (or better said teams work because there may be many teams involved) is all. Innovation has always been primarily collective, even if certain individuals may stand out for making decisive contributions, but it is especially so in the life sciences of the present day.

One interesting aspect of our enhanced understanding of how cellular malfunctions lead to disease is that some disease categories are fragmenting so that out of what was once considered a single disease arise several or many diseases. And out of one patient population sharing the same disease and being given the same treatment there may become several patient populations being treated for different, albeit closely related, diseases; or even patient subpopulations forming a larger population all having the same diseases but who need to be given either a completely different treatment, or a different dose or formulation of the same one.

This latter observation leads to another form of complexification: the differences between people having the same disease in terms of how they respond to the same treatment. Human bodies are complex enough without looking into individual human variation. Investigating the latter alone has cost vast amounts of money and generated staggering amounts of data out of which bioinformaticians and others strive to glean patterns, connections and insights that can contribute to better healthcare. But we have a long way to go. It is well known that a powerful drug in a particular dosage can cure one person with minor or no side-effects, whereas the same dosage may not work on another person, cause harmful side-effects, or even endanger a patient's life. Being able to accurately predict how an individual will respond to a drug treatment and what the optimal mode of delivery should be for that person is very important, but it requires further research including the gathering of data large enough in volume and breadth of coverage to generate findings enabling precise and individualized treatment including the discovery of biomarkers. Personalized medicine 'refers to the tailoring of medical treatment to the individual characteristics of each patient; to classify individuals into subpopulations that differ in their susceptibility to a particular disease or their response to a specific treatment so that preventive or therapeutic interventions can then be concentrated on those who will benefit, sparing expense and side-effects for those who will not' (President's Council of Advisors on Science and Technology, 2008).

The word 'biomarker' is increasingly part of the discourse of modern medical research including personalized medicine, denoting 'a broad subcategory of medical signs – that is, objective indications of medical state observed from outside the patient – which can be measured accurately and reproducibly' (Strimbu and Tavel, 2010). It is a somewhat broad term. Traditional physiological biomarkers include blood pressure and pulse rate, but nowadays the most valuable ones to discover (and patent) are molecular, including proteins, metabolites and gene sequences. They are hugely significant:

> Biomarkers and molecular individualized medicine are replacing the traditional 'one size fits all' medicine.... The essence of personalized oncology lies in the use of biomarkers.... Three different types of biomarkers are of particular importance: predictive, prognostic and early response ... Molecular diagnostics identify individual cancer patients who are more likely to respond positively to targeted chemotherapies. Molecular diagnostics include testing for genes, gene expression, proteins and metabolites.
>
> Kalia (2013)

In this context it is worth noting that some 4000 diseases are associated with single-gene defects.

These include cystic fibrosis and sickle-cell disorder. Additionally, numerous health disorders have polygenic (involving more than one gene) and multifactorial causes in which genes, lifestyle and environment are involved. These include diabetes, heart disease, cancers and hypertension. Molecular data analysis can thus provide not only a massive volume of health-related information, but also a diversity in the kinds of information we can derive, from responsiveness to drugs, to likelihood of contracting a particular disease, and ways to prevent or reduce risk of certain diseases later in life. No doubt both volume and variety stand to increase tremendously in the coming years.

Due to personalization and to the sheer amount of molecular data being generated, modern day healthcare is becoming increasingly information intensive, including at the personal level. As a prominent medical scientist in the USA was quoted as saying:

> The patient is an enormous repository of information that needs to be harvested as a partnership not only in clinical care but in discovery... The ability to stratify the phenotypic expression of wellness and disease will ultimately lead to better validation of human therapeutic targets for drug discovery.
>
> Elenco *et al.* (2015)

Much of this data is of course collected, stored and analysed in digital form. The role of the drug is still quite central to a great deal of patient care. But prescribing a drug with a generalized instruction as to how much to take and when, except for minor ailments like colds, is becoming insufficient. It is not just healthcare in the broad sense that is moving on to computer screens; medicine is becoming digital as much as it is chemical, especially when treatment concerns itself more and more with disease prediction, diagnosis, prognosis, and monitoring of sickness, health and treatment effects and side-effects, and of course with personalization. Personalized medicine is frequently held to be a very promising development that could dramatically enhance the health of millions of people. It seems fair to mention, however, that some doubt that the anticipated transformative effects of personalized medicine on healthcare will all come to pass, and that there is a certain amount of hype (Joyner and Paneth, 2015). Of course, not all of this turn to big data is convenient for industry. There is a genuine possibility, for example, that it can reveal links between a commercially successful drug and health risks for some people that would otherwise be very hard to detect. This could result in shrinkage of the product's patient base, and possibly even the withdrawal of the product from the market.

A third area of complexification is in relation to the drug itself. Until quite recently drugs were almost by definition small-molecule products. Historically, these were typically isolated plant metabolites or microbial products, or synthetic versions, perhaps with modifications, made in a lab. Others were synthetic without being modelled on ones found in nature. There were also animal products such as hormones. In recent decades there has been a big growth in biopharmaceuticals (or biologics), often therapeutic proteins like insulin, erythropoietin and various monoclonal antibodies (MAbs). These cannot be synthesized and are produced in living organisms or cells. With the early ones, bacteria could be used. MAbs are very large complex chemicals and can only be made in animal cells. The processes used to make them are immensely challenging. To accurately mass-produce these large-molecule products with a minimum of variability within and between batches and without impurities is technically difficult and expensive. As for making copies of such drugs, regulators have struggled to find a way to approve generic versions, which can never be considered to be exact reproductions as is possible with the active ingredients of small-molecule drugs. So called biosimilars have now been approved in some countries, but are comparatively small in number compared to small-molecule drugs even when we take into account that the first biologics only became available in the 1980s (Smith Hughes, 2011; Hoffman and Furcht, 2014; Rasmussen, 2014).

The driver we leave to last is that of fragmentation. But this is not to understate its importance for IP. Fragmentation follows from the point made earlier that the discovery and development of drugs and other health products such as vaccines and diagnostic tests increasingly requires the expertise of numerous individual scientists and teams having quite diverse disciplinary backgrounds that are unlikely to be found under the same roof. Indeed, they may be spread across countries and continents and work in companies, universities, hospitals, government health research institutes and health charities. Apart from these entities, organized patient groups are sometimes actively engaged in the search for new treatments as well. These are likely to have different missions, interests and ways of getting their work funded. And increasingly, the private-sector side contains not just the traditional giants ('Big Pharma') but highly specialized small firms, often spun out of universities who have useful services and possibly quite unique skills to offer, and perhaps a few highly valuable patents, but who have no products at all to sell. Building networks that satisfy all parties is no easy task. IPRs can help or get in the way, depending on who acquires them and how they are managed. The aggressive assertion of rights can be highly divisive and counter-productive in terms of stimulating innovation and ensuring that products are made available to all of those who need them and under fair terms and conditions. This in turn raises questions about the imperative to maximize taxpayer value (as opposed just to shareholder value) where much of the work is government supported, and the availability of public interest safeguards in patent law.

There is much talk in this context nowadays about 'open innovation'. According to proponents of the concept, business sectors whose firms typically tried to do everything in-house, and were thus practising 'closed innovation', are now collaborating with others, sharing each other's knowledge and materials in the hope that by widening participation in this way, the speed of innovation will increase. The life science industries appear to be embracing the concept. One must be cautious about all this. First, to suggest that pharmaceutical innovation was closed in the past but is finally now opening up is inaccurate. There is no closed versus open dichotomy, at least in this industry. University scientists and drug companies have collaborated as far back as the 19th century, and at no time was this ever abnormal, though it has often been controversial. Furthermore, one must be careful not to equate the 'open' in open innovation with absence of IP. There is no reduction in patent filing. None the less, it is true that the industry is becoming more pro-active and strategic about research networking and knowledge management. Whether 'open innovation' is an accurate label for the various practices they are adopting to further such collaborating is a matter for debate.

The discussion so far in this chapter suggests that medical science innovation is multifaceted and diverse in terms of what is being done, who is doing it and where, and it is constantly evolving. As for patenting, the subject matter of what is nowadays being claimed is hugely challenging and not necessarily easy to accommodate within the traditional product, process and use claim categories, nor to justify in view of the function of the patent system to promote invention in the public interest.

Several years ago while Steve Jobs was not working at Apple, he was interviewed and said he did not begrudge Bill Gates his billions – he just wished his company did not make such mediocre products. I would hazard a guess that while a lot of people are reasonably satisfied with Microsoft's products, many people would have agreed. One could say the same about the pharmaceutical industry. For every Glivec, which is a highly effective precision medicine, there are numerous products that do little for the majority of people who take them. According to a recent study not one of the ten best-selling drugs in the USA helps the majority of patients who are given them. In fact, they benefit only 'between 1 in 25 and 1 in 4 of the people who take them. For some drugs, such as statins – routinely used to lower cholesterol – as few as 1 in 50 may benefit' (Schork, 2015). That is really pretty staggering. The industry is raking in the cash for pretty useless products.

But is this really a low quality issue? Or is it more a problem of insufficient information? Here such ignorance is of two kinds. First, that of not knowing anything relevant about the individual patient that would enable the tailoring of the treatment to the person. The second is that of not knowing enough about the 'disease' for which the treatment is intended. A single disease may in fact be better subdivided into a series of related diseases each of which requires different therapeutic responses whether in the form of radically different drugs or alternative dosages or formulations of the same ones. If this is true, is the off-the-peg style of drug development now outmoded and ready for replacement by made-to-measure drugs? And if so, can drugs for the same disease still be cut from the same cloth or is it going to be a case of diverse treatment for different people having the same disease? As we will see in this chapter, the questions go beyond these rather simplistic ones. Note that I am not just talking about the science. How can firms make money from treatments that will be given to only some patients out of a much bigger population having a particular disease? And stratifying and personalizing patients by drug responsiveness is not the end of it. It is no good giving the right sort of drug in the right amount to the right sort of patient if it is for the wrong sort of the disease. This is a serious issue in cancer, which researchers over the years have been splitting into ever more cancers. Again, how can businesses make money from patients having a highly specific disease affecting very few people that is part of a 'family' of diseases previously thought to be one single disease?

Before going further it is important to place personalized medicine in context. Personalized medicine is in fact one aspect of the efforts currently being made by biomedical scientists and industry to enhance targeting of disease to achieve better health outcomes for more people. Personalized medicine deals with the tailoring of treatments in a way that responds to the variability of human beings, and to the fact that single diseases may really be families of sub-diseases. It implies individualization of medical attention, but whereas it does involve the testing of people for certain biomarkers conveying diagnostic or therapy-related information, those biomarkers are typically ones shared with other people. Thus some prefer the term stratified medicine, which makes it plainer that the focus is on those shared traits that are meaningful insofar as medical intervention goes. Precision medicine has emerged in the USA as another way of saying stratified medicine. Of course, the word 'precision' seems to imply also the idea of the 'magic bullet', the smart drug that forcefully hits the right target head on and without collateral damage. It is in this sense that Glivec is a prime example of a precision drug.

16.2 Patents and Commercial Biomedicine

What do pharmaceutical companies patent? Nowadays, it is increasingly simplistic if not erroneous to think merely in terms of product and process claims. What types of product and process claims are we talking about? What about uses? Clearly there is much more to patenting than claiming the active ingredient of a drug and the means by which it is produced.

A treatment for disease *x* may turn out to be an excellent treatment for similar diseases *y* and *z*, or even for more distantly related (or unrelated) diseases *d* and *h*. This is why research continues to be done on old drugs, both successful ones and failed ones, that is, natural or synthetic chemicals that were discovered, produced and tested but which failed to work well against drug targets known about at the time. As better knowledge of cell biology, genomics and proteomics increases the number of drug targets, there is a lot of interest in testing known substances against them in the hope of discovering new therapeutic applications. Nowadays this is referred to as drug repurposing (Mullard, 2012; Cragg *et al.*, 2014). Patent systems in many jurisdictions encourage such research by allowing claims on newly discovered medical uses of old chemical substances. We will look into how such claims are accepted below.

Another phenomenon driving such 'drug repurposing' is personalized or stratified medicine, which offers the possibility to 'save' rejected chemicals found to have some therapeutic effects across the whole spectrum of individuals suffering from a particular disease but not enough to justify further development (Allison, 2008). On the face of it, it seems counter-intuitive that personalized medicine is commercially attractive for industry. If we keep on splitting diseases and stratifying and personalizing patients how can the industry make money? Until recently, the industry has made its biggest profits from blockbuster drugs that are prescribed to all (or most) patients having the disease or disorder for which the drug is intended. But if the disease is in fact several diseases and a drug works effectively only against one or two of these *and* not all sufferers can be given the treatment anyway, the size of the market in terms of the numbers taking the drug will be much smaller. That might suggest it would not be worth developing and selling the drug. In fact, the possibility still exists to make large profits, and the patent system can potentially offer some strong legal protection for reasons that will now be explained, starting with a discussion on the scope of what can be claimed.

16.3 Beyond Product, Process and Use: What Drug Companies Actually Claim and What Patent Law Permits

For all the reasons given earlier, the pharmaceutical industry has seen a need – and been given the opportunity – to claim a wide diversity of 'subjects' in their attempts to protect their investments and secure legal monopoly protection with as much comprehensiveness as the law allows.

Determining whether a product is new or not for the purpose of granting or refusing a patent would appear to be a fairly straightforward calculation. But this is far from being the case with medicines. Apart from the fact that the concept of novelty in patent law is in large part a matter of description, specialized knowledge and public availability as opposed to whether the invention had no prior existence in any absolute sense, chemistry does present complications other types of invention do not necessarily share. A coffee lid with a useful slide to unlock mechanism aiding the retention of heat and reducing the risk of the drinker getting scalded while dashing to catch the train is novel if no such device on a coffee lid was known to exist before. Much invention involves some kind of mimicry of nature, but we can be certain that nobody will dig a hole in the ground, dive over a coral reef or trek through a rainforest and stumble across a natural coffee lid. Only other coffee lids form the prior art and these are all manufactured items that are easily and publicly available. The identity and novelty of a chemical composition is relatively far less clear-cut and the patent system can apply novelty in a very strict way accommodating some variability or else it can be applied more narrowly. By being strict, the system might demand some additional novelty requirement so that a small or trivial variation is disqualified because it is essentially the same thing as the disclosed or prior art chemical it most closely relates to. This is done in India (see below). If narrow, the system will accommodate only very limited variability: similar but different is different. Take two substances A and B. There is some difference of significance between them notwithstanding that they vary by as little as one being merely purer than the other but otherwise identical, or as one being a racemic mixture comprising both left- and right-handed versions (enantiomers) of the molecule, and the other being one or other of the enantiomers. Either way, A and B can be patented as separate inventions in Europe and in the USA. Accordingly, slightly different things are treated as being something else entirely if some new difference of therapeutic or other significance is disclosed (whether or not it is proven). Minor modifications to existing medicaments such as enantiomers, combinations and minor variants of existing products may well be deemed patentable.

Generally, the system has become relaxed about novelty with respect to pharmaceuticals and this of course suits industry (albeit not the generics producers) as it was intended. Purified versions of existing substances can be patented on either side of the North Atlantic and in other jurisdictions too. There are old precedents for this in the USA[1] and in the UK where purified insulin was patented in the 1920s. Selection inventions are also possible. The door may also be open to the patenting of naturally occurring drug metabolites which are basically 'made' by the human body (see below).

There is of course a difference between the active pharmaceutical ingredient (API) and the drug product in that the tablet (if that is the mode of delivery) contains a mixture of the API and other non-active chemicals called excipients, which may perform certain functions such as protecting the API on its journey through the body, controlling its rate of absorption, or enabling more convenient modes of delivery. For example, it may be possible for the same API to be administered in liquid rather than solid form, or can be made to be chewable so it does not have to be swallowed whole. New dosage forms, such as a pill with a known API having a new coating offering some sort of advantage, genuinely therapeutic or otherwise, are patentable.

European patent law is generally accommodating towards the pharmaceutical industry, arguably excessively so. The industry is highly organized and spends large sums of money on promoting its interests. Notwithstanding its power and influence, there are certain exclusions in terms of methods claims. As we shall see, though, these can largely be circumvented by claiming new or additional uses. There are no such statutory exceptions in the USA. European Patent Convention (EPC) Article 53(c) excludes:

> (c) methods for treatment of the human or animal body by surgery or therapy and diagnostic methods practised on the human or animal body; this provision shall not apply to products, in particular substances or compositions, for use in any of these methods.

This is somewhat similar to the allowable exclusions of 'diagnostic, therapeutic and surgical methods for the treatment of humans or animals' in the TRIPS Agreement. The methods for treatments and diagnostic methods exclusions require some explanation in terms of their meaning, purpose and scope of application. In terms of its rationale, Article 53(c) seeks to immunize physicians, surgeons and veterinarians from patent infringement suits concerning their adoption of methods. As such it is grounded in ethics and public health concerns. This interpretation has been affirmed in a series of European Patent Office (EPO) cases.[2] However, the exclusion does not extend to drugs or equipment that may be under patent. As far as therapy goes, a line of EPO appeal board decisions has distinguished between methods intended to benefit the health of patients in a range of possible ways which fall within the exclusion, from treatments that are cosmetic in nature that do not. Cosmetic treatments are quite broad. EPO cases have found them to include methods or treatments relating to weight control, baldness and hair removal, snoring and contraception to name a few (Sterckx and Cockbain, 2012). Also falling outside the exclusion are treatments intended to kill humanely the recipient of the substance, for example euthanasia treatments for pets.[3]

With respect to surgery, EPO cases over the years have been less consistent. But surgery is now deemed to constitute physical intervention on the body involving the application of professional medical skill and which entails substantial health risk to the subject person or animal. Sterckx and Cockbain[4] helpfully sum up the situation by suggesting that 'interventions such as massage, tattooing, tanning, shaving, ear-piercing, blood-drawing, and routine injecting or catheterisation will not classify as excluded methods of surgery', whereas 'bone-setting, castration, embryo implantation and cosmetic surgery involving anaesthetics seem likely to be included' (Sterckx and Cockbain, 2012).

Diagnosis of necessity involves at least two steps including data gathering or comparison followed by the act of diagnosis itself. These are all primarily mental, thus non-technical, acts that are outside of normal practice done on the body. But if diagnosis is inherently non-technical, diagnoses consist of 'methods for performing mental acts' and are therefore not inventions anyway under EPC Article 52(2)(c). This would render the exclusion unnecessary. However, the Enlarged Board of Appeal decision in G0001/04 identified technical steps: those implying 'an interaction with the human or animal body, necessitating the presence of the latter.'[5] This seems like a questionable *ex post facto* justification for the existence of a redundant provision.

The European Patent Convention allows known chemicals to be patented for previously unknown uses in the above-mentioned unpatentable methods. Providing for this (under Article 54(4)) is really a special exemption to the novelty provision, which reflects the fact that much pharmaceutical discovery concerns extant substances, not just ones brought into existence by pharmaceutical chemists and designed with a specific application in mind. As we will see, this is not the only relaxation of the novelty criterion provided in the EPC in favour of the pharmaceutical industry.

The question arose of what to do about discoveries of new medical uses of substances that were already being used as medicines for something else; that is, second medical uses. Given the genuine possibility of drugs turning out to be useful for treating diseases other than the ones for which they were initially indicated, the industry obviously had an interest in the patent system protecting second medical indications (note: 'indication' in European patent law practice is more general in meaning and not confined to a specific disease or medical condition). Given that this is not product protection being sought, the obstacle was the wording of Article 53(c). The solution arrived at was devised by the Swiss Patent Office and became commonly known in consequence as the 'Swiss-type claim'. Accordingly, a second medical indication could side-step the methods exclusions under Article 53(c) as long as the following claim language was adopted: 'Use of a substance or composition X for the manufacture of a medicament for therapeutic application Z'. This explanatory formula for making claims was affirmed and construed broadly in a 1984 Enlarged Board of Appeal case.[6] However, as from 29 January 2011 such claims are no longer accepted for new patent applications. Instead, a more straightforward claim is to be made: 'Product X for use in the treatment of Z'. This is a consequence of a clause added to the revised 2000 version of the EPC. Article 54(5) now states that the novelty requirement:

> shall also not exclude the patentability of any substance or composition referred to in paragraph 4 for any specific use in any method referred to in Article 53(c), provided that such use is not comprised in the state of the art.

That second and follow-on use claims may be under patent after the initial patent on the drug itself has expired gives rise to the following rather curious scenario: a generic version of a drug can enter the market, subject to regulatory approval, for the original indication. But if marketed for the second use, its manufacture and sale would be patent infringing. This potentially raises difficulties for physicians and pharmacists, especially in jurisdictions in which doctors are encouraged to write prescriptions using International Non-proprietary Names (i.e. generic names) and where indications may not necessarily be entered on the prescription note. It also means that generic drug-makers must be careful about the indications they mention on the label.

Whereas many second use claims are targeted towards diseases and other health problems that were not previously indicated, there are further types of claim that may be allowed in some jurisdictions where the disease is actually the same as before. This is where the earlier discussion on personalized medicine becomes particularly relevant. Following the EISAI decision, claims may be allowed on further uses and methods for optimized or personalized dosage regimes, that is, for a specific schedule of doses such as take one 200 mg tablet every 6 h. This rather generous interpretation of European patent law concerning second and further use claims was reaffirmed by the TBA in *Genentech/Method of administration of IFG-1*.[7] As an aside, it is important to differentiate between the terms 'formulation' when applied to a specific drug mixture in whatever type it is prepared for delivery, and a 'dosage form'. The latter tends to refer to the physical form of the drug. Dosage forms include tablets, capsules, injectable solutions, ointments or powders to give some examples. However, the term is also used to apply to the chemical composition of the drug. Thus, there is a certain overlap in the usage of the two terms, which can be a little confusing. As we will see below, new formulations such as different mixtures containing known APIs that might offer certain advantages to patients are patentable. Examples of such patient benefits include its enabling a more convenient dosage regimen (e.g. a one-a-day version of a drug previously to be taken every 4 h), or a reduction in side-effects. Dosage forms presented as novel and unobvious routes of administration for known drugs can also be patented (Gupta *et al.*, 2010).

In 2008, the British Court of Appeal made an important ruling on a medical product called Finasteride, which had first been patented in 1978 as a treatment for enlarged prostate. A decade later, the same company, Merck filed a patent application on the same product but for its use to treat male baldness at a daily dosage of over 5 mg using the Swiss form of claim. In 1996 a new patent relating to the same product was granted to Merck claiming its use 'for the preparation of a medicament for oral administration useful for the treatment of androgenic alopecia [male baldness] in a person and wherein the dosage amount is about 0.05 to 1.0 mg'. The court, taking relevant EPO decisions into account, reversed its earlier revocation, finding the new dosage regime to be novel: 'A claim to a pill containing 1 mg dose of finasteride would be a claim to a new thing. No-one had made or proposed such a thing, so why should it not be novel?' It was found non-obvious because there was deemed a low expectation of success.[8] Ironically, it was shortly after that the EPO decided in Decision No. G0002/08 - Dosage regime/ABBOT RESPIRATORY that use of Swiss-type claims should cease in light of the new language inserted into the EPC from 2000. This decision clarified two important points as follows:

1. Where it is already known to use a medicament to treat an illness, Article 54(5) EPC does not exclude that this medicament be patented for *use in a different treatment* by therapy of the *same illness*.
2. Such patenting is also not excluded where *a dosage regime is the only feature claimed* which is not comprised in the state of the art. [emphasis added]

Concerning personalized medicine and considering how far claims may be allowed for methods of use for something old, the following question arises: Is it possible to claim uses or methods relating to a patient population identified by a shared biomarker that forms part of a wider population of patients having the same disease *and who may already have received the medicine*? (Parker and Hall, 2014). Surely, if the drug has already been prescribed for that disease including to those members of the smaller population, there is no novelty. This is a tricky issue, but the EPO has allowed patents for uses targeted to clearly defined sub-populations. In T1399/04 (Combination therapy HCV/SCHERING), the TBA stated that prior public use is not necessarily novelty-destroying. Moreover, defining a group of patients medically where there is overlap with a population already receiving the same drug for the same purpose *can* lead to a patentable invention. As stated by the Board:

> If the use of a compound was known in the treatment or diagnosis of a disease of a particular group of subjects, the treatment or diagnosis of the same disease with the same compound could nevertheless represent a novel therapeutic or diagnostic application, provided that it is carried out on a new group of subjects which is distinguished from the former by its **physiological or pathological** status.
>
> According to the established case law of the Boards of appeal ..., the subject-matter ... represents a new therapeutic application as the patient group concerned is distinguishable ... by its **physiological and pathological** status. [emphasis in original]

16.4 Gaming the System? The Patenting of Enantiomers and Other Variants

Often there is a degree of variability among a class of things that many would assume still to be identical to each other. Is a form or variant of a chemical the same thing as that chemical? Or is it something else entirely? What about drugs? While you ponder this last question, bear in mind that a pill is not just a package comprising multiple numbers of the specified active ingredient, but a collection of those plus various additional chemicals, formulated to enhance the effectiveness of the drug product after it has entered the body. Paradoxically, it is both a pure substance and a mixture.

Let us consider one potentially significant type of variability in the pharmaceutical context: stereoisomerism. Two or more compounds may consist of the same elements in the same atomic proportions but be a different shape. The basic chemical formula is the same but each variant may interact with the human body in different ways that give rise to different effects. Isomers that are mirror images of each other are called enantiomers. A racemate is a chemical that may be one or other mirror image. Racemic mixtures tend to comprise both in equal measure, and many pharmaceutical products are such mixtures. Technically speaking, a racemate is a molecule of a certain type: it is chiral, meaning it comes in two opposite forms: left-handed and right-handed. Whether they are in the left-handed half of the mixture or in the right-handed one they do not differ in their atomic constituents. In terms of their relationship to the mixture and to their counterpart in the other half of the combination, they are referred to in chemistry-speak as enantiomers. Each may also be referred to as an optical isomer. It is possible that both enantiomers have the same therapeutic effect. But they may have very different therapeutic effects. One of the two may be toxic. In recent decades, separating enantiomers has become easier. Needless to say, the ability to carry out such a separation can provide genuine benefits for patients. It is much easier, as of course it should be, to acquire regulatory approval for an enantiomer than for a genuinely new chemical entity. Moreover, enantiomers are patentable. This might all appear to be reasonable, but the possibilities for 'gaming' the regulatory and patent systems are very much present. Chiral switching, that is, moving from the racemic mixture to an enantiomer, and filing a patent on the latter, is frequently attempted as the original patent covering the racemic mixture comes towards the end of its life.

Stereoisomerism is not the only type of variability that can be managed for commercial advantage. Prodrugs are chemically inactive drugs that are converted in the human body into the active ingredient by the action of enzymes. There are numerous reasons why prodrug forms can provide benefits, such as by enhancing in-body transportation or reducing toxicity effects. However, the actual metabolized product which causes the therapeutic effect can itself lead to an improved patient experience if delivered in that form instead. For example, the antihistamine prodrug terfenadine, marketed as Seldane, has cardio-toxic effects on some people. The metabolite, fexofenadine, subsequently marketed as Allegra (among various other names), is not. Metabolites of previously patented prodrugs can be patented too, as were both these two products. However, the situation is far from certain for companies. In the UK, the House of Lords revoked the patent on fexofenadine as the earlier patent on terfenadine had rendered it no longer novel even though the metabolite product had not been explicitly disclosed in the earlier patent.[9] But this did not necessarily make it impossible to patent a metabolite given the facts of the case were specific to the patents at issue. Indeed, in a 2003 US Court of Appeals for the Federal Circuit case, Judge Rader found a metabolite to be inherently anticipated (that is, not novel), but he went on to clarify that 'With proper claiming, patent protection is available for metabolites of known drugs'.[10] He suggested different ways to evade anticipation, including claiming an isolated and purified form of the metabolite or on a method of administering it to a patient.

Correa provides a useful summary of patenting targets chosen by companies to extend their monopolies on drugs. These include: polymorphs (crystalline forms of the active compound); pharmaceutical forms (i.e. new ways of administering the active compound); selective inventions (elements selected from a group that were not specifically named in earlier patents claiming the group); analogy processes; combinations of known products; optical isomers; active metabolites; prodrugs (inactive compounds that produce active metabolites when introduced into the body); new salts of known substances; variants of existing manufacturing processes; and new uses for old products (Correa, 2001).

Follow-on patents of these kinds that claim meaningful distinctions between similar things that in other contexts might be regarded as practically identical may promise benefits for patients. For example, whereas one patented minor molecular modification to a drug may make no difference to the average patient, another could lead to a much improved therapeutic effect on at least some patients. The former type would appear to be commercially valueless. If so, why file patents on it?

Notwithstanding possible patient benefits, patenting strategy here is clearly aimed to extend the market exclusivity of an existing drug or to support marketing aimed at switching patients to a supposedly upgraded version of a company's existing product. This is why they are controversial. Such incremental inventions may be regarded as examples of 'gaming the system' by acquiring extended or new patent monopolies that are not justified by the minor level of inventive contribution or the possibly negligible added benefit to the public. Such opportunistic practices are commonly referred to as 'evergreening' (Amin and Kesselheim, 2012). Whereas rigorous application of the novelty and inventive step criteria by patent offices and court decision may prevent the most egregious attempts to unfairly extend patent monopolies, India's patent law actually has an anti-evergreening provision. Accordingly, Article 3(d) of the Patents (Amendment) Act, 2005 states as follows:

> The following are not inventions within the meaning of this Act, ...
>
> *(d)* the mere discovery of a new form of a known substance which does not result in the enhancement of the known efficacy of that substance or the mere discovery of any new property or new use for a known substance or of the mere use of a known process, machine or apparatus unless such known process results in a new product or employs at least one new reactant.
>
> *Explanation.* – For the purposes of this clause, salts, esters, ethers, polymorphs, metabolites, pure form, particle size, isomers, mixtures of isomers, complexes, combinations and other derivatives of known substance shall be considered to be the same substance, unless they differ significantly in properties with regard to efficacy.

This measure was the subject of a highly publicized Supreme Court case which decided on the patentability of a new form of the anticancer drug Glivec and found it not to be, interpreting 'efficacy' for a medicine of necessity to mean 'therapeutic efficacy'. Novartis failed to prove that the new form had an enhanced and significantly different efficacy as compared to the earlier version of Glivec.[11]

Elsewhere, with good marketing it may be possible to shift patients from an about to become off-patent drug on to a new version marketed under a different name that is really no better. If the original drug was successful and profitable while the patent was valid, such product switching could well generate a lot of money. Some have alleged that this tactic was used by AstraZeneca. When its highly successful anti-ulcer drug omeprazole (sold as Losec, or Prilosec) was coming to the end of its patent life, and attempts to evergreen its monopoly were thwarted, the company sought to switch users to esomeprazole, branded as Nexium, a product that was both chemically and therapeutically virtually identical to Losec, but was ten times more expensive than the former drug. They did this by deploying aggressive marketing tactics claiming that it was both newer and better. Admittedly, the fact that it comprises optically pure salts of omeprazole enabled it to pass the novelty test in the key jurisdictions. But novelty in this context is of course a legal fiction. Like more than half of the drugs currently on the market, the active ingredient of Losec is a racemic mixture: a 50–50 mix of molecules that are mirror images of each other. Nexium is the so-called (S) enantiomer of omeprazole, the one that is therapeutically active. Putting it another way, esomeprazole is one of two optically pure salts of omeprazole. Therefore esomeprazole was part of the contents of omeprazole. In other words it was contained in the API of Losec.

This effort cost hundreds of million dollars in marketing, but it was worth it. 'About 40% of Prilosec users made the switch to Nexium earning the drug over $3 billion in 2003 and almost US$5 billion in 2004' (Manners, 2006).[12] So *was* Nexium any better than Losec to which it is chemically very closely related? It appears that the improvement was modest to say the least (Angell, 2004; Goozner, 2004; Law, 2006). Apart from needing less of it to have the same therapeutic effect by dint of the fact that the other enantiomer is inactive, both Losec and Nexium are prodrugs. Prodrugs are those which are converted by the human body into the substance that is the real active ingredient. Both Losec and Nexium are converted into the same compound, sulfenic acid, which only has a single-handed form: it is achiral, not chiral. Therefore, whether omeprazole or esomeprazole are taken, an optically and pharmacologically identical API is formed *in vivo* which goes about its business in a way that does not depend on which of the two products it is derived from.

The courts are not the place to provide a definitive answer to a scientific question, but in July 2015 the Canadian Federal Court of Appeal provided an answer of sorts. The court upheld a Federal Court decision invalidating AstraZeneca's Nexium patent. The reason was not for lack of novelty or for obviousness, but the absence of utility. As the court put it: 'while the patent promised that its compounds provide improved pharmacokinetic and metabolic properties which will give an improved therapeutic profile, such as a lower degree of inter-individual variation, this promise was neither demonstrated nor soundly predicted at the time the patent was filed.'[13] One should note in passing that this ruling came towards the end of the patent life so AstraZeneca had had plenty of time to enjoy the benefits of its rather questionable marketing practices and its ownership of a patent in that country.

16.5 Conclusion

This chapter has set out various approaches to fine-tuning biomedical research in ways that take account of human variability at the molecular level, and that enable a drug design methodology that 'homes in' on a target with greater accuracy. The patent system has generally been very supportive, among other ways by accommodating chemical, therapeutic and physiological variabilities to a very fine degree and by treating novelty as a legal fiction in some areas. If we accept that the incentives for investment in innovation provided by the patent system are genuine here as elsewhere in the biomedical field, there are good arguments to justify such expansive patentable subject matter practices. However, there are convincing counterarguments suggesting some reining in of the scope of patentability would almost certainly be desirable. The main reason is that the incentives can have negative consequences: legal monopolies on minor 'tweaks' that are unworthy of patent protection. The research-based pharmaceutical industry is a master at strategic patenting practices that can be used unjustifiably to extend market exclusivity over essential medicines well beyond the life of the original patent. The greater the number of ways companies can file additional patents surrounding this or that product, the greater is the opportunity to adopt business practices that reduce competition for ever longer periods of time. Obviously this has negative effects on patients in terms of access; it also appears to encourage investment in minor incremental innovation rather than in the discovery and development of breakthrough products that really make a difference. Competition is essential for reducing prices. In this regard India has set an excellent example, which other countries might wish to follow.

Notes

1 *Parke Davis and Co.* v. *H.K. Mulford and Co.*, 189 Fed. 95 (SDNY 1911) affirmed, 196 Fed. 496 (2nd Cir. 1912).
2 Most notably, see G0005/83 *Second medical indication/EISAI* (1984), G0001/04 *Diagnostic methods* (2005), G0001/07 *Treatment by surgery/MEDI-PHYSICS* (2011).
3 T-866/01 Euthanasia composition/MICHIGAN STATE UNIVERSITY, 2005.
4 G0001/07 *Treatment by surgery/MEDI-PHYSICS*, 2010.
5 G0001/04 *Diagnostic methods*, 2005.
6 G0005/83 *Second medical indication*, 1984
7 T1020/03, *Method of administration of IGF-I/ Genentech Inc.*, 2004.
8 *Actavis UK Ltd* v. *Merck & Co Inc* (2008) EWCA Civ 444.
9 *Merrell Dow* v. *HN Norton* (1996), *Intellectual Property Reports* 33, 1–14.
10 *Schering* v. *Geneva Pharmaceutical*, 339 F.3d 1373 (FCCA, 2003).
11 *Novartis* v. *Union of India*, Civil Appeal No. 2706-2716 of 2013 before the Supreme Court of India (2013). For commentaries on the case, see Thikkavarapu (2014) and Thambisetty (2014).
12 In Europe, the company sought also to delay generic competition with Losec by means that the European Commission found in June 2005 to be an abuse of its dominant position. AstraZeneca was fined €60 million.
13 *AstraZeneca Canada* v. *Apotex*. 2015 FCA 158.

References

Allison, M. (2008) Is personalized medicine finally arriving? *Nature Biotechnology* 26(5), 509–517.

Amin, T. and Kesselheim, A.S. (2012) Secondary patenting of branded pharmaceuticals: A case study of how patents on two HIV drugs could be extended for decades. *Health Affairs* 31(10), 2286–2294.

Angell, M. (2004) *The Truth about the Drug Companies: How They Deceive Us and What to Do about It.* Random House, New York.

Correa, C.M. (2001) *Trends in Drug Patenting: Case Studies.* Corregidor, Buenos Aires.

Cragg, G.M., Grothaus, P.G. and Newman, D.J. (2014) New horizons for old drugs and drug leads. *Journal of Natural Products* 77(3), 703–723.

Elenco, E., Underwood, L. and Zohar, D. (2015) Defining digital medicine. *Nature Biotechnology* 33(5), 460–461.

Goozner, M. (2004) *The $800 Million Pill: The Truth behind the Cost of New Drugs*. University of California Press, Berkeley, California.

Gupta, H., Kumar, S., Roy, S.K. and Gaud, R.S. (2010) Patent protection strategies. *Journal of Pharmacy and Bio Allied Sciences* 2(1), 2–7.

Hoffman, W. and Furcht, L. (2014) *The Biologist's Imagination: Innovation in the Biosciences*. Oxford University Press, New York.

Joyner, M.J. and Paneth, N. (2015) Seven questions for personalized medicine. *Journal of the American Medical Association* 314(10), 999–1000.

Kalia, M. (2013) Personalized oncology: recent advances and future challenges. *Metabolism* 62, S11–14.

Law, J. (2006) *Big Pharma: How the World's Biggest Drug Companies Control Illness*. Constable, London.

Manners, S. (2006) *Super Pills: The Prescription Drugs We Love to Take*. Raincoast Books, Vancouver.

Mullard, A. (2012) Drug purposing programmes get lift off. *Nature Reviews Drug Discovery* 11, 1–2.

Parker, S. and Hall, B. (2014) Patenting personalized medicines in the UK, Europe and USA. *Pharmaceutical Patent Analyst* 3(2), 163–169.

President's Council of Advisors on Science and Technology (2008) *Priorities for Personalized Medicine*, Executive Office of the President of the United States.

Rasmussen, N. (2014) *Gene Jockeys: Life Science and the Rise of Biotech Enterprise.* Johns Hopkins University Press, Baltimore, Maryland.

Schork, N. (2015) Personalized medicine: time for one-person trials. *Nature* 520, 609–611.

Smith Hughes, S. (2011) *Genentech: The Beginnings of Biotech.* Chicago University Press, Chicago, Illinois.

Sterckx, S. and Cockbain, J. (2012) *Exclusions from Patentability: How far has the European Patent Office Eroded Boundaries?* Cambridge University Press, Cambridge, UK.

Strimbu, K. and Tavel, J.A. (2010) What are biomarkers? *Current Opinion in HIV and AIDS* 5(6), 463–466.

Thambisetty, S. (2014) Novartis v Union of India and the person skilled in the art: A missed opportunity. *Queen Mary Journal of Intellectual Property* 4(1), 79–94.

Thikkavarapu, P.R. (2014) The Indian Supreme Court's judgment in the case of Glivec® – the uncertain future of pharmaceutical patents in India. *Pharmaceutical Patent Analyst* 3(2), 117–119.

17 Patentability of Human Embryo Stem Cells: A Comparative Analysis of Case WARF in the United States of America and Europe

JIANG LI*

Kenneth Wang School of Law, Soochow University, Su Zhou, China

17.1 Introduction

Human embryonic stem cells are potentially of great therapeutic value in a number of areas including spinal cord injuries, Parkinson's disease, stroke and transplantation therapy (Thomson, 1998). The patentability of human embryonic stem cells has raised worldwide controversy and dispute in the last two decades, especially in the USA and Europe.

Following the basic principle of 'anything under the sun that is made by man can be patented',[1] US patents on human embryonic stem cells have been granted. The US Patent and Trademark Office (USPTO) granted a broad patent on primate embryonic stem cells (ESC) in December 1998 and a second patent on human embryonic stem cells (HESC) in March 2001 (Loring and Campbell, 2006). However, the government appeared not to carry out the supporting policies to embryo research. In August 2001, President Bush imposed a ban on federal funding of HESC research (Passeggio, 2002).

However, invention regarding ESC is still contentious in Europe. The ethics of stem cells patentability is evaluated by the European Group on Ethics in Science and New Technologies to the European Commission (EGE, 2002). Unlike the US government, EGE had not required to prohibit public funding for embryo research at the community level. In its opinion, EGE suggested isolated and cultured stem cells without modification should not be patented (EGE, 2002).

The recent case WARF (Wisconsin Alumni Research Foundation) reflected well the difference between the USA and Europe. Patents related to ESC were granted by USPTO whereas the EPO rejected to grant any patents involving the destruction of human embryos. This chapter mainly focuses on commenting on the WARF case in the USA and Europe. Simultaneously, it will analyse the patentability of human ESCs.

17.2 What is an Embryonic Stem Cell?

Before looking at the case, it is necessary to give some explanation about ESC. An ESC is a type of undifferentiated cell, which stems from human early embryo and is cultured *in vitro*. These cells are characterized by a powerful self-renewal capability. They are also potentially able to differentiate into various kinds of tissue and organ. Other than ESCs, adult-specific tissues are another source of stem cell (Crespi, 2006). The ESC, which can survive a long time in an undifferentiated state, is a valuable research tool for medicine and drug screening.

*Corresponding author, e-mail: jelly3636@hotmail.com

17.3 WARF in the USA

WARF was granted three patents by USPTO. The first patent No.5,843,780 on primate ESCs was issued on 1 December 1998.[2] On 13 March 2001, the second granted patent No.6,200,806 directed attention to human ESCs.[3] Then on 28 April 2006, the third patent No.7,029,913 was granted, relating to the culture and discovery of human or primate ESCs.[4]

These three patents aroused much controversy in the scientific community. Many scientists argued that these patents were too broad. As Tim Friend (2001) indicated for patent 6200806, however, WARF is claiming intellectual property rights (IPR) in every human ESC line that qualifies for federal funding under President Bush's plan (Friend, 2001). It is no exaggeration to say that patent 6200806 and 5843780 reference almost the whole field of stem cell research (Beardsley, 2007). In July 2007, the Foundation for Taxpayer and Consumer Rights[5] (FTCR) and the Public Patent Foundation (PUBPAT) applied to re-examine the patents on the basis of prior art and obviousness. There were also some notions that these patents were too broad and might become the obstacle to stem cell research (Washburn, 2006).

The key point of obviousness examines whether the prior art which applied to species such as mouse and pig can obviously extend to humans. According to plentiful previous experiments, the inference is that the method applied to an animal cannot certainly suit a human. The complexities of living organisms lead to the uncertainty of the outcome. Thus, I support the viewpoint that these patents are non-obvious.

The main argument for stifling stem cell research is that the monopoly of WARF is driving both research and money out of the USA. However, WARF had not prohibited the application of these patents in stem cell research. WARF has changed its policies and distributes HESC at a low price to researchers (Madison, 2007). Perhaps the bad funding policies of the government should be responsible for the flow.

After revoking three patents on 3 April 2007, the USPTO finally upheld them in 2008. The WARF stem cell patents were ruled non-obvious under the light of prior art. Despite that some claims were revised and minor amendments were made, some still disagree with the amended claims and think the patent is obvious (McDermott, 2008).

17.4 WARF in Europe

Compared with the controversies of case WARF in the USA, the situation in Europe is very different and mainly focuses on moral issues. WARF made application in 1995 and was refused by the examining division of the European Patent Office (EPO). Finally, on 25 November 2008 the Enlarged Board of Appeal rejected the appeal and 'exclude patentability of any biotechnological inventions that concern use of human embryos for industrial or commercial purposes'.[6]

The decision mainly answered four questions referred by Technical Board of Appeal. The first question, whether Rule 28(c) (formerly 23d(c)) EPC apply to applications filed before the effect of the rule, is indisputable. The Appellant, the President of the EPO and the Enlarged Board of Appeal all agreed.[7]

The second question addressed the core of the debate. Six alleged arguments were put forward to defend that the Rule 28(c) was not applied in this case. The board replied to all arguments with contrary claims. First, the appellant explained 'the embryo of 14 days or older'. In the medical field, the pre-14 day embryos have no central nervous system therefore are allowed for research in some countries such as the UK. Human embryo, in WARF case, refers to embryos of 14 days or older. The board stated that there is neither explicit definition nor restrictive interpretation to the term 'embryo'. Second, the appellant argued the claims did not contain 'the use of human embryos'. The board pointed out this rule stipulated 'invention' instead of 'claim', therefore the destruction of the human embryo is one part of invention falling within Rule 28(c). Third, the appellant contended that it was too extreme to examine all the steps that whether they might be excluded by Rule 28(c) or not. However, the board noted that it is irrelevant. Fourth, the appellant claimed that the destruction of human embryo was aimed at obtaining the claimed HESC. This could not characterize 'a use for industrial or commercial purpose'.[8] The board rejected the opinion that the manufacturing process that involves destruction is 'an integral and essential part of claimed invention'.[9] Fifth, the appellant quoted the history of the directive as an evidence of narrowing of the provision. The board denied any tightening and emphasized the distinction between 'the uses of human embryos for industrial or commercial purposes'.[10] Last, the appellant assumed that Rule 28(c) *ultra vires* if this case was within the scope of exclusion. The board disapproved it and indicated that 'the performing of the invention including a step'[11] pertains to commercial exploitation.

After affirmation of the second question, the third question is unnecessary to answer. Moreover, the fourth question is not of relevance to the patentability. To sum up, the Enlarged Board of Appeal held that the application of patent involving destruction of human embryos is rejected.

However, the rejection decision seems not fully clarified by the law. According to Mullally (2010), 'It's very specific in what it says, which means it's very non-specific in what it doesn't say. At the time when Warf were filing there was no way they could carry out research without harming the embryo' (Sigrid and Julian, 2010). He further indicated that the Enlarged Board of Appeal had not given clear answer to the circumstance in which the applicants use stem cells as the starting point without themselves destroying an embryo. What is worse, these situations such as derived from a stem cell line are very common among the applications. The uncertainty of law, which is the greatest enemy to patent law, might be inevitable due to this grey area (Mullally, 2010).

Another challenging phenomenon is that most patent applicants file not only at the EU level but also the national level. It is a well-known fact that Europe reveals the ethical diversity. The standards of patentability of HESC vary greatly from state to state. For instance, the UK Patent Office tends to narrowly explain their counterpart of 28(c) EPC (Crespi, 2006). In the UK, HESC derivation and nuclear transfer of human somatic cells are both permitted and licensed by the Human Fertilisation and Embryology Authority (HFEA) (Abdalla, 2008). On the contrary, in a civil law system such as Germany, the majority point of the research on human stem cell is immoral. Scientists are merely allowed to use HESC lines derived from the labs in a foreign country, and human embryonic stem cell lines could only be obtained before 2002 outside Germany (Russo, 2005). Obviously, different perspectives on HESC research between countries lead to the inconsistent interpretations of the moral provision. Thus, the decision had not eliminated the uncertainty of the scope of moral exclusion.

Currently, ESC research has become one of the key areas in biotechnology. The concern that the moral rejection would hinder the development of stem cell research in Europe is reasonable. It might lead to Europe lagging behind other economic areas such as the USA and Japan.

17.5 Conclusion

From the discussions above, a significant distinction can be concluded. The USPTO made a decision mainly through examining the patents on technical grounds, whereas the EPO focused on moral grounds. Because the biotechnological invention was not distinguished from other technological invention in the USA, the moral opposition had little effect on patentability (Sorenson, 2005). However, the rules of moral exception can be found in most national regulations and the European documents (Porter *et al.*, 2006).

I personally support the decision of the Enlarged Board of Appeal. First, although the decision did not thoroughly clarify the scope of moral exception, it did unambiguously preclude the patent related to the destruction of the human embryo. So many homologous applications will not be protected. This prohibition will press scientists to draw more attention to non-destructive ESC research. To some extent, the destruction of the human embryo could be reduced. Second, the decision is in favour of stem cell research in Europe. The claims of WARF were relevant to the basis of stem cell research. If the relevant basic work patents were granted to the inventors, expensive royalties and unnecessary restraints would hinder subsequent stem cell research in this area. With the increased cost of research and development, investment in this area will correspondingly decrease. Obviously, much research on disease based on the ESC could be influenced by the declining investment. Third, the sanctity of human embryos that can be developed into human life should be respected. In spite of the fact that many researchers use human embryos that are discarded as medical waste from IVF (*in vitro fertilization*) clinics, those embryos still have the right to be respectfully treated instead of being destroyed. The use of HESC damages the integrity of life and is a violation of human dignity. The consequences of abuse of human embryos would be inconceivable.

For the above reasons, I strongly advocate the decision of the Enlarged Board of Appeal. From my point of view, the USPTO should not simply examine the application on technical grounds. Biotechnological inventions, which correlate with science, ethics and law, should be specifically treated. Moral consideration is not excessive but indispensable for patentability of the HESC.

Notes

1 *Diamond* v. *Chakrabarty*, 447 US 303 (1980).
2 *Primate Embryonic Stem Cell*, US Patent No. 5,843,780 (filed 18 January 1996) (issued 1 December 1998).
3 *Primate Embryonic Stem Cell*, US Patent No. 6,200,806 (filed 26 January 1998) (issued 13 March 2001).
4 *Primate Embryonic Stem Cell*, US Patent No. 7,029,913 (filed 18 October 2001) (issued 18 April 2006).
5 Foundation for Taxpayer and Consumer Rights (FTCR). Now renamed as 'Consumer Watchdog': http://consumerwatchdog.org
6 G-02/06 of the Enlarged Board of Appeal of the European Patent Office 18.
7 G-02/06 of the Enlarged Board of Appeal of the European Patent Office 19.
8 G-02/06 of the Enlarged Board of Appeal of the European Patent Office 24.
9 G-02/06 of the Enlarged Board of Appeal of the European Patent Office 24.
10 G-02/06 of the Enlarged Board of Appeal of the European Patent Office 25.
11 G-02/06 of the Enlarged Board of Appeal of the European Patent Office 26.

References

Abdalla, H.I. (2008) The role of the Human Fertilisation and Embryology Authority. *Obstetrics, Gynaecology and Reproductive Medicine* 19, 51–56.

Beardsley, D. (2007) A two-front assault on the stem cell patents. *The John Marshall Review of Intellectual Property Law* 6, 501.

Crespi, R.S. (2006) The human embryo and patent law – a major challenge ahead? *European Intellectual Property Review* 28, 569.

European Group on Ethics in Science and New Technologies (EGE) (2002) Opinion on the ethical aspects of patenting inventions involving human stem cells. Available at: http://www.unal.edu.co/bioetica/documentos/6_4.pdf (accessed 9 August 2015).

Friend, T. (2001) *Free stem-cell lines will be offered to researcher. USA TODAY* D10.

Loring, J.F. and Campbell, C. (2006) Intellectual property and human embryonic stem cell research. *Science* 311, 1716–1717.

Madison, W.I. (2007) Wisconsin alumni research foundation changes stem cell policies to encourage greater collaboration. Willcell Research Institution. Available at: http://www.wicell.org/home/about-wicell/news-room/12207/12207.cmsx (accessed 10 October 2015).

McDermott, E. (2008) USPTO backs WARF stem cell patents. *Managing Intellectual Property* 178, 62.

Mullally, K.C. (2010) Legal uncertainty, legal process and patent law. *Loyola of Los Angeles Law Review* 43, 1109–1162.

Passeggio, E.R. (2002) Embryonic stem cell research: shifting availability of federal funds. *Boston University Journal of Science and Technology Law* 8, 347.

Porter, G., Denning, C., Plomer, A., Sinden, J. and Torremans, P. (2006) The patentability of human embryonic stem cells in Europe. *Nature Biotechnology* 2, 653.

Russo, E. (2005) Follow the money – the politics of embryonic stem cell research. *PLoS Biol* 3, 234.

Sigrid, S. and Julian, C. (2010) Assessing the morality of the commercial exploitation of inventions concerning uses of human embryos and the relevance of moral complicit: comments on the EPO's WARF decision. *SCRIPTed - A Journal of Law, Technology & Society* 7(1), 83–103.

Sorenson, C.W. (2005) Stem cell research and human cloning: where do we draw the line? *New England Law Review* 39, 477.

Thomson, J.A. (1998) Embryonic stem cell lines derived from human blastocysts. *Science* 282, 1145.

Washburn, J. (2006) The legal lock on stem cells. *L.A. Times* 13.

18 Innovation and Intellectual Property Issues in the 'Decade of Vaccines': A Brazilian Perspective

CRISTINA POSSAS,[1,3]* ADELAIDE ANTUNES,[2,4] FLAVIA M.L. MENDES,[2] REINALDO M. MARTINS[1] AND AKIRA HOMMA[1]

[1]Bio-Manguinhos, Oswaldo Cruz Foundation, Rio de Janeiro; [2]School of Chemistry, Federal University of Rio de Janeiro; [3]Evandro Chagas National Institute of Infectious Diseases, Oswaldo Cruz Foundation, Rio de Janeiro; [4]National Institute of Industrial Property, Rio de Janeiro, Brazil

18.1 Introduction

Human and veterinary vaccines are recognized as the most powerful preventive, low-risk and cost-effective interventions, with an extensive impact on global health and livestock production. This chapter examines vaccine innovation and intellectual property rights (IPR) issues in biotechnology, from the Brazilian perspective. It discusses how vaccine innovation policies and intellectual property (IP) regimes can affect countries in particular ways, according to their different stages of development. From this perspective, it proposes a comprehensive approach to vaccine innovation in the developing world, examining the global scenario in IPR and how it impacts on policies and access of the population to new health products.

It also stresses the crucial role of the state in formulating policies, sustainable long-term financing and regulation, aiming at innovation, competitiveness and social welfare. It examines conflicts involving IPR in drug and vaccine innovation, the increasing global demand for new preventive and therapeutic vaccines, the promising role of immunotherapy in diverse cure strategies, such as HIV, cancer and other chronic and degenerative diseases. It discusses the globalization of IP in the health sector with the World Trade Organization (WTO) Agreement on Trade-Related Aspects of Intellectual Property Rights (TRIPS), signed in 1995, and its implications for vaccine innovation and development.

From a market perspective, it discusses IPR constraints in technology transfer agreements and the impact of bilateral and multilateral trade agreements on IPR and on prices of pharmaceutical products. The need for a scientific and technological cross-fertilization between the fields of human and veterinary vaccines is also stressed. Final considerations are made on future scenarios in the new developments of the 'Decade of Vaccines', a World Health Organization (WHO) Global Vaccine Action Plan 2011–2020 (WHO, 2013), discussing alternative incentives to innovation, aiming to place vaccines in the core of health innovation systems in the developing world.

18.2 The Global Epidemiological Scenario: Challenges for Vaccines

In a complex global epidemiological scenario, infectious diseases remain one of the leading causes of morbidity and mortality in the contemporary world. In spite of progress in the global control of infectious diseases and of the social gains resulting from development and urbanization, in the last five

*Corresponding author, e-mail: cristina.possas@bio.fiocruz.br

decades new conditions related to environmental change and to international population mobility have led to the unexpected phenomenon of emergence and re-emergence of infectious diseases not preventable by vaccination and of drug-resistant diseases. Moreover, available data indicate that around 60% of infectious and parasitic diseases are zoonose-related (CDC, 2013), a scenario aggravated in recent decades by environmental change and mobility of vectors and wild animals to urban areas. These changing conditions have favoured the global emergence and resurgence of several infectious diseases with complex dynamic cycles, such as HIV/AIDS, influenza, tuberculosis, dengue, yellow fever, malaria, leishmaniasis, leptospirosis, hantavirus pulmonary syndrome, Ebola and many others. The rapid spread of these diseases worldwide challenges the national health systems, particularly affecting developing countries like Brazil, a nation plagued by social exclusion and environmental degradation and with a rapidly ageing population.

This epidemiological complexity poses new challenges and a need for an integrated view of emerging and re-emerging infectious diseases, including zoonoses, based on evolution theory and on eco-social and anthropological perspectives, supporting alternative strategies to anticipate risks and to prevent the amplification of risk conditions.

Finally, from this perspective, new global scientific and technological initiatives supporting the development of new biotechnology-based vaccines will be key to deal with emerging and re-emerging infectious diseases and to support global surveillance.

18.3 Biotechnology: The New Global Vaccine Scenario

The international scenario in science and technology has been marked by accelerated advancement in diverse aspects of biotechnology, in different fields of human activity. The new biotechnology, resulting from advances in molecular biology and genetic engineering, plays a crucial role in the dynamics of scientific and technological development in the contemporary world. In a global economic environment where competition is driven by knowledge and innovation, the new developments in the pharmaceutical industry, resulting from biotechnology, have been extraordinary. In this scenario, recombinant vaccines, biopharmaceuticals and drugs have emerged as scientific and technological breakthroughs, radically changing the strategies for prevention and treatment of a broad range of infectious, parasitic and chronic-degenerative diseases.

These extraordinary achievements in new biotechnology have emerged from radical changes in scientific and technological paradigms and also in scientific collaboration through international networks and regulatory procedures, increasingly submitted to harmonization. They have contributed to the rapid development of new innovative products with wide application in diverse economic and social sectors such as health, energy, agriculture, livestock, environment and others.

Drugs of synthetic origin, produced by organic synthesis, have been the main biotechnology products of the pharmaceutical industry in recent decades. Nevertheless, since the early 1980s, small start-up biotechnology enterprises began to conduct research on large molecules, derived from or modified from biological systems. New biotechnology-based products are currently the focus of many pharmaceutical companies, both in terms of R&D and product sales. The importance of these biotechnology products can also be illustrated by the fact that 190 biotech companies were acquired by large pharmaceutical companies in the last three decades, at a cost of US$393.3 billion (Evens and Kaitin, 2015).

The global pharmaceutical market for these products is spiralling. The pharmaceutical multinationals are increasingly investing in new biotechnology. In addition to high investments in innovation and technological development of new drugs and vaccines, these highly competitive companies are also rapidly increasing their market share in 'start-up' companies.

In developed countries, there is a clear division of labour related to position of the diverse actors in the flow from basic research to product development and production. The public sector is the main investor in basic science, generating new knowledge and discoveries, while the private sector concentrates in activities related to product development.

The big pharmaceutical companies invest 20% of net sales in product development (US$2 billion annually) and 50% of the drugs and vaccines developed in the world are now biotechnology-based. They are also expanding their market, investing in the acquisition of consolidated companies, in an oligopolistic trend: Pfizer bought Wyeth; GSK bought Novartis/vaccines and sold its Oncology component for Novartis, in a strategy to strengthen and consolidate new areas of expertise; Bayer bought from Merck & Co. the division of consumer goods and health for US$14.2 billion.

18.4 Human and Veterinary Vaccines: Innovations and Global Market

In this new biotechnology scenario, the global scientific and technological capacity for innovation in human vaccines against infectious diseases has advanced rapidly. Innovative strategies and new developments in biotechnology have resulted in new and effective vaccines against hepatitis B virus, *Haemophilus influenzae* type b, *Streptococcus pneumoniae*, *Neisseria meningitidis* and human papillomaviruses, as well as new combination vaccines, such as the pentavalent and hexavalent vaccines. The global vaccine scenario has changed and the international demand for vaccines has sharply increased (Homma, 2009; Homma and Possas, 2014). The poorest countries are now incorporating new vaccines, with financial support from the GAVI Alliance. Support from the Bill & Melinda Gates Foundation (BMGF) and other international organizations, such as WHO, UNICEF and governments from developed and developing countries, has contributed to significant advances in immunization and eradication strategies. Many countries have already attained high immunization coverage, saving the lives of millions of children and helping to reach Millennium Development Goals.

Moreover, in a global context of economic crisis, permeated by sharp social inequalities, there is increasing awareness among policy makers of the crucial role that vaccines can play in improving the living conditions of the poorest populations because of their high public health impact and cost-effectiveness. In fact, WHO estimates that immunization against vaccine-preventable diseases currently averts nearly 2.5 million deaths per year worldwide, and that an additional 2 million deaths could be prevented by improving access to vaccines. Successful international immunization strategies – such as the eradication of smallpox, the significant reduction of polio's incidence and the rapid decrease in measles deaths – illustrate well the global impact of vaccines (Homma *et al.*, 2013).

Veterinary vaccines, on the other hand, have been developed with wide applications in animal production and reproduction. Available literature indicates that veterinary vaccines have already resulted in extraordinary impacts not only in animal health but also on human health. A good example of the contribution of veterinary vaccine for human disease is a cattle vaccine against *Escherichia coli* 0157:117, which is a leading cause of food-borne disease in humans worldwide (Meeusen *et al.*, 2007).

18.5 Human Vaccines and their Global Market

According to WHO, the human vaccine markets are organized according to an oligopolistic structure, with about 80% of global human vaccine sales coming from five large multinational corporations, represented by the International Federation of Pharmaceutical Manufacturers and Associations (IFPMA). Besides concentrating on developed countries' markets, these corporations have also increased their participation in developing countries' markets, playing an important role in Global Health Initiatives, focusing mainly on joint-development and technological transfer projects.

On the other hand, emerging countries' manufacturers, represented by the Developing Countries Vaccine Manufacturers Network (DCVMN), have increased their participation in the global market and now play a crucial role in the supply of vaccines to developing countries, particularly for some basic and combination vaccines. Now they supply about 70% of UNICEF's vaccine procurement in volume of doses and their increased participation in the developing countries' vaccine market has contributed to lower these vaccines' prices, due to increased competition.

In spite of these advances, the limited number of vaccine suppliers leads to a critical situation in the human vaccine markets and requires constant negotiations and management between the main players, i.e. manufacturers, governments and international organizations, in order to assure sufficient global supply of vaccines, particularly in the developing world.

18.6 Veterinary Vaccine Market

The increasing international demand for food and meat-derived products explains the rapid evolution of the veterinary vaccine market. It is a highly competitive oligopolistic market, integrated by few multinational corporations, most of them related to the human vaccine market, which contributes to reduce the high R&D costs of vaccine development. In this economic sector, the capacity of innovation and market differentiation is key to assure entrepreneurial competitiveness.

The relevance of the agribusiness is a crucial component of some emerging economies, such as Brazil. Government authorities in these countries are aware of the need to follow strictly the recommendations of the World Organisation for Animal Health (OIE), an intergovernmental organization responsible for improving animal health worldwide. OIE has maintained since 1990 a list of countries free of certain animal diseases, such as foot-and-mouth disease (or aphthous fever), and this list has important implications for market decisions related to meat import.

18.7 International Strategies: the Global Vaccine Action Plan 2011–2020

International awareness of vaccines' impact and increased global demand for vaccines have highlighted the need for global strategies to strengthen local capacity for vaccine research and development in emerging economies and to increase access to vaccines through policies supporting free and universal access to vaccines.

In this context, the Decade of Vaccines (DoV) initiative was launched at the World Economic Forum in Davos in 2010, signed by international agencies, such as the World Health Organization (WHO), UNICEF, the US National Institute of Allergy and Infectious Diseases (NIAID) and the Bill & Melinda Gates Foundation, with the mission: 'to extend, by 2020 and beyond, the full benefits of immunization to all people, regardless of where they are born, who they are, or where they live'. This declaration was supported by a commitment by the Bill & Melinda Gates Foundation to donate US$10 billion to research and development and to delivering vaccines for the poorest countries.

The DoV initiative gained significant international support and visibility. Two years later, after consultations with DoV stakeholders, including industry groups, a Global Vaccine Action Plan (GVAP) was launched by the 194 member states of the 65th World Health Assembly in May 2012, aiming to deliver universal access to immunization.

Following the collaborative DoV strategies, the GVAP brought together multiple stakeholders to achieve the ambitious goals of the plan: the leadership of the Bill & Melinda Gates Foundation, GAVI Alliance, UNICEF, US NIAID and WHO, mobilizing many partners: governments, health professionals, academia, manufacturers, funding agencies, development partners, civil society, media and the private sector.

If the GVAP is translated into action and resources are mobilized, it is expected that between 24.6 and 25.8 million deaths could be averted by the end of the decade, with gains in billions of dollars in productivity.

Nevertheless, it is important to note that actions and resources will not be sufficient for the success of GVAP if the plan does not conceive a global strategy to support manufacturers in the developing world to overcome the main IPR and regulatory barriers that delay and hinder vaccine development and production.

18.8 Intellectual Property: Concept and Economic Rationale

The new vaccine products and processes resulting from biotechnology are challenging both developed and developing countries to the implementation of adequate regulations and IPR, raising several controversies.

IPR can be defined as private rights granted by the state with the main goal of protecting intellectual creations and innovations with a potential of economic application (Godinho, 2011). It is a private right that can be exerted by the state, through public enterprises, universities and research institutes.

The outcome of intellectual innovations is new knowledge, clearly distinct from previous existing knowledge. This new knowledge can be applicable or materialized in new processes or products that can meet social and economic demands.

IPR are granted by the state to individuals, enterprises or organizations under temporary monopolistic conditions (patents) in order to compensate them for the investments made in their creations/innovations. In industry, a patent is clearly an instrument to guarantee the returns of the investments on R&D through the commercialization of the patented products and through the payment of property rights.

The rationale for IPR and patents in technological inventions was conceived by economic theory six decades ago (Arrow, 1962). The main argument of the economists was at that time that in the absence of patent protection the market would act alone and a 'failure of the market' resulting from the lack of adequate incentives could lead to a deficit in technological innovation of processes and products necessary for social and economic life.

Patents, from this perspective, are viewed as a crucial incentive to innovation. Nevertheless, Arrow (1962) recognized in his pioneer theory that in spite of its advantages, the patent system creates a sub-optimal situation in economic terms: patents create a monopoly that restricts the diffusion and dissemination of innovation. The argument is that this restriction is temporary (after 20 years the patent protection 'falls' to public domain) and is compensated by the fact that the knowledge related to the patent is necessarily published at the moment the patent is granted.

18.9 TRIPS: Implications for Vaccines

The introduction of new regulatory barriers favouring imbalance in the international trade of life-saving drugs has been supported by an international system for patenting life and life-saving products, which has emerged after the Uruguay Round in 1984 and the creation of the World Trade Organization (WTO). The new IPR system introduced by TRIPS included pharmaceutical products as subject to patents, equivalent to any other product.

This change in the IPR scenario contrasted sharply with the previous more favourable international regulatory framework, where local firms in developing countries had been so far allowed to copy patented pharmaceuticals and vaccines and to produce generic drugs (Possas, 2013).

International IPR regulations and procedures have extended their scope far beyond the need to protect innovators and have become increasingly subordinated to oligopolistic and monopolistic market procedures. Multinational pharmaceutical enterprises are increasingly using these regulatory IP constraints as important economic barriers, limiting the access by the developing world manufacturers to favourable international trade conditions to drugs, vaccines and other health products, thus increasing the gap between the North and the South.

In addition, this trend towards increasingly strong IPR regimes related to international trade has been recently confirmed and aggravated by the so-called 'TRIPS Plus' bilateral commercial agreements, signed by the USA, Europe and other developed nations with developing countries, imposing on these countries even stronger IPR constraints related to trade. We discuss here the impacts of these international constraints by TRIPS and 'TRIPS Plus' agreements on the production of vaccines by manufacturers in developing nations. We examine IPR policy options in developing countries and how they have affected the access of their poorest populations to vaccines and drugs.

18.10 Vaccine Patents: An Overview

New vaccines against *Pneumococcus* conjugate, combination meningococcal A/C/W/Y conjugate, hepatitis B virus, *Haemophilus influenzae* type b, *Streptococcus pneumoniae*, *Neisseria meningitidis* and human papillomaviruses, as well as new combination vaccines, such as the pentavalent vaccine, DPT/HB/Hib and HPV are requiring from developing countries more effective governance strategies to accelerate the vaccine availability for their populations, through public–private partnerships and technology transfer agreements (Homma, 2009; Homma *et al.*, 2013). In this context, IP and vaccine patents emerge as a crucial issue for vaccine development in these countries (Mahoney *et al.*, 2004; Milstien et *al.*, 2007; Padmanabhan *et al.*, 2010).

A search in the Derwent Innovation Index retrieved 6060 patent deposit documents in the world from 2008 to 2013. Figure 18.1 provides the distribution of vaccine patent deposits in the top ten priority countries, with China (36%), the USA (33%) and the European Patent Office (8%) accounting for 77% of the world's vaccine patent deposits.

Figure 18.2 provides an overview of the deposits of vaccine patents in Brazil from residents and non-residents, which indicates the low number of deposits from Brazil and other emerging economies (Possas *et al.*, 2015).

Another important aspect of vaccine patent protection is that vaccines are increasingly being developed as immunotherapy strategies for non-infectious chronic and degenerative diseases such as cancer. 'Vaccines' rank fourth in a list of the most cited topics in patents deposits of biotechnological products related to cancer, after 'antibodies', 'therapeutic proteins' and 'gene therapy' (Von Ranke, 2015).

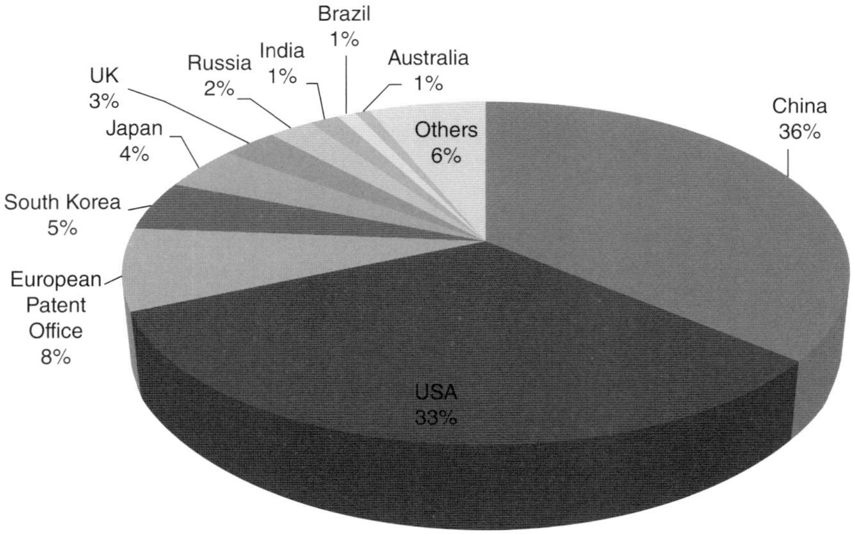

Fig. 18.1. Vaccine patent deposits in the world for the top ten priority countries, 2008–2013 (Federal University of Rio de Janeiro School of Chemistry Information System on the Chemical Industry (SIQUIM); Derwent Innovations Index, date of search: 17 June 2015).

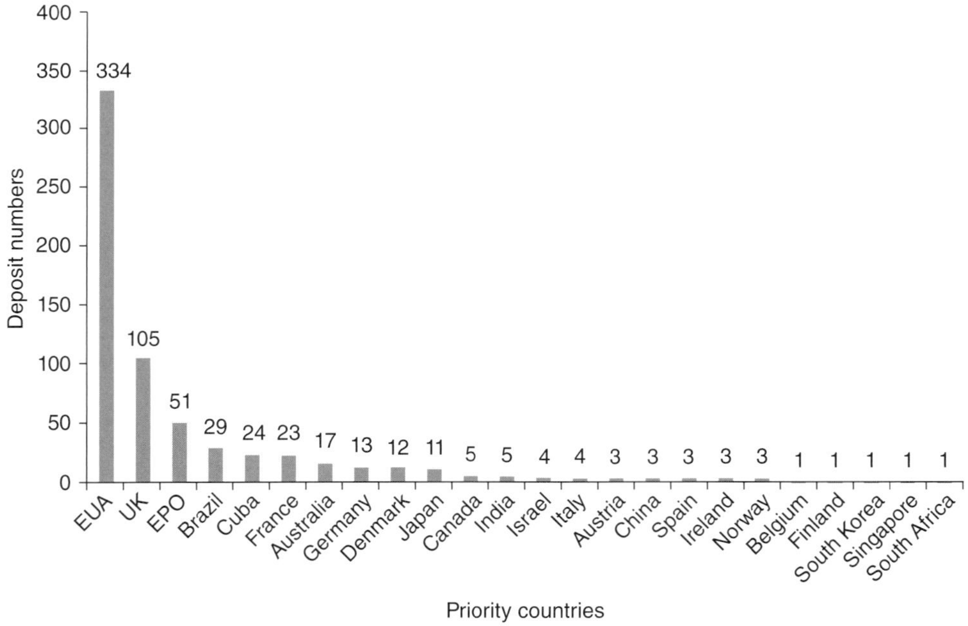

Fig. 18.2. Number of patent deposits per country of priority: vaccines against infectious diseases, Brazil 2000–2011 (Federal University of Rio de Janeiro School of Chemistry Information System on the Chemical Industry (SIQUIM); Derwent Innovations Index, in Possas *et al*., 2015).

The new global vaccine patent scenario indicates the need for new strategies based on comprehensive national surveys of vaccine patents in emerging economies. In these economies new incentives should be conceived to promote incorporation of patent information into technological prospect and into models to anticipate future scenarios for vaccine innovation and development.

Table 18.1 presents for selected countries a detailed description of patentable subject matters

Table 18.1. Patentable subject matters in biotechnology, selected countries, 2013 (IPEA, 2013).

		USA	Europe	Japan	China	India	Brazil
1	Microorganisms and associated biological materials isolated or extracted from nature: bacteria, fungi – including yeast, algae, protozoa, viruses and lichens	P	P	P	P	NP	NP
2	Processes that utilize microorganisms, enzymes or associated biological materials	P	P	P	P	P	P
3	Genetically modified microorganisms	P	P	P	P	P	P
4	Processes for the genetic modification of microorganisms	P	P	P	P	P	P
5	Genetically modified plants and non-human animals	P	P	P	NP	NP	NP
6	Processes for the genetic modification of plants and non-human animals	P	P	P	P	P	P
7	Processes used in the production of plant varieties	P	P	P	P	P	P
8	Plant varieties	P	NP	P	NP	NP	NP
9	Processes used in the production of non-human animal varieties	P	P	P	P	P	P
10	Non-human animal varieties	P	NP	P	NP	NP	NP
11	Genes or DNA fragments; RNA and chromosomes of humans or other animals isolated or extracted from nature	P	P	P	P	NP	NP
12	Genes or DNA fragments; RNA and chromosomes of microorganisms isolated or extracted from nature	P	P	P	P	NP	NP
13	Genes or DNA fragments; RNA and chromosomes of plants isolated or extracted from nature	P	P	P	P	NP	NP
14	Genes or DNA fragments; RNA and chromosomes of genetically modified humans or other animals	P	P	P	P	I	NP/P
15	Genes or DNA fragments; RNA and chromosomes of genetically modified microorganisms	P	P	P	P	P	P
16	Genes or DNA fragments; RNA and chromosomes of genetically modified plants	P	P	P	P	I	P
17	Embryonic stem cells	I	I	I	NP	NP	NP
18	Non-embryonic stem cells	P	I	P	P	NP	NP
19	Human cells	P	P	P	P	NP	NP
20	Animal (non-human) cells, tissues and organs	P	P	P	P	NP	NP
21	Plant cells, tissues and organs	P	P	P	P	NP	NP
22	Fused cells	P	P	P	P	I	P
23	Cell fusion process	P	P	P	P	P	P
24	Sequences of amino acids, polypeptides and enzymes isolated or extracted from nature	P	P	P	P	NP	NP
25	Sequences of genetically modified amino acids, polypeptides and enzymes	P	P	P	P	P	P
26	Polypeptide production processes	P	P	P	P	P	P
27	Monoclonal antibodies	P	P	P	P	I	P
28	Therapeutic method; surgical, diagnostic, and disease treatment methods for humans or other living animals	P	NP	NP	NP	NP	NP
29	Processes for cloning humans or other animals	I/P	NP/I	I/P	NP/P	NP	NP/P
30	Processes for modifying the genetic identity or germ cell lines of humans or other animals	I/P	NP/P	I/P	NP/P	NP/I	NP/P
31	Use of human or other animal embryos	I/P	I/P	I/P	NP/I	NP/I	NP/I

P, patentable; NP, non-patentable; I, unable to infer, undefined or under discussion.

in biotechnology, many of them related to vaccine research, development and production (IPEA, 2013).

18.11 Intellectual Property Rights: Barriers to Access Vaccine Technology

A global strategy to support a sustainable progress in innovation and access to new human and veterinary vaccines will require overcoming several challenges, including legal and regulatory barriers, such as IP, which provides incentives to innovators but may contribute to rising costs and delays in vaccine development and production. The IPR constraints for vaccine adjuvants illustrate well these barriers.

18.12 New Adjuvants: Intellectual Property Rights Barriers

Vaccine adjuvants are an important component in the development of new vaccines and play a crucial role in boosting the immune response necessary for vaccine protection (Sette and Rappuoli, 2010). Table 18.2 (Mbow *et al.*, 2010) provides a list of the main licensed adjuvants.

There seems to be a consensus in the international scientific and technological communities that access to patented vaccine adjuvants is a crucial issue and a major constraint, limiting the possibility of development of promising vaccines by developing countries' researchers and manufacturers. There is enough evidence indicating that some multinational companies have been unwilling to share information on these adjuvants, protected by patents with external investigators (Bregu *et al.*, 2011) in their international collaborations and even after technology transfer agreements.

This is a major barrier and contributes to a global impasse in vaccine development, since access to adjuvant information protected by patent could contribute to significantly accelerate vaccine development.

Patent barriers and lack of access to new vaccine technologies have also had adverse effects on vaccine development for the so-called 'diseases of the poor' (neglected diseases) for which there is limited commercial interest.

In an attempt to overcome these constraints, some non-profit organizations, with support from international funding agencies, have established a new Global Adjuvant Development Initiative (GADI) laboratory, located at the University of Lausanne, under the guidance of the WHO. GADI has been conceived to support new strategies and mechanisms to make adjuvants available to developing countries' vaccine manufacturers.

18.13 Cross-fertilization: Human and Veterinary Vaccines

There are many more veterinarian than human vaccines, but, surprisingly, contrasting with the huge economic impact of veterinary vaccines in the livestock market and meat imports, available data indicate that the profits of veterinary vaccines are 30 times lower than for human vaccines (Knight-Jones *et al.*, 2014). Veterinary vaccines must be very cheap as they are often paid for directly by farmers. For this reason, many of the veterinary vaccines available in the developing world rely on old technologies and many are of poor quality (Greenwood *et al.*, 2011). This is, probably, one of the reasons producers of veterinary vaccines are less competitive than the human vaccine producers. However, improved vaccines against several viral diseases of domestic animals are now being developed using newer technologies.

There are many reasons to believe that scientific and technological cooperation between these two fields should be mutually beneficial. This is clear from the early days of vaccination. It was common knowledge in rural England that cowpox conferred immunity to smallpox, but without acceptance by the medical community; Jenner's main merit was to

Table 18.2. Licensed vaccine adjuvants (Mbow *et al.*, 2010).

Adjuvant	Company	Class	Indications
Alum	Various	Mineral salts	Various
MF59	Novartis	O/w emulsion	Influenza (Fluad)/pandemic flu
ASO3	GSK	O/w emulsion+α-tocopherol	Pandemic flu (Pandemrix)
AS04	GSK	MPL+alum	HBV (Fendrix), HPV (Cervarix)
Liposomes	Crucell	O/w emulsion	HAV, Flu (EU)

give a scientific demonstration of it, by systematic observations of disease in animals and men and an experimental approach. After Jenner's paper (Jenner, 1798), inoculation of cowpox to prevent smallpox became widely accepted and practised around the world.

A second example is that, in 1878, Pasteur received, from the veterinarian Henry Toussaint, a strain of bacteria (now *Pasteurella multocida*) that caused chicken cholera. He infected chicken with food contaminated with this bacterium and many chickens died, but some recovered and were resistant by a challenge with the same bacteria. Continuing his researches in the laboratory, subjecting the cultures to several stressing conditions, he was able to obtain two strains of the bacteria, one virulent and another non-virulent, and so he could protect chickens against cholera by vaccination with the attenuated strain. This was the first attenuated vaccine developed in the laboratory, and the results of these experiments were communicated to the 'Académie des Sciences' in Paris, in 1880 (Bazin, 2011). Work followed with vaccines against anthrax, basically a veterinarian vaccine, and rabies, a veterinarian and human vaccine.

A third example is the collaboration between Albert Calmette, a physician, and Camille Guérin, a veterinarian and microbiologist, for the development of BCG vaccine, beginning in 1897. Calmette needed the cooperation of Guérin for the animal studies. They discovered that animals infected with low virulence bacilli resisted a virulent reinfection, and this was the starting point. From a strain of *Mycobacterium bovis* transferred to their laboratory, in 1904, they performed 239 passages in culture media from 1906 to 1919, and obtained an attenuated strain that could protect several animal species against bovine and human tubercle bacilli. This strain has been used as the BCG vaccine against human tuberculosis since 1921 (Gheorgiu *et al.*, 2010; Gheorgiu, 2011).

The last example, but many more could be easily found, was the development of the technique of inoculation in eggs, developed by Ernest Goodpasture, then a pathologist studying fowlpox. This technique was employed for the development of human influenza and mumps vaccines (Offit, 2007).

The difficulty of convincing human vaccine producers to enter the field of veterinarian vaccines is long-standing. Maurice Hilleman developed a vaccine for Marek's disease, a viral disease that causes serious illness in chicken, including cancer, but it was difficult to convince Merck to enter into this business. This was the first vaccine against cancer. In 1974, Merck bought Hubbard Farms, which raised chicken, and soon became the biggest chicken and egg producer in the world, because these vaccinated flocks had much greater productivity (Offit, 2007).

Nevertheless, it is important to note that scientific and technological collaborations between the fields of human and veterinary vaccines have significantly increased in the last four decades and now there are several common grounds in vaccine development and production. As a matter of fact, the area of veterinary vaccines has provided many important contributions for the development of human vaccinology, especially for the production of vaccines at an industrial scale.

Several scientific and economic advantages may result from partnerships between the human and veterinarian vaccine researchers and businessmen. A remarkable example is the development of a vaccine against *Clostridium chauvei*, an important pathogen for cattle in Brazil and responsible for huge economic losses, by Alcides Godoy, a Brazilian scientist, in 1906, at the Oswaldo Cruz Institute. The financial returns of this vaccine were fundamental for the maintenance of the institute and investments, and the prohibition of its commercialization by the institute has led to an institutional crisis (Benchimol, 2001).

18.14 Accelerating Vaccine Development in Emerging Countries

Manufacturers in emerging countries, such as Brazil and other DCVMN countries, have reasonable national capacity for the production of vaccines. In Brazil, Bio-Manguinhos at Fiocruz Rio de Janeiro and the Institute Butantan in São Paulo meet 90% of the demand of vaccines from the National Health System (SUS). This was due to the significant support from the government by the National Self-Sufficiency Program for Immunobiologicals (PASNI), created in the mid-1980s, and for over 10 years it has supported the modernization of public manufacturing laboratories, resulting in quality products. In this context, Bio-Manguinhos has been investing in the development and production of vaccines, kits for diagnosis and monitoring and has provided, for the country's blood banks, the NAT kit (nuclear acid testing), consisting of rapid tests for diagnosis of HIV and other infectious diseases. However, major challenges still remain in the country, both in the areas of innovation, technological development and in production. This challenge was clearly pointed out by a study by the Bio-Minas Foundation. A survey of around 300 biotechnology companies in Brazil indicates that in spite of recent advances, most of them are still small. Another study by CEBRAP found, in the same direction, that the biotechnology enterprises in the country are still embryonic, which raises concerns about national competitiveness, considering the strategic economic role of this sector. This situation, among other barriers, has contributed to increase the deficit of the country's trade balance, which in 2012 reached US$12.2 billion (Homma and Possas, 2014). Public–private partnerships in biotechnology are therefore a crucial issue for Brazil (Rezaie *et al.*, 2008).

Only 4% of all pharmaceutical products (including vaccines) purchased by the Brazilian Ministry of Health are biotechnology-based, but they account for 43% of all national expenditure with pharmaceuticals, around US$3 billion, demonstrating the fragility of technological and industrial capacity of the country in this area (Homma and Possas, 2014). This scenario exposes the population to the possibility of shortages and creates a highly vulnerable situation. Recent government policy, creating agreements between public laboratories and private enterprises (Partnerships for Product Development – PDPs) as an attempt to overcome this critical situation, form an important strategy in national capacity building in this area. The country has now 104 PDPs for development and production of 97 health products (drugs, vaccines and other products), linking 19 public laboratories and 60 private enterprises, 30 of them multinational and 30 national. However, there are several challenges in the organization of PDPs, such as difficulties in defining innovative partners and especially the adequate identification of the partner that can transfer the technology. In this context, it will be necessary that the receiver of the technology transfer has adequate knowledge of the technological aspects involved in vaccine production, particularly those related to IP, information on processes and production procedures; pre-clinical and clinical studies and post-marketing surveillance, which assures the quality of the product. Local capacity and experience in vaccine development, production, technology transfer agreements, marketing and commercialization is also very important.

It is crucial to ensure the strategic management implementation of PDPs, such as organizing groups with capacity for technological prospect and vaccine patent analysis, with specific focus on strategic search results. Competence and speed are necessary to create, organize, structure, strengthen or build new facilities production in the institutions involved, especially in technology recipient institutions. It requires the preparation, on an urgent basis, of a large number of specialized professionals in the fields of innovation, technology development, production and management. This national strategy will also require sustainable medium- and long-term financing, creating significant incentives for innovation and technological development of the new biotechnology processes and products (Homma and Possas, 2014).

18.15 Final Considerations

Vaccine discovery and innovations require, particularly in the developing countries, increasing investments in high-throughput screening of new vaccine antigens and candidates and access to new vaccine adjuvants. New platforms will be necessary, supported by multi-stakeholder, manufacturing facilities capable of rapid production of new human and veterinary vaccine products.

From this perspective, a more flexible global IPR legislation is necessary to accelerate the flow of vaccine development from laboratory to production and delivery in the developing nations, supported by other incentives to innovation, such as patent pools, providing rapid access of population to the new scientific advances.

The new global initiatives, in the context of the Global Vaccine Action Plan (GVAP), following the Decade of Vaccines, will be crucial to accelerate vaccine development and to overcome current IPR constraints affecting manufacturers in developing countries. Considering the complexity of the issues involved in vaccine innovation, development and patent protection, it is urgent, from a problem-solving approach, to intensify the existing collaborations between academy, industry, research agencies, regulators and government representatives and developing countries' manufacturers.

References

Arrow, K. (1962) Economic welfare and the allocation of economic resources for invention. In: Universities-National Bureau Committee for Economic Research (ed.) *The Rate and Direction of Inventive Activity: Economic and Social Factors*. Princeton University Press, Princeton, New Jersey, pp. 609–626.

Bazin, H. (2011) Pasteur and the birth of vaccines made in the laboratory. In: Plotkin, S.A. (ed.) *History of Vaccine Development*. Springer, New York, pp. 33–45.

Benchimol, J.L. (2001) *Febre amarela: a doença e a vacina, uma história inacabada* [Yellow Fever: The Disease and the Vaccine, an Unfinished History]. Editora Fiocruz (ed.), Rio de Janeiro, Brazil.

Bregu, B., Draper, S.J., Hill, A.V.S. and Greenwood, B.M. (2011) Accelerating vaccine development and deployment: report of a Royal Society satellite meeting. *Philosophical Transactions of the Royal Society of London. Series B, Biological Sciences* 366, 2841–2849.

CDC (Centers for Disease Control and Prevention, US Government) (2013) *Zoonotic diseases.* Available at: http://www.cdc.gov/onehealth/zoonotic-diseases.html (accessed 20 September 2015).

Evens, R. and Kaitin, K. (2015) The evolution of biotechnology and its impact on health care. *Health Affairs* 34(2), 201–219.

Gheorgiu, M. (2011) Antituberculosis BCG vaccine: lessons from the past. In: Plotkin, S.A. (ed.) *History of Vaccine Development*. Springer, New York, pp. 47–55.

Gheorgiu, M., Lagranderie, M. and Balazuc, A.M. (2010) Tuberculosis and BCG. In: Artenstein, A.W. (ed.) *Vaccines: A Biography*. Springer, New York, pp. 125–140.

Godinho, M.M. (2011) Propriedade Intelectual: Racional de Utilização e Desafios Futuros em Países de Língua Portuguesa [Intellectual Property: Rationale of Utilization and Future Challenges in Portuguese Speaking Countries]. In: Castro, A.C., Possas, C. and Godinho, M.M. (eds) *Propriedade Intelectual nos Países de Língua Portuguesa* [Intellectual Property in Portuguese Speaking Countries: Topics and Perspectives]. E-papers, Rio de Janeiro, Brazil, pp. 55–74.

Greenwood, B., Salisbury, D. and Hill, A.V.S. (2011) Vaccines and global health. *Philosophical Transactions of the Royal Society of London. Series B, Biological Sciences* 366(1579), 2733–2742.

Homma, A. (2009) The Brazilian vaccine manufacturers' perspective and its current status. *Biologicals* 37(3), 173–176.

Homma, A. and Possas, C. (2014) Panorama da Biotecnologia em Saúde [An overview of Health Biotechnology]. *Facto Abifina*, 40, ano VIII.

Homma, A., Tanuri, A., Duarte, A.J., Marques, E., de Almeida, A., Martins, R. and Possas, C. (2013) Vaccine research, development, and innovation in Brazil: a translational science perspective. *Vaccine* 31S, B54–B60.

IPEA (2013) Propriedade Intelectual e Aspectos regulatórios em Biotecnologia [Intellectual Property and Regulatory Aspects in Biotecnology]. Zucoloto, G.F. and Freitas, R.E (org.) IPEA, Rio de Janeiro, Brazil.

Jenner, E. (1798) *An Inquiry into the Causes and Effects of Variolae Vaccinae*, xth edn. Sampson Low, London.

Knight-Jones, T.J.D., Edmond, K., Gubbins, S. and Paton, D.J. (2014) Veterinary and human vaccine evaluation methods. *Proceedings of the Royal Society B* 281, 2013–2839.

Mahoney, R., Pablos-Mendez, A. and Ramachandran, S. (2004) The introduction of new vaccines in developing countries. III. The role of intellectual property. *Vaccine* 22(5–6), 786–792.

Mbow, M.L., De Gregorio, E., Valiante, N.M. and Rappuoli, R. (2010) New adjuvants for human vaccines. *Current Opinion in Immunology* 22(3), 411–416.

Meeusen, E.N., Walker, J., Peters, A., Pastoret, P.P. and Jungersen, G. (2007) Current status of veterinary vaccines. *Clinical Microbiology Reviews* 20(3), 489–510.

Milstien, J., Gaule, G. and Kaddar, M. (2007) Access to vaccine technologies in developing countries: Brazil and India. *Vaccine* 25(44), 7610–7619.

Offit, P.A. (2007) *Vaccinated: One Man's Quest to Defeat the World's Deadliest Diseases*, xth edn. Smithsonian Books, New York.

Padmanabhan, S., Amin, T., Sampat, B., Cook-Deegan, R. and Chandrasekharan, S.S. (2010) Intellectual property, technology transfer and manufacture of low-cost HPV vaccines in India. *Nature Biotechnology* 28, 671–678.

Possas, C.A. (2013) Propriété intellectuelle et le SIDA dans les pays em développement: innovation et accès aux produits pharmaceutiques. In: Possas, C. and Larouzé, B. (eds) Propriété intellectuelle et politiques publiques pour l'accès aux antirétroviraux dans les pays du Sud. ANRS, Paris, pp. 37–50.

Possas, C., Antunes, A.M.S., Mendes, F.M.L., Schumacher, S.O.R., Martins, R.M. and Homma, A. (2015) Access to new technologies in multi-patented vaccines: challenges for Brazil. *Nature Biotechnology* 33, 599–603.

Rezaie, R., Frew, S.E., Sammut, S.M., Maliakkal, M.R., Daar, A.S. and Singer, P.A. (2008) Brazilian health biotech-fostering crosstalk between public and private sectors. *Nature Biotechnology* 26(6), 627–644.

Sette, A. and Rappuoli, R. (2010) Reverse vaccinology: developing vaccines in the era of genomics. *Immunity* 33, 530–541.

Von Ranke, N.L. (2015) Prospecção Tecnológica na Base Integrity em Documentos de Patentes na Área de Biotecnologia com Foco em Câncer [Technology Foresight in Patent Documents using Integrity Database in the Biotechnology area with Focus on Cancer]. MSc thesis. Brazilian Institute of Industrial Property (INPI), Rio de Janeiro, Brazil.

World Health Organization (2013) *Global Vaccine Action Plan, 2011–2020*. WHO, Geneva.

19 Promoting Access to Healthcare Through Biosimilars: Addressing Intellectual Property Rights and Regulatory Barriers

KANIKARAM SATYANARAYANA[1,2]* AND SADHANA SRIVASTAVA[1]

[1]*Intellectual Property Rights Unit, Indian Council of Medical Research, Ansari Nagar, New Delhi;* [2]*Biosafety Unit, Department of Biotechnology, New Delhi*

19.1 Introduction

Biopharmaceutical drugs have become an important and integral part of modern healthcare system all over the world. Biological product is itself defined by the US Food and Drug Administration (US FDA, 2015) as:

> a wide range of products including vaccines, blood and blood components, allergenics, somatic cells, gene therapy, tissues, and proteins. Unlike most traditional, small-molecule prescription drugs that are made through chemical processes, biological products are generally made from human and/or animal materials. Biological products are usually larger than and have a more complex structure than small-molecule prescription drugs. Such products may be manufactured through biotechnology, derived from natural sources, or, in some cases, produced synthetically

Biological products are both therapeutically and economically important. In contrast to other therapeutic agents such as chemotherapies, biological products, particularly monoclonal antibodies, are highly selective, offering effective treatments against dreaded diseases with fewer side-effects (So and Sachs, 2012) A biotherapeutic is a large structurally complex molecular structure (acetylsalicylic acid: 21 atoms and IgG1 antibody >20,000 atoms) typically derived from living cells and used in the treatment, diagnosis or prevention of disease. By understanding the mechanisms of diseases, such as multiple sclerosis, biologic medicines can be developed to target and modify underlying causes of disease, potentially altering the course of disease rather than simply treating symptoms (Hartung, 2004). From the first recombinant insulin introduced during the early 1980s, biotherapeutic medicines now benefit more than 350 million patients worldwide. Biotherapeutics have been proven effective for diseases that chemical drugs could not treat and are a critical part of treatment regimens for diseases such as cancer, hepatitis C, chronic renal failure, rheumatoid arthritis, psoriasis, Crohn's disease, ulcerative colitis, haemophilia, Fabry's disease, growth deficiency, multiple sclerosis etc. (Grewal, 2009; IFPMA, 2013).

19.2 Definition of Biosimilars/Biogenerics

Biogenerics are also called 'follow-on-biologics' or 'biosimilars' and 'similar biologics' in various countries (Table 19.1).

*Corresponding author, e-mail: kanikaram_s@yahoo.com

Table 19.1. Definitions of biological medicines according to regional regulatory agencies (Health Canada, 2010; Ministry of Food and Drug Safety, South Korea (MFDS), 2010; Food and Drug Administration, China, 2015).

Nomenclature	Agency	Definition
Similar biotherapeutic product	WHO (2009)	A biotherapeutic product which is similar in terms of quality, safety, and efficacy to an already licensed reference biotherapeutic product.
Similar biological medicinal product	EMA (2005)	'Biosimilar', is a product which is similar to a biological medicine that has already been authorized, the so-called 'reference medicinal product'. The active substance of a biosimilar medicine is a known biological active substance and similar to the one of the reference medicinal product. A similar biological medicinal product and its reference medicinal product are expected to have the same safety and efficacy profile and are generally used to treat the same conditions.
Biological product/ biosimilar	US FDA (2012a)	The term 'biosimilar' or 'biosimilarity' in reference to a biological product as a biological product that is the subject of an application under section (k), means: (a) that the biological product is highly similar to the reference product notwithstanding minor differences in clinically inactive compounds; and (b) there are no clinically meaningful differences between the biological product in terms of the safety, purity and potency of the product.
Biosimilar	NIFDSE, South Korea (2009)	A biological product that is comparable to an already marketed reference product in terms of quality, safety and efficacy.
Subsequent entry biologic (SEB)	Canada (2010)	A biologic drug that enters the market subsequent to a version previously authorized in Canada, and with demonstrated similarity to a reference biologic drug. An SEB relies in part on prior information regarding safety and efficacy that is deemed relevant due to the demonstration of similarity to the reference biologic drug and which influences the amount and type of original data required.
Similar biologics	DBT, India (2012)	A biological drug produced by generic engineering techniques and claimed to be 'similar' in terms of safety, efficacy and quality to a reference biologic, which has been granted marketing authorization in India by the DCGI on the basis of a complete dossier, and with a history of safe use in India.
Follow-on biologic	MHLW, Japan (2009)	A 'follow-on' biologic is a biotechnological product developed to be comparable in regard to quality, safety and efficacy to an already approved biotechnology-derived product ('original biologic' of a different company.
Biosimilar	FDA, China (2015)	A biosimilar is defined as a therapeutic biologic that is similar to a reference product approved in China or elsewhere in quality, safety and efficacy; the biosimilar should in principle have the same amino acid sequence as the reference product.

19.2.1 The World Health Organization

The WHO's definition is a biotherapeutic product which is similar in terms of quality, safety and efficacy to an already licensed reference biotherapeutic product (WHO, 2009). The WHO refers to biosimilars or 'similar biologics' as 'similar biotherapeutic products' (SBPs) (WHO, 2009).

19.2.2 The European Medicines Agency

The European Medicines Agency defines 'similar biological' or 'biosimilar' medicine as one 'that is similar to another biological medicine that has already been authorized for use. Biological medicines are medicines that are made by or derived from a biological source, such as a bacterium or yeast. They can consist of relatively small molecules such as human insulin or erythropoietin, or complex molecules such as monoclonal antibodies' (EMA, 2013a). A biosimilar is a biological medicinal product that contains a version of the active substance of an already authorized original biological medicinal product (reference medicinal product). A biosimilar demonstrates similarity to the reference product in terms of quality characteristics, biological activity, safety and efficacy based on a comprehensive comparability exercise (EMA, 2012, 2013a).

Eventually, 7 years after the release of initial regulatory guidelines (EMA, 2005), the European Medicine Agency (EMA) included in 2012 a definition of a 'biosimilar' in an EMA procedural guidance document (EMA, 2013b):

> A similar biological medicinal product, also known as 'Biosimilar', is a product which is similar to a biological medicine that has already been authorised, the so-called 'reference medicinal product'. The active substance of a biosimilar medicine is a known biological active substance and similar to the one of the reference medicinal product. A similar biological medicinal product and its reference medicinal product are expected to have the same safety and efficacy profile and are generally used to treat the same conditions.

19.2.3 The United States Food and Drug Administration

The FDA defines a biosimilar as a biological product that is highly similar to a US-licensed reference biological product notwithstanding minor differences in clinically inactive components, and for which there are no clinically meaningful differences between the biological product and the reference product in terms of the safety, purity and potency of the product (US FDA, 2012a).

However, the FDA does not consider biosimilars as generic versions of biological products and 'the term "generic" applies only to small-molecule drugs that are the same as, and bioequivalent to, an already-approved small-molecule drug regulated under the FD&C Act'. This is unlike the definition of a generic drug by the FDA:

> A generic drug is the same as a brand name drug in dosage, safety, strength, how it is taken, quality, performance, and intended use. Before approving a generic drug product, FDA mandates many rigorous tests and procedures to assure that the generic drug can be substituted for the brand name drug. The FDA bases evaluations of substitutability, or 'therapeutic equivalence,' of generic drugs on scientific evaluations. By law, a generic drug product must contain the identical amounts of the same active ingredient(s) as the brand name product. Drug products evaluated as 'therapeutically equivalent' can be expected to have equal effect and no difference when substituted for the brand name product.
>
> US FDA, 2012b

19.2.4 India

In India, a 'similar biologic' is defined as: 'A biological product/drug produced by genetic engineering techniques and claimed to be "similar" in terms of safety, efficacy and quality to a reference biologic, which has been granted a marketing authorization in India by the Drug Controller General of India (DCGI) on the basis of a complete dossier, and with a history of safe use in India' (DBT, 2012).

'Biosimilars' are also called 'follow-on pharmaceuticals' in the USA and Japan, 'subsequent entry biologics' in Canada and 'biocomparables' in Mexico.

There is another term 'biobetters' that refers to enhanced or 'better' versions of biosimilars, or the innovator drugs. Typically, this therapeutic molecule is modified from the innovator drug or a biosimilar using platform technologies to create a biological product with superior features in terms of improved efficacy, different mode of delivery, reduced toxicity, longer shelf life etc., overall aiming to provide increased treatment gains for patients. Unlike biosimilars, which aim to establish similarity to a known biological, biobetters seek superiority in one or more aspects of their clinical profile (Dolinar and Reilly, 2013). For example, while working against the same target protein, biobetters may have structural changes, bi-functional targeting (with or without a biosimilar core) or an improved formulation that may result in an expected improvement in safety and/or efficacy (Dolinar and Reilly, 2013). One example of a biobetter could be Roche's obinutuzumab (Gazyvara), an anti-CD20 monoclonal antibody, which has shown superior efficacy in the treatment of chronic lymphocytic leukaemia (CLL) as compared to its 'originator' rituximab (MabThera, Roche) (Anour, 2014). Biobetters straddle the space between biosimilars and new biological entities (NBEs). Also, a biobetter is treated as an NBE by the regulator and has to meet the stiff regulatory stipulations (Anour, 2014). Biobetters of improved profiles can qualify for patent protection, making their low business-risk proposition even more attractive. NBEs, by nature, garner multiple product and process patents covering ideation, technology, clinical application and a host of other areas, amplifying the returns on innovation investments. Often, biobetters are used as a defence strategy by originator companies, to protect their market against potential competing biosimilars (Dolinar and Reilly, 2013; Anour, 2014).

In this chapter, the term biosimilar includes: similar biotherapeutic product, similar biological medicinal product, biological product, subsequent entry biologic, similar biologics and follow-on biologic.

19.3 Importance of Biotherapeutics in Healthcare

Biological products including biosimilars are both therapeutically and economically important as, unlike conventional therapeutics, biologicals like monoclonal antibodies are highly selective and specific as treatment options. There are over 150 'reference product' biologics approved for marketing in the USA. Over 370 biopharmaceutical drug products and vaccines are in various stages of clinical trials focusing on about 200 diseases including cancer, Alzheimer's disease, heart disease, multiple sclerosis, AIDS, arthritis etc. (IFPMA, 2012, 2013). There is a trend towards increasing use of biologicals, which is expected to continue with biological products increasingly replacing 'small molecule' products and, in some disease conditions, as the preferred therapeutic method (Krishnan *et al.*, 2015). Monoclonal antibodies (mAbs)-based therapies are claimed to have ushered in a marked change in treatment from conventional therapeutic regimens, as the protein-based biologics offer novel and more efficient options for disease care as they bind to specific target cells (IFPMA, 2012, 2013). Global sales of biosimilars are expected to reach between US$1.9–2.6 billion (2015) and US$25 billion in 2020, with an estimated annual compound growth rate of 7.7% (Bourgoin and Nuskey, 2013). Among different therapeutic classes, like erythropoietins, growth hormones, interferons, stimulating factors, low molecular weight heparins, insulins, monoclonal antibodies and Fc fusion proteins, monoclonal antibodies have been established as a major product class and occupy a large sector of the biosimilar market space (IFPMA, 2012, 2013).

19.4 Intellectual Property Rights Issues

Intellectual property rights (IPRs), especially patents, play an important role in pharmaceutical innovation as patent protection encourages the creation, dissemination and efficient exploitation of technology and leads to products for public use. Patents give inventors a geographically distinct, limited term legal monopoly on their inventions with the right to exclude all others from the scope of protected invention. There is little doubt that intellectual property (IP) protection incentivizes innovators (and their assignees, typically pharma companies) to create new health products through research and development with significant investments. IP protection thus creates a system of monopoly wherein some or many of the newer products may be priced so high that they remain out of reach of many patients. In addition, the patent owners like to extend the monopoly (typically 20 years) through several means to prevent the entry of generic manufacturers into the market after the expiry of patent protection of the innovator product. However, there are arguments that strong patent protection is needed for the development of biosimilars (Grabowski *et al.*, 2015).

The market for biologicals is quite large and therefore very lucrative for the manufacturers of biotherapeutics. According to estimates, four of the top ten biologics may account for over US$60 billion in revenue while the rest contribute to about US$30 billion (Gal, 2011; BCC Research, 2012). At the top of that list is Humira, which goes off patent protection in 2016, which alone accounts for about US$10 billion followed by Remicade, which loses protection in Europe in early 2015 and in the USA by 2018. In 2010 to 2014, a number of blockbusters are projected to go off patent, representing more than US$209 billion in annual drug sales (Carroll, 2014). This trend is projected to result in a decrease in the sale of innovator products with a decrease in the sale of branded products of about US$113 billion (Paul *et al.*, 2010). IMS Health estimates that biologicals comprising US$64 billion worth of products will be off patent in the US and European markets by 2015 (Greer, 2013).

Even biosimilars garner significant sales. According to estimates by Frost and Sullivan and Allied Marketing Research, the global biosimilar companies earned a revenue of about US$172 million in 2010, which is expected to reach US$1.3 billion by 2013 and US$3.987 billion by 2017. Estimates also suggest that the overall annual compounded growth of innovator biologicals and biosimilars is set to touch an annual growth rate of 56.7% between 2010 and 2017. The total revenues are expected to swell to US$35 billion by 2020 as newer products undergoing trials penetrate the market in North America, Europe and Asia (GABI Online, 2014). Biologicals are also significant drivers of growth and revenue source as the healthcare market is expected to expand due to the increasing ageing population coupled with the shift in the disease profile to non-communicable and chronic diseases not easily amenable to treatment by conventional chemical drugs.

In view of the large revenues, innovator pharma companies try to extend their monopoly of their inventions by several means, including patent evergreening, filing divisional application, patent linkage and data exclusivity etc.

19.4.1 Patent evergreening

'Patent evergreening' generally refers to the strategy of obtaining more than one patent covering diverse aspects of the same product, typically by obtaining patents on improved versions of existing product(s) to unjustly prolong the patent monopoly of existing products (Roox *et al.*, 2008; Schacht and Thomas, 2009). The global IP protection system permits the grant of new patents for improvement of an existing invention and there are always attempts by innovator companies to file patents for even minor improvement or change. Such patent applications, also called follow-on patents, are generally weak and could be trivial (Schacht and Thomas, 2009). More rigorous examination by patent examiners would show that the inventive step is not novel enough for justifying granting of a new patent (Schacht and Thomas, 2009). In India, some patent protection for innovator products has been refused recently by Indian courts as the innovation component of these patents was considered too trivial.

According to the European Generic Manufacturers Association (Sheppard, 2009) the following are some methods used by innovator companies for evergreening to build a patent thicket:

1. First patent obtained for the active pharmaceutical ingredient itself, typically the first patent in the thicket. The specific claim may be for just one molecule, or the particular molecule might be disclosed as a component of a broader family.
2. A polymorph or hydrated form of the active substance.
3. A simple salt of the active pharmaceutical substance.
4. An isomeric form of the drug.
5. A substantially pure form of the drug.
6. An impurity inherent in an already disclosed process of making the drug.
7. Formulations, whether in solution or in solid form.
8. Concentrations in dosage forms.
9. Second medical use of the patent, especially where that use is in a related field to the one originally disclosed in the first patent or a variation of the dosage regime already disclosed.

19.4.2 Filing multiple divisional patent applications

A patent application filed later for an independent or distinct invention, carved out of a pending patent application and disclosing and claiming only subject matter disclosed in the earlier or parent application, is known as a divisional application or 'division'. A divisional application thus divides a single (original) parent patent application into two or more applications. Generally, the subsequent applications are filed with a time gap to attempt extension of patent protection period. The divisional application must only contain subject matter in the application from which it is divided (its parent), but retains the filing and priority date of that parent. A properly filed divisional retains the filing and priority date of the parent application but thereafter is treated independently, which ensures the company owning the original patent extension of monopoly with the divisional application. Generally, a divisional application is considered useful if a unity of invention objection is issued, in which case the second invention can be protected as a divisional application. But in actual practice divisional applications are typically filed to delay the entry of generics into the market. Generic companies will be reluctant to enter the market as the pending (divisional) patent application of the innovator company could pose them a constant risk, because if the pending divisional patent is granted it could well result in another 20-year patent protection. Some innovator companies resort to this method by keeping a series of pending divisional applications on file to maintain a stage of uncertainty for the generic companies (EMA, 2014a; Shepherd, 2015).

19.4.3 Patent linkage

This is another unfair practice and is considered abuse and exploitation of procedural loopholes.

Patent linkage is the practice of linking either market approval, the pricing and reimbursement status or any regulatory approval for a generic medicine to the patent status of the originator reference product. Patent linkage is essentially an administrative and/or regulatory process under which the regulator and/or other responsible agency in a country will not grant the relevant authorization for a generic medicine to any party not authorized by the patent holder for that drug product until either: (i) the expiry of all relevant patents covering the drug product; or (ii) the determination by a governmental body, either executive or judicial, that the relevant patents are not being infringed, or are invalid or unenforceable (EMA, 2014a).

19.4.4 Data protection/data exclusivity

As a condition for registering pharmaceutical products, countries normally require applicants to submit (disclose) data relating to quality, safety and efficacy (so called 'test data'), as well as all other relevant information on the physical, chemical characteristics and composition of the product for marketing approval (Satyanarayana *et al.*, 2006). The innovator companies which approach the regulator for marketing approval of their drug submit their test data file. Several countries offer such data protection for varying periods. This system of 'data exclusivity' virtually extends the monopoly of the innovator company during which period the regulator cannot approve any biosimilar (or generic) versions of a drug, granting the original biologic an effective monopoly regardless of developed biosimilars (Roth, 2013).

In the case of biosimilars, data exclusivity refers to the period during which a biosimilar is not permitted to use a reference drug's safety information to file a truncated or expedited application for regulatory approval in the USA where data exclusivity is permitted. During this period of data exclusivity, a proposed biosimilar applicant must pay the costs – in time and capital – to secure FDA approval as though it were an original biologic. Innovator pharma companies defend data exclusivity claiming that the 20-year patent protection and regulatory data protection, while appearing to be complementary, serve two distinct purposes. Patent protection is granted for any new invention that fulfils the criteria of a patent as defined under national laws. Data exclusivity is a special privilege granted by some countries to innovator companies to further incentivize biomedical innovations and innovator companies to invest resources and time for research and development necessary to establish that the new drug or biological product is safe and effective for introduction into the healthcare system. Data exclusivity therefore essentially prevents third parties (generic companies) from approaching the regulator for marketing approval of a same or similar product and is a major barrier for generic companies from entering the market even after the expiry of patent protection. This is because typically the generic companies depend on the preclinical and clinical data submitted by the innovator company and they need not redo clinical studies for gathering evidence on the safety and efficacy of the biologic and would typically need to do 'bridging' studies to prove that the product is similar to the innovator products in terms of safety and efficacy. Data exclusivity ensures that the preclinical and clinical evidence submitted by the innovator company to the regulator cannot be accessed by the generic company for marketing purposes, extending the monopoly beyond the 20-year patent protection period.

The US Patient Protection and Affordable Care Act (2009) proposes 12 years of data exclusivity (including 4 years of market exclusivity) for biologics, which is 7 more years than conventional, small-molecule drugs. Some analysts feel that the additional 7 years is not warranted as there appears to be little difference in the R&D costs to bring a biologic to market (US$1.2 billion) as compared to a conventional drug (US$1.318 billion) (PHRMA, 2009). The justification provided by the innovator pharma is that the complexity of the developmental phase of biologics (as compared to conventional chemical molecules) necessitates more rigorous clinical testing to prove safety and efficacy, which is very expensive. Biosimilar manufacturers argue that they also need to invest more time and money to establish bioequivalence of biologicals as compared to conventional drugs and therefore the barrier is unreasonably high. In fact, recognizing this higher barrier to generic competition, the US Federal Trade Commission has recommended that biologics may not be given any period of data exclusivity in the USA (Federal Trade Commission, 2009).

The 10-year data exclusivity given by the EU has since been revised at the latest review of pharmaceutical EU legislation to the following: 10 years if the reference product is centrally approved or application to the centralized procedure has been made before 20 November 2005; or 8 years' data exclusivity + 2 + 1 formula if a full dossier is submitted on or after 30 October 2005 via national procedure or 20 November 2005 via the centralized procedure (ECID, 2012). The 12-year data exclusivity given to innovator biologic companies in the USA is an important contentious issue in the negotiations of the Trans-Pacific Partnership, a free-trade agreement between (and led by) the USA and several nations (Bourgoin and Nuskey, 2013).

After detailed analysis, Harvard economist Kotlikoff has opined that there is no justification for the exclusivity proposed under the Biologics Price Competition and Innovation Act (BPCIA) (Kotlikoff, 2008, 2009). Comparative economic analysis of the R&D of small molecules with biosimilars shows that while small-molecule drugs are generally manufactured in labs using chemical synthesis, the production of most biologics requires cell-culture facilities that can take 3–5 years to build, using materials 20–100 times more expensive than those for facilities used to manufacture molecule-based drugs, at a cost of US$200 million or more (Shapiro, 2008). These factors also could extend the process of approval for many biogenerics, with one analyst estimating that developing a generic biologic, producing it and gaining FDA approval could take 5–8 years: 1–2 years to carry out the cell biology, 1 year for process analysis and 2–4 years for regulatory approval (Shapiro, 2008).

19.5 The Patent Dance

The patent dance is a procedure for resolving patent disputes between a biosimilar applicant and the innovator company on applications for biosimilars submitted to the FDA under the Biologics Price Competition and Innovation Act, 2009 (Fogel and Hanna, 2014; Mann and Mahinka, 2014). This procedure envisages exchange of information and data between the biosimilar applicant and the innovator company with a strict time limit after submission of the applicant for marketing approval by the biosimilar company.

Within 20 days after the FDA has accepted the application, the biosimilar applicant must provide the innovator company with confidential access to the biosimilar application as well as other relevant manufacturing details. Within 60 days of receiving these materials, the innovator company must provide to the biosimilar applicant: (i) a list of patents it believes are being infringed; and (ii) identify which, if any, of these patents the innovator company would be willing to license to the biosimilar applicant. Within 60 days of receipt of the patent list, the biosimilar applicant must respond to the innovator company with a statement indicating, on a claim-by-claim basis, the factual and legal basis as to why each patent of the innovator company is invalid, unenforceable and/or not infringed. In addition, during this period, the biosimilar applicant may also provide to the innovator company a counter list of patents that the biosimilar applicant believes could be subject to a claim of patent infringement. Within 60 days of receiving these materials, the innovator company must provide a reciprocal statement describing, on a claim-by-claim basis, the factual and legal basis that each patent will be infringed, as well as a response to any statement regarding validity and enforceability.

Both the parties then have up to 15 days to negotiate in good faith to arrive at a list of patents, if any, that should be subject to a patent infringement action. If both the parties reach agreement, then the innovator company must bring an infringement action within 30 days for each patent on the negotiated list. On the other hand, if the parties fail to reach an agreement, the biosimilar applicant must notify the innovator company the number of patents it will provide in a second list. Both the parties then simultaneously exchange within 5 days of this notice a list of patents that each party believes should be the subject of the infringement litigation. Within 30 days after this exchange, the reference product sponsor must bring an infringement action on all the patents on the simultaneously exchanged lists.

While there is no automatic regulatory stay of approval of the biosimilar application during this period of patent litigation by the regulator, the biosimilar applicant is always at risk of legal action on patent infringement launched by the innovator company. A proactive innovator company could thus potentially delay the launch of a biosimilar through this complex process. Not surprisingly, there are several cases pending before US courts on the interpretation of the patent dance system applicable to the biosimilar applicants under the BPCI Act (Fogel and Hanna, 2014). These initial patent challenges under the BPCIA and the resultant uncertainty may result in further delays with respect to approval of biosimilars and their marketing (Mann and Mahinka, 2014). There are hundreds of cases pending before US courts on the issue of patent dance (Mann and Mahinka, 2014).

19.6 New IP Challenges

In addition to IP and regulatory barriers, innovative pharma companies have been devising newer strategies to deter biosimilar competition, such as frequently changing delivery devices (e.g. switching from vials to prefilled syringes to autoinjectors) and changing of presentations (from liquid to lyophilized). In addition, often several older drugs and administrative devices are withdrawn merely to increase the market for newer products at a higher price. This strategy puts the biosimilar companies at a great disadvantage as the original reference standard that was used at the time of marketing approval to the regulator is no longer valid (Krishnan *et al.*, 2015). They need to collect the data for the biosimilar all over again, resulting in significant cost and time overrun and delaying access of an affordable drug to the needy. Another strategy for extending monopoly is patenting of analytical methods by the innovator companies that are used to evaluate critical quality attributes of the marketed biologic, which limits the biosimilar applicant in accurately determining the potency of their product, as the patent life of the method is much beyond the patent life of the original product. Such frequent process changes of reference standards are well documented for Enbrel, Rituxan, Mabthera and Aranesp (Schiestl *et al.*, 2011).

Compounding the complexity of patenting in the area of biologics are some recent decisions of the US Supreme Court on patent protection of human biological materials, such as the invalidation of some granted patents on isolated genes and some diagnostic methods, holding that they are not patentable under Section 101 of the US Patent Act.[1] These decisions of the US Supreme Court are likely to impact the IP protection of biologicals as about 8000 patents were invalidated in the Myriad decision alone.[2] Therefore, the validity of several other granted patents involving biological samples for the purposes of drug screening, mechanisms of action, diagnostic and treatment methods (Mann and Mahinka, 2014) is under the glare of the US Supreme Court. Thus, IP issues in respect of biologicals are likely to become more complex as the patents are essentially based on living matter (Mann and Mahinka, 2014). To some extent, this ruling of the US courts may well help the biosimilar manufacturers in managing the 'patent cliff'. It appears that the patented methods of analysing biologics may be available to biosimilar manufacturers under the safe harbour provision of Section 271(e)(1)[3] of the patent code, and it is therefore possible that biosimilars will be immunized against infringement of highly valuable patents directed to methods of analysing biologics by the same provision (Mann and Mahinka, 2014).

The future patent challenges for the approval for biosimilars is going to be very messy if the current spate of litigations between innovator majors and biosimilar companies is any indication (Mann and Mahinka, 2014). Analysts predict that the complete process of IP litigation is yet to be completely tested in the US legal system. This complex, tedious, unclear regulatory path for approval of biosimilars in the patent dance system has created a climate of uncertainty in the minds of new biosimilar entrants into the US market (Mann and Mahinka, 2014). The patent exchange process in the patent dance system has created some panic among biosimilar companies as they have to share confidential and sensitive information on manufacturing and testing systems with the innovator company, which is its competitor trying to block the introduction of the biosimilar (Mann and Mahinka, 2014). This is an entirely new experience for generic companies accustomed to the rather clear system of paragraph IV filings and patent challenges (Mann and Mahinka, 2014). The number of patent challenges between the innovator biological manufacturers and generic applicant reflects the complexity of the process (Mann and Mahinka, 2014). Until December 2012, biosimilar company Teva is battling 169 patent challenges followed by Actavis, current partner with innovator major Amgen with 132. Mylan has 131 patent challenges while Novartis, the parent company of Sandoz, is in fourth place with 125 patent litigations (Mann and Mahinka, 2014).

Compounding the complex and messy patent landscape of biologicals in general and biosimilar in particular, the newly emerging biosimilar companies hold process patents for innovator biologics, viz. rituximab, aetanercept, trastuzumab and adalimumab. Two Indian companies, Avesthagen and Zydus-Cadila Healthcare Ltd, hold process patents for trastuzumab, and for etanercept and rituximab (respectively). Another company Therapeutic Proteins International, Chicago processes patents for rituximab, trastuzmab and etanercept while Sandoz holds a process patent for rituximab (Mann and Mahinka, 2014). The impact of this is yet unclear.

19.7 Regulatory Issues

Safety is a priority for all medicines, and biologicals especially raise significant safety concerns as compared to chemical drugs. This is because biologics are more structurally complex where even the slightest modification of their manufacturing process may lead to undetected changes in the biological composition of the product with impact on the safety and efficacy of the biological. Biologics are typically made by genetically engineering living cells that are used as miniature factories to produce the desired protein. Being living cells, they are inherently variable and susceptible to even slight changes in their environment that can significantly alter the composition of the protein being produced. Just like no two living cell lines are identical, no two biologic manufacturing processes can have identical starting materials or proceed in the same way (Krishnan *et al.*, 2015). Biosimilar manufacturers will use different starting materials and a different process to develop a product that is different from the innovative product. Immunogenicity is a critical concern for the safety of patients for all biologicals more so for biosimilars, as our bodies treat a protein (biological) as a foreign substance and try to attack the protein with antibodies. Therefore, all biological drugs can potentially stimulate antibody production in patients and such responses are highly unpredictable. The antibodies produced in response to a biological may have no effect, may render the biological ineffective or could even cause serious health problems later (Gershon *et al.*, 2002). Recognizing this, both manufacturers and regulators associated with biologicals exercise extreme caution while assessing their safety and efficacy. Due to their inherent structural complexity, it is not possible with the current technology to establish that the same biological product (biosimilar and reference product) are exactly identical and would behave exactly the same way in two different patients. The process of clinical trials is done only to establish that the biosimilar product is safe and effective in all patients for the intended intervention, just like the innovator product.

19.8 WHO Regulatory Framework

To ensure safety, efficacy and quality of biotherapeutic products, the WHO drafted guidelines based upon some core principles considered mandatory for approval of a biosimilar product (WHO, 2009).

The following are some basic principles.

19.8.1 Reference product

The reference biotherapeutic product (RBP), the innovator product, under consideration should be approved by the regulator based on full quality, safety and efficacy data and should be authorized for marketing in the country or region in question. However, for countries that do not have nationally licensed RBPs, additional criteria (like the product under consideration must have been licensed and widely marketed in another jurisdiction) may be applied.

19.8.2 Quality

It is essential to submit comprehensive data on quality and heterogeneity of the similar biologic product (SBP) that includes head-to-head comparison with the RBP. If the composition of the RBP is not known, the comparative studies could be done with the commercial product. However, it must be ensured that the excipients of the biosimilar do not cause any interference in the analytical test methods. If such an analysis is not possible, extraction of drug substance needs to be done. In such cases, it must be ensured that the product heterogeneity and relevant attributes of the active moiety are not impacted by the extraction procedure. Wherever possible, the biosimilar product must be tested with and without manipulation. A full quality dossier for both SBP and RBP drug substance and drug product is necessary. The protocol for manufacture of a SBP must be robust and conform to good manufacturing practices, modern quality control and assurance procedures, in-process controls and process validation. State-of-the-art biochemical, biophysical and biological analytical methods must be employed for the characterization of both RBP and SBP. Details of the active ingredient(s) should be provided on primary and higher-order structure, post-translational modifications, biological activity, purity, impurities, product-related (active) substances (variants) and immunochemical properties, where relevant.

19.8.3 Non-clinical data

The non-clinical component of the data includes the pharmaco-toxicological assessment of the SBP with the safety and efficacy established through the some non-clinical studies with the final formulation of the SBP. The non-clinical evaluation of a

SBP includes broad spectrum pharmacodynamic (PD), pharmacokinetic (PK) and comparative repeat-dose toxicity studies in a relevant species. The selection of PK study design could be single-dose studies, steady-state studies, or repeated determination of PK parameters, all of which need to be adequately justified by the applicant. The pharmacokinetics of the SBP and RBP must be compared to assess absorption, bioavailability and elimination characteristics. Clinically relevant pharmacodynamic markers should be selected and may be investigated in the context of combined PK/PD studies.

19.8.4 Clinical studies

All critical clinical data should be generated using the SBP derived from the final manufacturing process for which marketing authorization is being sought. If there are any deviations from this procedure, these must either be properly justified or additional data, such as PK bridging studies, comparing the PK profiles of the products from the earlier and final product formulations, be provided. Similarity of the efficacy of the SBP and the RBP must be demonstrated in adequately powered, randomized and controlled clinical trial(s). Although equivalence study designs are the preferred choice for the comparison of efficacy and safety of the SBP with the RBP, non-inferiority designs may also be considered with suitable justification. The immunogenicity studies of the RBP in sufficient number of humans must always be conducted before pre-authorization and must be done even if the efficacy and safety of a SBP and RBP have been shown to be similar, as immunogenicity could still be different between SBP and RBP.

19.8.5 Pharmacovigilance

In common with all biologicals, data from pre-authorization clinical studies are usually too limited to identify all potential unwanted effects of a SBP, especially rare adverse events that are unlikely to be encountered in the limited clinical trial populations (WHO, 2015). Therefore, further close monitoring of the clinical safety of these products in all approved indications is warranted with a continued benefit–risk assessment during the post-marketing phase. A comprehensive safety specification and pharmacovigilance plan must be submitted by the manufacturer. Relevant guidelines prescribed in documents as ICH E2E (International Conference on Harmonisation of Technical Requirements For Registration of Pharmaceuticals For Human Use) may be followed for planning the pharmacovigilance. Wherever relevant, risk minimization measures such as educational material for patients and/or treating physicians will help promote the SBP.

19.9 EU EMA Guidelines

The European Medicine Agency (EMA) was the first to recognize the potential of biosimilars and has brought out the first guidelines for this purpose and has since issued a series of guidelines, the most recent one in October 2014 (Table 19.2). The European guidelines cover similar biological with biotechnology-derived proteins as active substance with well laid down guidance on considerations of quality for a biotherapeutic product with claims of similarity with an already marketed product. The European guidelines address the requirements regarding manufacturing processes, the comparability exercise for quality, the choice of reference medicinal product, analytical methods, physicochemical characterization, biological activity, purity and quality attributes for relevant specifications of the similar biological medicinal product. Only such products will be considered which already have reference biological product that has been granted market authorization in Europe. The proposed product must have active substance similar in terms of physicochemical and biological terms with the active ingredient of the reference biological product.

19.9.1 Manufacturing process

The pathway for the manufacturing process should conform to: (i) the product profile in terms of molecular characteristics and quality attributes of the proposed biosimilar must be comparable to the reference biological; and (ii) there must be consistency in the manufacturing process to ensure consistency of the product.

A biosimilar must be manufactured using state-of-the-art methods and detailed information may be provided on the impact of the manufacturing process on the product characteristics. If any novel expression systems are employed, their impact on potential risk such as atypical glycosylation pattern, higher variability or even a different impurity profile, as compared to the reference biological

Table 19.2. European Medicines Agency guidelines relevant to biosimilar development and approval.

EMEA/CPMP/ BWP/3207/2000/ Revision 1	Guideline on comparability of medicinal products containing biotechnology-derived proteins as active substance; quality issues.	http://www.ema.europa.eu/docs/ en_GB/documentlibrary/Scientific_ guideline/2009/09/WC500003573.pdf
EMEA/CPMP/ 3097/2002	Guideline on comparability of medicinal products containing biotechnology-derived proteins as active substance; non-clinical and clinical issues.	http://www.ema.europa.eu/docs/ en_GB/document_library/Scientific_ guideline/2009/09/WC500003963.pdf
CHMP/437/2004/ Revision 1	Guideline on similar biological medicinal products.	http://www.ema.europa.eu/docs/ en_GB/document_library/Scientific_ guideline/2013/05/WC500142978.pdf
EMA/CHMP/ BWP/247713/2012/ Revision 1	Guideline on similar biological medicinal products containing biotechnology-derived proteins as active substance; quality issues.	http://www.ema.europa.eu/docs/ en_GB/document_library/Scientific_ guideline/2012/05/WC500127960.pdf
EMA/CHMP/ BMWP/572828/2011	Concept paper on the revision of the guideline on similar biological medicinal products containing biotechnology-derived proteins as active substance; non-clinical and clinical issues.	http://www.ema.europa.eu/docs/ en_GB/document_library/Scientific_ guideline/2011/10/WC500115611.pdf
EMA/275542/2013	Concept paper on the revision of the guideline on immunogenicity assessment of biotechnology-derived therapeutic proteins.	http://www.ema.europa.eu/docs/en_ GB/document_library/Scientific_ guideline/2014/03/WC500163623.pdf
EMEA/CHMP/ BWP/157653/2007	Guideline on development, production, characterisation, and specifications for monoclonal antibodies and related products.	http://www.ema.europa.eu/docs/en_ GB/document_library/Scientific_ guideline/2009/09/WC500003074.pdf
EMA/CHMP/ BMWP/86289/2012	Guideline on immunogenicity assessment of monoclonal antibodies intended for in vivo clinical use.	http://www.ema.europa.eu/docs/en_ GB/document_library/Scientific_ guideline/2012/06/WC500128688.pdf
EMA/CHMP/ BMWP/403543/2010	Guideline on similar biological medicinal products containing monoclonal antibodies.	http://www.ema.europa.eu/docs/en_ GB/document_library/Scientific_ guideline/2010/11/WC500099361.pdf

BMWP, Biosimilar Medicines Working Party; BWP, Biologics Working Party; CHMP, Committee for Medicinal Products for Human Use; CPMP, Committee for Proprietary Medicinal Products; EMA, European Medicines Agency; EMEA, European Medicines Agency.

must be considered and recorded. While the formulation of the biosimilar need not be identical to the reference biological, its suitability in respect of stability, compatibility (i.e. interaction with excipients, diluents and packaging materials), integrity, activity and strength of the active substance with the reference biological must be demonstrated. Biosimilars are to be manufactured with quality control following relevant ICH Q5C and CHMP guidelines. If a different formulation and/or container/closure system to the reference biological is selected its potential impact on the safety and efficacy should be appropriately justified. If any changes are introduced in the manufacturing process (in active substance and/or finished product) a comparability assessment study must be done (as per ICH Q5E) *vis-à-vis* the reference biological and all details must be provided.

19.9.2 Comparability with reference medicinal product

Several different batches of the reference medicinal product should be used to provide a robust analysis and to generate a representative quality profile. The relative age of the different batches of reference medicinal product should also be considered when establishing the target quality profile.

Extensive comparability studies are required to demonstrate that the biosimilar has a highly similar quality profile with the reference biological, which includes comprehensive side-by-side analyses using sensitive and orthogonal methods. If any change is detected, justification on its potential impact of quality, safety and efficacy must be provided. As the biosimilar will be used to treat patients, all quality attributes of the biosimilar must be identical and if any minor differences are noticed they must not have any bearing on clinical efficacy and these must be detailed with adequate justification. The comparability exercise should take into account quantitative limits with an adequate number of reference medicinal product lots tested with details of test methods used. The data should be collected with statistically sound study design. In these comparability exercises, all materials used must be clearly identified – for both the reference biological and the biosimilar (e.g. brand name, pharmaceutical form, formulation, strength, origin of the reference medicinal product, number of batches, lot number, age of batches, use).

19.9.3 Analytical considerations

All analytical studies must be carried out using state-of-the-art methods to provide robust evidence that the biosimilar is comparable to the reference biological in terms of quality of active substance etc. The methods adapted by the biosimilar manufacturers must be sensitive enough to detect the slightest difference with the reference biological. It is the responsibility of the applicant to demonstrate that the selected methods used in the appropriate international standards must be used for all analytical studies with justification for the methods used.

19.9.4 Physicochemical properties

Physicochemical evaluation for structural identification of product-related substances and impurities must include determination of the composition, physical properties, primary and higher order structures of the biosimilar. The target amino acid sequence of the biosimilar must be the same as in the reference biological. If post-translational modified forms are detected, they must be appropriately characterized and if any variations are found with the reference biological, detailed justification must be provided that these are not a cause of concern.

19.9.5 Biological activity

Assessment of biological activity done through a comparability analysis is critical for demonstrating that the biosimilar is able to achieve the specific and defined biological effect of the reference biological. Different assay formats, viz. ligand or receptor binding assays, enzymatic assays or cell-based assays may be used. The assay systems must follow a complementary approach to ensure sensitivity and specificity to be able to detect changes, if any, in biological activity between the products and comply with the stipulated guidances provided in European Pharmacopoeia.

19.9.6 Immunochemical properties

All biologicals have the potential to cause immune reactions in the body. Therefore comparative immunological studies for monoclonal antibodies or related substances (e.g. fusion proteins based on IgG Fc) must include comparative affinity of both the products on the intended target.

19.9.7 Purity and impurities

The biosimilar must be thoroughly investigated for its purity and impurity profiles of the active substance with both comparative qualitative and quantitative studies with a combination of analytical procedures. The methodology must be state-of-the-art and must take into account specific degradation pathways (e.g. oxidation, de-amidation, aggregation) of the biosimilar product and potential post-translational modifications of the proteins. Details on the age/shelf life of the reference biological at the time of testing should be mentioned with its potential impact on the quality. Other factors that impact quality, such as selected time points for testing or storage conditions (e.g. accelerated or stress conditions), must be indicated to demonstrate that the degradation pathways of the reference biological and the biosimilar are similar. There could be the presence of process-related impurities such as host cell proteins, host cell DNA, reagents, downstream impurities, etc. that would differ qualitatively between processes, therefore comparison between both the biologicals may be relevant. The analytic methodology for this assessment must be state-of-the-art, and if there are any newly identified impurities that may cause immunogenicity, these must be appropriately documented and justified.

19.9.8 Specifications

The choice of the battery of tests for determining specifications (or control strategy) for both the reference biological and biosimilar must be as per ICH Q6B: 'Note For Guidance on Specifications: Test Procedures and Acceptance Criteria for Biotechnological/Biological Products'. The rationale for establishing the proposed range of acceptance criteria should be described with each of the criterion established and justified. These include data obtained from different lots from non-clinical and/or clinical studies, demonstration of manufacturing consistency, stability studies with all other relevant development data necessary for establishing biosimilarity (quality, safety and efficacy).

With its proactive strategies, Europe eventually led the way by setting up the first science-driven regulatory regime for the approval of biosimilars for consumer use in 2004 (EMA, 2004). The first biosimilar drug – Omnitrope, a human growth hormone – was approved by the Commission in April 2006, following scientific advice from the European Medicines Agency. The EU has since approved 18 biosimilar medicines for use in Europe thus far (Table 19.3).

Significantly, a proposed biosimilar under the EU guidance does not have to demonstrate bioequivalence, but merely biosimilarity to the innovator's product. This distinction rectifies Hatch-Waxman's failure with biosimilars by loosening the contextually impossible bioequivalence standard. EU guidelines are widely considered to be one of the most stringent regulatory norms for biosimilar product development (Krishnan *et al.*, 2015). The chemistry, manufacturing and controls package and other clinical requirements are extensive, which ensures safety and efficacy of biosimilars.

Europe's faster biosimilar approval pathways have resulted in substantial benefits to patients, including lower drug costs and wider biologic availability in the EU. The regulatory framework of the EU has helped many countries around the world, e.g. Australia, Canada, Japan, Turkey, Singapore, South Africa, Taiwan, USA, India etc. to develop their own guidelines.

19.10 United States Food and Drug Administration

The Affordable Care Act introduced a pathway for FDA approval of 'abbreviated' applications for biological products that are shown to be 'biosimilar' to an already-approved biologic ('reference product'). This abbreviated pathway part of the law is called the Biologics Price Competition and Innovation Act (BPCIA, 2009) and provides the framework for the applicant to the FDA and the eventual patent litigation that may result from a biosimilar application.

The BPCIA is part of the Patient Protection and Affordable Care Act (2010), which created a new licensure pathway for biosimilars of the FDA.

Table 19.3. EMA-approved biosimilars.

Product	Active substance	Authorization date	Manufacturer/Company
Omnitrope	Somatropin	2006	Sandoz GmbH
Valtropin	Somatropin	2006 (withdrawn May 2012)	BioPartners GmbH
Abseamed	Epoetinalfa	2007	MediceArzneimittelPütter GmbH
Binocrit	Epoetinalfa	2007	Sandoz GmbH
Epoetinalfahexal	Epoetinalfa	2007	Hexal AG
Retacrit	Epoetin zeta	2007	Hospira UK Limited
Silapo	Epoetin zeta	2007	StadaArzneimittel AG
Biograstim	Filgrastim	2008	AbZ-Pharma GmbH
Ratiograstim	Filgrastim	2008	Ratiopharm GmbH
Filgrastimratiopharm	Filgrastim	2008 (withdrawn 20 July 2011)	Ratiopharm GmbH
Tevagrastim	Filgrastim	2008	Teva GmbH
Filgrastimhexal	Filgrastim	2009	Hexal AG
Zarzio	Filgrastim	2009	Sandoz GmbH
Nivestim	Filgrastim	2010	Hospira UK Limited
SomatropinBiopartners	Somatropin	2013	BioPartners GmbH
Inflectra	Infliximab	2013	Hospira UK Limited
Remsima	Infliximab	2013	Celltrion Healthcare Hungary Kft
Ovaleap	Follitropinalfa	2013	TevaPharma BV
Grastofil	Filgrastim	2013	ApotexEurope BV
Bemfola	Follitropinalfa	2014	Finox Biotech AG

Biosimilar applications are to be filed under Section 351(k) of the Public Health Service Act. The act defines the basic requirement of a demonstration of biosimilarity between the reference and biosimilar products. According to the act, the biosimilar must be shown to be interchangeable with the reference product. The framework for approval is quite comprehensive and includes a system for resolving patent disputes between the biosimilar applicant and the innovator company.

Since the BPCI Act was signed into law on 23 March 2010, the FDA has issued six guidances relating to biosimilars or reference biological products:. These guidances are:

1. Guidance for Industry on Biosimilars: Q & A's Regarding Implementation of the BPCI Act (2009) – issued 9 February 2012 (US FDA, 2012c).
2. Scientific Considerations in Demonstrating Biosimilarity to a Reference Product – issued 9 February 2012 (US FDA, 2012b).
3. Quality Considerations in Demonstrating Biosimilarity to a Reference Protein Product – issued 9 February 2012 (US FDA, 2012a).
4. Formal Meetings Between the FDA and Biosimilar Biological Product Sponsors or Applicants – issued 29 March 2013 (US FDA, 2013).
5. Clinical Pharmacology Data to Support a Demonstration of Biosimilarity to a Reference Product (Clinical Pharmacology Guidance) – issued 13 May 2014 (US FDA, 2014b).
6. Reference Product Exclusivity for Biological Products Filed Under Section 351(a) of the PHS Act (Exclusivity Guidance) – issued 4 August 2014 (US FDA, 2014c).

A step-wise assessment of biosimilarity through extensive and comparative structural and functional studies (such as, bioassays, binding assays and studies of enzyme kinetics) is recommended to assess whether the proposed biosimilar product and the reference product are 'similar' under the following four categories: (i) the first is 'highly similar' with fingerprint-like similarity. This is the gold standard; (ii) highly similar, where the product meets the statutory standard for analytical similarity; (iii) similar, where data are needed to determine that the proposed biosimilar product is highly similar to the reference product; and (iv) not similar, where the FDA considers the product is not similar to the innovator's product.

The Scientific Considerations guidance calls for the demonstration of biosimilarity between the reference and biosimilar in terms of safety, potency and purity, with no clinically meaningful differences. The guidance prescribes the following main procedures:

- Analytical studies: demonstration of the proposed biosimilar product as highly similar to the reference product notwithstanding minor clinically irrelevant differences;
- Animal studies: including toxicity studies, PK and PD measurements, and immunogenicity studies; and
- Human clinical studies: including PK and PD studies to generate immunogenicity results, and safety and efficacy data and equivalence studies to demonstrate comparability on both upper and lower margins.

The quality considerations guidance focuses more on the physicochemical characterization issues. Among the factors the US FDA considers important include the manufacturing process consistency, structural similarity, potency assays relevant to mechanism of action, PD studies extending the understanding of the mechanism of action, comparative PK and immunogenicity. In physicochemical characterization, there is emphasis on display of primary and higher order structure and post-translational and chemical modifications. In data on biological characterization, comparative potency and receptor binding assays are sought. The commercial history and experience of the sponsor would be helpful. An important aspect of the BPCIA is the interchangeability of biosimilars. The sponsor is expected to demonstrate that the biosimilar drug produces the same clinical effects as the innovator product in clinical studies. This will ensure automatic interchangeability of the biosimilars with the reference originator product, without the need for clinician's discretion. Once approved or licensed for marketing, the clinician could switch between the innovator and biosimilar.

As in other countries, biosimilars present unique challenges for FDA review and approval due to the complexities in the manufacture of biological products. The guidelines also recommend sponsors of the proposed biosimilar product to set up an early meeting with US FDA officials with preliminary data and product development plans to help the FDA understand the requirements of submission and draw up a plan for the process for approval.

19.11 India's Biosimilar Guidelines

In India, biosimilar biologics are regulated as per the Drugs and Cosmetics Act (1940), the Drugs and Cosmetics Rules (1945) (as amended from time to time) and the Rules for the Manufacture, Use, Import, Export and Storage of Hazardous Microorganisms and Genetically Engineered Organisms or Cells 1989 (Rules 1989) notified under the Environment (Protection) Act (1986). The Central Drugs Standard Control Organization (CDSCO) and Department of Biotechnology (DBT), through the Review Committee on Genetic Manipulation (RCGM), approves the preclinical studies of biosimilars using an abbreviated version of the pathway applicable to new drugs. CDSCO is the apex regulatory authority in India for ensuring the safety and efficacy of similar biological products and establishing appropriate regulatory pathways. CDSCO, in collaboration with the Department of Biotechnology (DBT), issued the Guidelines on Similar Biologics: Regulatory Requirements for Marketing Authorization in India in 2012 (DBT, 2012). These guidelines detail the regulatory requirements, such as data requirements for production process, characterization, preclinical studies and clinical trials, for the marketing authorization of biosimilar biologics.

One of the most important considerations in the regulation of biosimilars in India is that a similar biologic to be considered must be against an authorized reference biologic already marketed in India. If the reference biologic is not authorized for use in India, this product must have been licensed and marketed for a minimum of 4 years in the innovator's jurisdiction having a well-established regulatory framework with significant systems for monitoring safety and efficacy. Only when these essential criteria are fulfilled can the multi-step approval process be initiated. The biosimilar manufacturer has to submit adequate data demonstrating similarity with the innovator product. These data include product characterization and preclinical and clinical studies carried out as per the stipulated guidelines.

In case no medicine or only palliative therapy is available, or in the case of a national healthcare emergency, this period of 4 years may be reduced or waived. The same reference biologic should be used throughout the development programme and when conducting comparative studies to support the safety, efficacy and quality of the similar biologic product. The dosage form, strength and route of administration of the similar biologic should be the same as that of the reference biologic. The active substance (active ingredient) of the reference biologic and that of the similar biologic must be shown to be similar.

Unlike the US FDA system where the regulator determines on a case-to-case basis the extent of animal and clinical data required, the Indian guidelines provide guidance about animal studies to be conducted with the approval of the Institutional Animal Ethics Committee (IAEC) for studying the immunological characterization of the proposed biosimilar product. The preclinical toxicology studies must have at least one repeat-dose toxicity study in a relevant species for a duration of generally not less than 28 days followed by a 14-day recovery period. The dosage and other parameters for preclinical studies, however, could vary depending upon the product under review. The biosimilar applicant must provide adequate scientific justification on the choice of animal models to be used. However, if the relevant animal species is unavailable the applicant can seek RCGM approval to conduct these toxicity studies with two animal models, viz. a rodent and one non-rodent species as stipulated under the Schedule Y of the Drugs and Cosmetics Act (1940). These pre-approval safety data are primarily intended to provide assurance to the regulator of the absence of any unexpected safety concerns.

After the completion of preclinical studies, the reports are submitted to RCGM for review and consideration. Based on the evaluation of preclinical study reports, including demonstrations of consistency of the process and product, product characterization, product specifications and similarity to reference biologics, RCGM will recommend the applicant to approach DCGI to conduct the appropriate phase of clinical trial as per the CDSCO requirements.

The DCGI gives permission to conduct clinical trials after due diligence of the data on preclinical studies approved by the RCGM. The manufacturer is required to perform comparative PK and PD studies in healthy volunteers or patients within the therapeutic dose range of reference biologic to demonstrate the similarities in PK and PD characteristics between the biosimilar and the product. Typically, multiple-dose, comparative, parallel arm steady-state PK studies are mandated where markedly higher or lower concentrations are expected at steady-state than that expected from single-dose data PK measurements and to rule out time- and dose-dependent PK parameters. The PD studies must be comparative, parallel arm or cross-over, in the most relevant population (patients or healthy volunteers) to detect differences, if any, between the reference and innovator products with appropriate dose ranges to establish the response and efficacy of the reference product.

Comparative clinical trials with the reference product and innovator product are to be conducted to demonstrate the similarity in safety and efficacy in a larger population. The nature, severity and frequency of adverse events should be compared between the similar biologic and reference biologic, and should be based on safety data from a sufficient number of patients treated for an acceptable period of time. The manufacturer should ensure that comparative clinical studies have a sufficient number of patients treated for an acceptable period of time, in order to allow the detection of significant differences in safety between similar biologic and reference biologic.

The confirmatory clinical safety and efficacy studies can be waived by the regulator if the applicant of the biosimilar is able to establish that: (i) the structural and functional comparability of similar biologic and reference biologic can be reliably mapped out through appropriate methods; (ii) the preclinical data are quite comparable between both the products; (iii) PK and PD studies have established comparability typically in-patient setting, with safety shown for an adequate period through safety efficacy data; and (iv) a robust post-marketing risk management strategy is in place that will on a systematic and continuous basis assess safety data with a focus on immunogenicity. Most of these regulatory stipulations are in conformity with the EU and US guidelines.

Significantly, Indian guidelines permit the biosimilar use only for the indication for which approval is accorded by the regulator. This is unlike the US FDA guidances that permit potential extrapolation to allow the biological product to be used for more than one indication for licensing once the data from clinical studies are considered adequate to establish safety, purity and potency in an appropriate condition of use without substantial or any additional clinical testing.

In respect of manufacturing, the Indian guidelines stipulate a set of quality standards for cell lines to be used to ensure that the process is highly consistent and robust. There are also standards outlined for the biosimilar manufacturing process, fermentation and downstream process, analytical methods, product characterization, stability and other specifications. The applicant is expected to submit at least three standardized batches to demonstrate consistency in the manufacturing process. The guidelines also stipulate head-to-head characterization studies to compare the biosimilar and the innovator product for quality comparison through analytical methods robust and sensitive enough to detect the smallest changes in the biosimilar and a list of tests employed for the comparative exercise (Ohly, 2012).

For considerations of safety of a product it is important to have a system in place to track and trace all biologics throughout the product life cycle, especially since the biologicals, unlike chemical generics, are more prone to cause adverse events in patients. The treating physicians should also be informed about any adverse events linked to treatments to enable them to exercise precaution if warranted. The biosimilar applicant must propose a system for proper product identification, accurate record keeping and attribution of adverse events and system of pharmacovigilance and adverse event recording and reporting.

India's new guidelines describe a regulatory pathway that addresses the 'similarity exercise' and the quality considerations of the biosimilar. These include requirements on manufacturing processes and quality control. The Indian regulator's requirements are similar in many respects to the regulatory stipulations of the USA and the EU. The Indian guidelines define a 'similar biologic' as a 'biological product/drug produced by genetic engineering techniques and claimed to be "similar" in terms of safety, efficacy and quality to a reference biologic, which has been granted a marketing authorization in India by DCGI on the basis of a complete dossier, and with a history of safe use in India.' In addition, the Indian law, unlike the US law and in common with the EU law, does not provide any market or data exclusivity to the innovator company. The biosimilar applicant has to show data to demonstrate that the biosimilar can be substituted with the reference product without increasing 'the risk of using the reference product without such alternation or switch' (Ohly, 2012). Several biosimilars have been approved in India (Table 19.4).

Table 19.4. 'Similar biologics' approved and marketed in India (from: CDSCO, New Delhi, India, company websites and other sources).

Product name[a]	Active substance	Therapeutic area[b]	Approval/ launch in India[c]	Company
AbcixiRel	abciximab	Angina Cardiac ischaemia	2013	Reliance Life Sciences
Actorise	darbepoetin alfa	Anaemia Cancer Chronic kidney failure	2014	Cipla/Hetero
Basalog	insulin glargine	Diabetes	2009	Biocon
Biovac-B	hepatitis B vaccine	Hepatitis B	2000	Wockhardt
CanMab	trastuzumab	Breast cancer	2013	Biocon
Ceriton	epoetin alfa	Anaemia Cancer Chronic kidney failure	NR	Ranbaxy
Choriorel	chorionic gonadotrophin hormone r-hCG	Female infertility	2011	Reliance Life Sciences
Cresp	darbepoetin alfa	Anaemia Cancer Chronic kidney failure	2010	Dr Reddy's Laboratories
Darbatitor	darbepoetin alfa	Anaemia Cancer Chronic kidney failure	2014	Torrent Pharmaceuticals
Emgrast	filgrastim	Cancer Neutropenia	2010	Gennova Biopharmaceuticals (Emcure)
Epofer	epoetin alfa	Anaemia Cancer Chronic kidney failure	NR	Emcure
Epofit/Erykine	epoetin alfa	Anaemia Cancer Chronic kidney failure	2005	Intas Pharmaceuticals
Eporec	erythropoietin	Anaemia Chronic kidney failure	2011	Bioviz Technologies
Epotin	epoetin alfa	Anaemia Cancer Chronic kidney failure	NR	Claris Lifesciences
Erypro	epoetin alfa	Anaemia Cancer Chronic kidney failure	NR	Biocon
Etacept	etanercept	Ankylosing spondylitis Rheumatoid arthritis Psoriatic arthritis Psoriasis Juvenile rheumatoid arthritis	2013 [5]	Cipla
Fegrast	filgrastim	Cancer Haematopoietic stem cell transplantation Neutropenia	NR	Claris Lifesciences
Filgrastim	filgrastim	Neutropenia	2013	Cadila Pharmaceutical
Filgrastim	filgrastim	Neutropenia	2013	Lupin
Filgrastim	filgrastim	Neutropenia	2013	USV
Folisurge	follitropin alfa (follicle stimulating hormone)	Female infertility Spermatogenesis in men	2013	Intas Pharmaceuticals

Continued

Table 19.4. Continued.

Product name[a]	Active substance	Therapeutic area[b]	Approval/ launch in India[c]	Company
FostiRel	follitropin beta (follicle stimulating hormone)	Female infertility	2010	Reliance Life Sciences
Glaritus	insulin glargine	Diabetes mellitus	Mar 2009	Wockhardt
Grafeel	filgrastim	Neutropenia Haematopoietic stem cell transplantation Cancer	NR	Dr Reddy's Laboratories
Infimab	infliximab	Ankylosing spondylitis Crohn's disease Psoriasis Psoriatic arthritis Rheumatoid arthritis Ulcerative colitis	2014	Epirus Biopharmaceuticals
Insugen	human insulin	Diabetes mellitus	NR	Biocon
Insulin	insulin	Diabetes mellitus	2011	Gland Pharma
Intacept	etanercept	Ankylosing spondylitis Juvenile idiopathic arthritis Psoriasis Psoriatic arthritis Rheumatoid arthritis	2015	Intas Pharmaceuticals
Intalfa	interferon alfa-2b	Carcinoid tumour Chronic hepatitis B Chronic hepatitis C Hairy cell leukaemia Chronic myelogenous leukaemia BCR-ABL positive Follicular lymphoma Malignant melanoma Multiple myeloma	2007	Intas Pharmaceuticals
MabTas	rituximab	Lymphoma Non-Hodgkin's lymphoma	2013	Intas Pharmaceuticals
Molgramostim	Recombinant human granulocyte macrophage colony stimulating factor (molgramostim)	Neutropenia	2013	Zenotech Laboratories
Mirel	reteplase (tissue plasminogen activator)	Myocardial infarction	2009	Reliance Life Sciences
Myokinase	streptokinase	Acute myocardial infarction Deep venous thrombosis Acute pulmonary embolism	NR	Biocon
Neukine	filgrastim	Neutropenia Hematopoietic stem cell transplantation Cancer	2004	Intas Pharmaceuticals
Neupeg	peg-filgrastim	Cancer neutropenia	2007	Intas Pharmaceuticals
Nufil	filgrastim	Cancer neutropenia	NR	Biocon
Pegex	peg-filgrastim	Cancer neutropenia	2010	Gennova Biopharmaceuticals (Emcure)
Peg-filgrastim	peg-filgrastim	Cancer neutropenia	2013	Lupin

Continued

Table 19.4. Continued.

Product name[a]	Active substance	Therapeutic area[b]	Approval/ launch in India[c]	Company
Peg-interferon alfa 2b	Pegylated recombinant human interferon alfa-2b	Chronic hepatitis B Chronic hepatitis C	2013	Intas Pharmaceuticals
Platelet derived growth factor	rh-PDGF-BB + β-TCP	Peridontal defect Gingival recession	2010	Virchow Biotech
Rasburicase	rasburicase	Malignancy-associated hyperuricaemia	2012	Virchow Biotech
Razumab	ranibizumab	Wet macular degeneration Macular oedema Degenerative myopia Diabetes complications	2015	Intas Pharmaceuticals
Reditux	rituximab	Leukaemia Lymphoma Rheumatoid arthritis	2007	Dr Reddy's Laboratories
Relibeta	interferon beta-1a	Multiple sclerosis	2011	Reliance Life Sciences
Reliferon	interferon alfa-2b	BCR-ABL positive Carcinoid tumour Chronic hepatitis B Chronic hepatitis C Chronic myelogenous leukaemia Follicular lymphoma Hairy cell leukaemia Melanoma Multiple myeloma	2008	Reliance Life Sciences
Religrast	filgrastim	Neutropenia	2008	Reliance Life Sciences
Relipoietin	epoetin alpha	Anaemia Autologous blood transfusion Chronic kidney failure HIV	2008	Reliance Life Sciences
Repoitin	erythropoietin	Anaemia Chronic kidney failure	2011	Serum Institute of India
Rituximab	rituximab	Non-Hodgkin's lymphoma	2013	Zenotech Laboratories
Shankinase	streptokinase	Arterial occlusions Deep vein thrombosis Pulmonary embolism	2004	Shantha Biotechnics/ Merieux Alliance
Shanferon	interferon alfa-2b	BCR-ABL positive Carcinoid tumour Chronic hepatitis B Chronic hepatitis C Chronic myelogenous leukaemia Follicular lymphoma Hairy cell leukaemia Melanoma Multiple myeloma	2002	Shantha Biotechnics/ Merieux Alliance
Shanpoietin	erythropoetin	Anaemia Chronic kidney failure	2005	Shantha Biotechnics/ Merieux Alliance

Continued

Table 19.4. Continued.

Product name[a]	Active substance	Therapeutic area[b]	Approval/ launch in India[c]	Company
Terifrac	teriparatide (parathyroid hormone)	Post-menopausal women with osteoporosis who are at high risk for fracture	2010	Intas Pharmaceuticals
Teriparatide	teriparatide (parathyroid hormone)	Post-menopausal women with osteoporosis who are at high risk for fracture	2012	Cadila Healthcare
Teriparatide	teriparatide (parathyroid hormone)	Post-menopausal women with osteoporosis who are at high risk for fracture	2012	USV
Wepox	epoetin alfa	Anaemia Cancer Chronic kidney failure	2001	Wockhardt
Wosulin	human insulin	Diabetes mellitus	2003	Wockhardt
Zavinex	interferon alfa-2b	Chronic hepatitis B Chronic hepatitis C	2011	Cadila Healthcare
Zyrop	erythropoietin	Chronic kidney failure	2010	Cadila Healthcare

[a] Where brand-name is not known active substance name is given.
[b] Therapeutic area taken from company information, from originator product information on EMA website or from CDSCO information; NR: not reported.
[c] 'Similar biologics' launched in India before the Indian 'similar biologics' guideline came into effect on 15 September 2012, were approved using an ad-hoc abbreviated procedure on a case-by-case basis.

19.12 Nomenclature of Biologicals

A good pharmacovigilance post-approval safety surveillance programme must have accurate adverse event (AE) tracking and tracing capabilities for detecting and evaluating safety signals and ensure efficient and coordinated responses (e.g. recalls and alerts) (Casadevall *et al.*, 2002). Accurate traceability requires the medicinal products to have a system of rational nomenclature for each biosimilar with a unique non-proprietary name, Healthcare Common Procedure Coding System/NDC and lot number (Casadevall *et al.*, 2002). The pharmacovigilance systems should have processes that can be easily used by the prescribing practitioner. Any complex pharmacovigilance systems with more than the needed identifiers would lead to an inconsistent system that would delay accurate identification, thereby undermining the accuracy and reliability of the pharmacovigilance system as well as exposing patients to avoidable risks (Casadevall *et al.*, 2002).

As biosimilars are not exactly identical with respect to their reference products, if a substituted biosimilar results in a problem there should be a system of tracking in place to enable prevention of future episodes, especially since a biological can lead to averse immunogenic reactions. It is important to know exactly which product has led to the adverse reaction; confusion in tracing the biological product may be serious as some may cause delayed onset of adverse event, and the patient may receive both the innovator and the biosimilar product, which makes it difficult to identify the exact product (Casedevall *et al.*, 2002; Gershon *et al.*, 2002; Mellstedt *et al.*, 2008; EuropaBio, 2011). Regulatory authorities in Europe have recognized the risks of substitution for biologic medicines and have brought out guidelines for this purpose (EMA, 2012, 2013a, b).

A system of distinguishable names for all biologics will minimize the chances of inadvertent and inappropriate product switching and strengthen the accuracy of the prescription practices. The FDA concluded that the non-proprietary names for ziv-aflibercept and tbo-filgrastim should be different to their reference biologics, to avoid patients receiving the incorrect product and to reduce confusion among healthcare providers who may perceive them to be clinically the same, because they have the same non-proprietary name (CDER, 2012a, b). The FDA has also made the broader conclusion that the use of distinguishable non-proprietary names will help post-marketing safety monitoring, allowing better traceability of medicines in the case of an adverse event. In addition, the use of brand names alone was determined to be insufficient as brand names are often not used by healthcare professionals for prescribing, and many pharmacovigilance systems do not require them (CDER, 2012a, b).

The WHO introduced the concept of a biological qualifier (BQ) for naming biologicals in 2014 (WHO, 2014a). This proposes that the BQ be used in conjunction with the International Non-proprietary Name (INN) and would consist of a random alphabetic code, made up of four random consonants (WHO, 2014a). In the EU, however, both the brand name and INN are used to identify the specific biological product, whether it is a innovator's product or a biosimilar. The system has been followed for over 6 years and has been found to be effective. While they are marketed, each biosimilar with its own brand name would have other identifiers such as the batch number and other details for the purpose of tracking the product. In at least three product classes (filgrastim, epoetin, somatropin) the European regulators were successful in the nomenclature with over 96.2% product identification done in three product classes (GaBI Online, 2015). Though the US FDA has approved its first biosimilar, Zarxio, in 2015, the policy of naming the biosimilars is yet to be announced. The draft FDA guidance on naming biosimilars proposes that all biologicals and biosimilars have non-proprietary names with a four-letter suffix added to the names to distinguish them from each other. There would be four lowercase letters in the suffix that may not have any specific meaning. For example, the non-proprietary name of an innovator product could be replicamab-cznm and its biosimilar could be replicamab-hixf (GaBI Online, 2015). This naming of biologicals, both innovator products and biosimilars, has been criticized by the biosimilar companies as well as civil society representatives as this process is likely to cause problems both with the physicians and the users.

The regulatory pathways provided by BPCIA (for biologicals) and Hatch-Waxman (for chemical generics) represent a compromise on barriers to entry between original drug manufacturers and subsequent potential entrants, balancing entry concerns by permitting some barriers while reducing others (BPCIA, 2009). Original drug manufacturers claim – with some support – that the high research and development costs necessary to produce successful pharmaceuticals effectively require extensive monopoly profits merely to recoup their investments (Grabowski and DiMasi, 2009). Subsequent drug manufacturers, seeking profits and original manufacturers' market share, can safely and more affordably reproduce these expensive drugs, thus reducing the opportunity of original manufacturers to recoup costs (or amass profits) while passing savings on to consumers. Each side claims its position will increase innovation; original manufacturers contend that an extensive monopoly period is the only way to allow for research into new and often unproductive drugs, while subsequent manufacturers point out that competitive markets tend to spur innovation as subsequent firms jostle for market share and original manufacturers continue to innovate to stay ahead of the pack (Grabowski and DiMasi, 2009).

There is also broad criticism that, unlike the European guidelines, the broad scope of the FDA's draft guidance documents for biosimilar approval is found wanting on several counts (Bourgoin and Nuskey, 2013). The advice on the time and course of assay systems necessary for each product class for head-to-head conduct immunogenic studies is very inadequate (Bourgoin and Nuskey, 2013). Also, it does not contain adequate advice for the conduct of clinical studies (Bourgoin and Nuskey, 2013), which is the most expensive component of drug development. In the absence of such clarity the biosimilars applicants may run the risk of submitting data that the FDA may not find adequate or is incomplete, which would result in repetition of some clinical studies (Bourgoin and Nuskey, 2013). It appears that the regulator would rely heavily on the biosimilar applicant to anticipate necessary measures in protocol development rather than provide clarity of the expectations of the FDA (Bourgoin and Nuskey, 2013). In addition, it is widely believed that the success of the envisaged policy may be limited, especially since the FDA does not believe in a 'one-size-fits-all' assessment for various classes of biologicals (Bourgoin and Nuskey, 2013). Unlike its intended purpose, the 'totality-of-the-evidence' approach of the FDA does not guide the biosimilar company through clearly laid out product-specific trial design and implementation, which is quite unsatisfactory, unlike the EMA guidelines (Bourgoin and Nuskey, 2013).

The FDA draft guidance does permit bridging studies for non-US licensed products for marketing approval in the USA, but the guidelines in respect of bridging data requirements are unclear (Bourgoin and Nuskey, 2013).

The complex regulatory stipulation of the US FDA would mean that only large biotherapeutic companies with strong resources can hope to penetrate the US market (Bourgoin and Nuskey, 2013). Only companies that have substantial experience and exposure to big regulated markets like the EU can afford to take a potential risk for large investments. Not surprisingly, the list of companies currently conducting clinical trials on biosimilar molecules include Teva, Roche, Merck, Sandoz, Hospira and Celltrion. Smaller pharmaceutical companies are compelled to partner with the pharma majors to gain a foothold in the US biosimilars market. Indian biotech company Biocon has partnered with US-based Mylan in a US$33 billion deal to develop a biosimilar version of Roche's Rituxin, which is among the major biologicals coming off patent. Another blockbuster, Herceptin of Roche, is being targeted for biosimilar development by a consortium of Synthon, Amgen and Actavis with potential clinical trial support from Amgen and Actavis (Bourgoin and Nuskey, 2013).

This has since become part of a new strategy being attempted by big pharma companies for managing the risk in R&D for biotherapeutics including biosimilars. Both large and small companies engaged in the development of biologicals are partnering with Contract Research Organizations (CROs) and other entities that have experience in the design and conduct of biosimilar clinical trials. Hospira, a biotherapeutic major, has been in partnership with DaVita, Inc., and Fresenius Medical Care has conducted phase III clinical trials in over 200 haemodialysis centres across the USA in 2011 for its biosimilar erythropoietin. Samsung Biologics has joined hands with CRO giant Quintiles for conducting clinical trials for their biosimilar molecules (Bourgoin and Nuskey, 2013). Samsung Biologics along with Quintiles have entered into an agreement with Biogen Idec to complement business expertise to form Samsung Bioepis. Samsung Bioepis in turn has agreed to join the development and conduct of clinical trials with Merck Inc. for their new biosimilars (Bourgoin and Nuskey, 2013). In the fiercely competitive environment due to huge returns on investment for biosimilars and the complex and cumbersome procedures of the US FDA for biosimilars, only such companies that have global experience and expertise relating to approval of biosimilars in the USA and/or EMA are likely to succeed (Bourgoin and Nuskey, 2013). This is the reason why companies that have such experience are partnering with other entities with similar experience in order to overcome cost and experience barriers (Bourgoin and Nuskey, 2013). Big pharma companies such as Pfizer and Amgen also have handsome marketing budgets due to their successful introduction of innovator biologics, some of which have proven to be blockbuster drugs (Bourgoin and Nuskey, 2013). The strongest competitors will be able to bring experience in all areas of market experience. There are few that have all the attributes required to successfully overcome the marketing barrier, but deal activity could continue to create an opportunity for more marketing. The major innovator biologic company Amgen has joined with Actavis to tap its understanding of the marketing for biosimilars (Bourgoin and Nuskey, 2013). These new initiatives are likely to significantly reduce competition as only big companies will be able to compete eliminating small and medium companies which are the primary reason for the competition and the resultant cost cutting in chemical generics. Experts opine that a well-crafted abbreviated regulatory pathway for biosimilars in the USA may help hasten the introduction of expensive biotherapeutics into the domestic and international healthcare systems besides encouraging the discovery of new biologics through increased competition (Kotlikoff, 2008).

19.13 Access to Biotherapeutics

Access to healthcare includes availability, affordability and acceptability of health products. It is essential that access to medicines is maintained on a sustainable basis, viz. a system to be able to function over prolonged periods (EMA, 2015). In any healthcare system, generic medicines play an crucial role in promoting the accessibility and affordability due to their cost (Don't trade our lives away, 2014). At least two barriers have been identified that impact access to affordable healthcare: (i) intellectual property rights (IPR; patents, trade secrets etc.) that create monopolistic regimes; and (ii) an unduly complex regulatory framework that delays the introduction of cheaper generic medicines. It is therefore essential to address these two issues to promote access to biotherapeutic products without compromising the safety of patients through generics (GPHA, 2013). It is essential to address this issue as healthcare costs are going up all over the world. In the USA, for example, the national health expenditure grew 5.3% to US$3.0 trillion in 2014, which means US$9,523 per person accounting for 17.5% of Gross Domestic Product (Centers for Medicare and Medicaid, 2014). It is projected to touch 20% of GDP by 2017 of about US$4.3 trillion (Bouggy, 2010).

Other barriers to access include any legal, economic or practical limitations that prevent firms from offering products in a given market; such barriers necessarily increase the likelihood that an entity will hold a monopoly and charge monopoly prices (Shepherd, 2015). In a practical setting, these are costs that would be borne by firms not currently serving a market, but that are *not* currently felt by firms in the market; in other words, costs that only affect outsiders looking in on a market (Shepherd, 2015). Economic barriers are of scale with cost advantage, where a company obtains a certain size of market while the competitors face the cost barrier. Some barriers to entry are purely legal, such as data and market exclusivity periods (included in part in both Hatch-Waxman and the BPCIA), which make it difficult for new entities from offering products at a competitive price (Generics and Biosimilars Initiative, 2011). Compounding the problem in the USA is the requirement of large patient co-insurance for speciality drugs, like biologics, and most prescription drug insurance plans do not fully reimburse these huge costs (Purvis, 2014), which results in many patients being deprived of these life-saving drugs.

The legal manoeuvres employed by pharmaceutical and biologics manufacturers take many forms, from delaying approval of safe biosimilars, to abrogating previous commitments to international drug-naming protocols, to circumventing Congressional intent for biosimilar substitution. These policies reduce competition in the market for biologic drugs, impede drug innovation, increase drug costs and limit patient access to these important medications, thus frustrating the Affordable Care Act's (BPCIA) goals of increasing healthcare availability while controlling healthcare costs (Shepherd, 2015). Also as is common in the pharma sector, especially in the USA, innovator companies that have a monopoly of market due to IP protection also continuously lobby with local and national policy makers, international agencies and others to prevent the entry of biosimilar companies that would offer similar products at a lower price (Pollack, 2013).

Biologic drugs are currently prohibitively expensive for many consumers. The average cost of a biologic drug is 22 times greater than a traditional drug (So and Katz, 2010). The average annual cost of a biologic drug is an estimated US$34,550, but annual costs for many drugs exceed US$200,000 (Megerlin *et al.*, 2013); for example, Roche-Genentech's Herceptin, a common drug used to treat breast cancer, costs upwards of US$48,000, and Johnson & Johnson's rheumatoid arthritis treatment Remicade is priced at US$20,000 annually. The annual increase in the price of biologics is 36% as compared to other medicines (Musselwhite, 2009), keeping these lifesaving medications out of reach for many. Humura, one of the highest selling drugs in the world, costs about US$1200 per month. A biosimilar Exemptia marketed by Cadila, an Indian company, will cost one-fifth of Humira's US price at US$200 a vial (Helfand, 2014).

Several patients' groups, especially cancer patients, are hoping for wider access to biologicals through biosimilars, as for several cancers such drugs are the only hope. Biotherapeutics for cancers are prohibitively expensive: researchers have calculated that treatment of metastatic colorectal cancer with bevacizumab (Avastin) costs about US$75,000 per year of life gained (Shankaran *et al.*, 2014). Generic competition is the single best proven method of bringing drug prices down. It is estimated that introduction of biosimilars in the US market would bring down the costs by about 20–40% with an estimated saving of over US$71 billion over the next decade (Musselwhite, 2009). A report by the RAND Corporation, California, estimated that biosimilars could save users US$44.2 billion by 2024 (Ledford, 2015).

In general, generic medicines: (i) provide an affordable, high quality option for several disease conditions; (ii) help expand the patient groups that require these medicines in view of the cost; (iii) trigger healthy competition with the innovator companies in terms of pricing of products; (iv) reduce considerably healthcare costs of people and governments; (v) save cost of treatments even during hospitalization, as a significant share of cost is due to medicines; and (vi) provide scope for further reduction of the cost of medicines due to competition among the generic manufacturers.

The EMA has greatly promoted the access of cheaper medicines to patient groups (IMS Health, 2011). According to some estimates, savings due to these generic medicines for the European population could well be in excess of €33.4 billion (US$45.5 billion) by 2020, with over €20.4 billion in savings from biosimilar antibody drugs alone (Haustein *et al.*, 2011; Hirschler, 2013). Yet another study suggests that the faster regulatory approval process for biosimilars in Europe saves patients as much as 60% after 4 years of market penetration (Bongers and Carradinha, 2009; Hirschler, 2013; Shepherd, 2015).

Any policy framework that discourages or delays the entry of generic medicines into the market would have a very adverse impact on the overall healthcare costs as patients, governments and others need to pay more. A study by the EGA Health Economics Committee underscores the overall significance of increasing patient access to generic medicines and the benefits accrued thereof (Sheppard, 2009). According to estimates by the EGA, the total savings until now due to generic medicines in the EU are over €30 billion, which is set almost to double with the expansion of the EU to 27 member states (Sheppard, 2009). Another study by the IGES institute charts a pathway to improve the use of biosimilars to combat the increasing costs of the European healthcare system due to the rising cost of biologicals (Sandoz, 2011). A quicker shift from reference drugs to biosimilars in eight countries of the EU, viz. Germany, France, the UK, Italy, Spain, Sweden, Poland and Romania, for only three classes of biologics – erythropoetin alfa (EPO), granulocyte colony stimulating factors (G-CSF) and monoclonal antibodies (mAbs) – could lead to potential savings for the EU of between €11.8 billion and €33.4 billion by 2020. Although cost reductions for biosimilars probably will not be so great, the US Federal Trade Commission predicts that the availability of biosimilars in the USA may not significantly reduce the cost of biologics and increase their accessibility (FTC, 2009). The FTC is expected to scrutinize any agreements between biosimilar applicants and the innovator product sponsor to monitor delay in marketing of the biosimilars.[4]

The IGES report also recognizes that, currently, many developing countries such as the USA and Europe are increasingly importing generic medicines from countries like India that have a strong and vibrant generic manufacturing system. These manufacturing systems are being closely monitored by regulatory agencies. The US FDA, for example, has set up a monitoring unit in India to ensure that adequate standards are maintained for safety and efficacy. The setting up of such systems of quality-monitoring generic medicines in the country of manufacture would ensure the availability of quality medicines at affordable cost.

19.14 Research and Development on Biosimilars

More work and cost is involved for biosimilar companies as compared with chemical generic companies. The development of a biosimilar, unlike a small molecule generic, requires substantial R&D investments for both preclinical and clinical study trials with hundreds of patients, with an estimated overall timeline of 7–8 years and investment of between US$100 million and US$250 million (Sandoz, 2011) compared to about less than half the cost for chemical generics. Recognizing the increasing markets, most biosimilar companies have begun investing in R&D (Pharmatching Portrays, 2012), especially for innovation towards improved formulations and better delivery systems to promote better patient compliance. At a higher level, to encourage the increasing use of biosimilars the IGES study calls for major role of national governments including the European authorities to create a policy climate through such systems as providing ready access to data on biosimilars of current and past users as part of the post-marketing adverse event monitoring, which is currently a very contentious issue (Sandoz, 2011).

19.15 Impact of Hatch-Waxman Act on Access

It is well known that the US Hatch-Waxman Act[5] (1984) has successfully promoted the development of generic medicines with little impact on the overall innovation for new drugs in the USA. In fact R&D investments by innovator pharma companies have increased between three to six times (Congress Budget Office, 2006; PHRMA, 2011) and more than 8000 generics have been approved for marketing by the FDA since 1984 when Hatch-Waxman was enacted (Schacht and Thomas, 2009). By putting a clear legal and regulatory framework in place, this legislation has significantly reduced both the time and cost for generic manufacturers seeking FDA approval resulting in substantial share for generic medicines in the US market (Grabowski *et al.*, 2011). Generic medicines in the US market formed only 19% of all prescriptions filled prior to 1984, but currently form 84% of prescriptions filled (Dicken, 2012). This surge of cheaper generic medicines has resulted in significant savings for consumers; in the last decade alone, generic medicines have resulted in savings of US$1 trillion to the US healthcare system (Dicken, 2012).

There is thus adequate evidence to show that the Hatch-Waxman Act has indeed spurred innovation in the pharma sector, greatly increased competition, reduced costs for traditional chemical drugs and significantly improved access to several lifesaving medicines. Hatch-Waxman has created a climate for competition among the generic manufacturers themselves to come out with newer and better innovations, generic medicines with better and safer formulations and means of administration to greatly improve patient compliance with continued reduction in cost. Analysts predict that in common with the known impact of generic medicines on overall access to healthcare by the Hatch-Waxman Act, with policy support, the BPCIA would also spur innovation and competition in the market for biologics, making biotherapeutics affordable. According to Kotlikoff, the success of the Hatch-Waxman Act in balancing innovation with affordability, as well as the apparent absence of any material differences between the biologics and chemical medical industries in respect of R&D does not justify longer monopoly protection (Kotlikoff, 2008). He suggests that the US Congress must seriously consider the Hatch-Waxman model for data exclusivity rather than the arguments put forward by the innovator companies (Kotlikoff, 2008).

19.16 Some Strategies to Promote Access

19.16.1 Increase competition for biotherapeutics

It is now understood that competition would bring down prices of medicines and that the barriers of legal, economic, or practical limitations erected by innovator pharma companies have driven up the cost of medicines (Shepherd, 2015). Confidentiality along with IP protection are widely known to be tools for innovative pharma companies to keep generic companies from competing. According to Nicholson Price, a patent-law specialist at the University of New Hampshire School of Law in Concord, innovative biological companies often keep their manufacturing methods confidential (Ledford, 2015). As the manufacturing of biotherapeutics is quite complex, this strategy gives the innovator companies ample scope to file patents on manufacturing methods or ways of characterizing molecules minimizing the entry of biosimilar companies (Ledford, 2015).

While the Hatch-Waxman Act has successfully increased generic competition in the market for traditional drugs, it has not been successful in the development and marketing of biosimilars to increase competition. Biosimilar drugs and follow-on biologics as a class are required to have full, individual FDA testing and approval (Kanter and Feldman, 2012). This procedural asymmetry has made biotherapeutics broadly immune to the competitive pricing pressure that drives down their cost, unlike the chemical generics following the Hatch-Waxman (Kanter and Feldman, 2012). This asymmetry has been allowed to persist for several decades in the USA even while Europe surged ahead by putting in place science-driven regulatory regimes for approval of biosimilars for consumer use (Wiecek and Mikhail, 2006).

There is empirical evidence suggesting that the current anti-biosimilars policies have stifled competition, raising prices and limiting consumer access to lifesaving medications. The adoption of varying non-proprietary naming system successfully by the innovator companies has restricted biologic competition in Australia, Japan and Europe (McCamish, 2013). There is also evidence to show how state-level generic-unfriendly laws in several states in the USA have increased the burden on patients by denying them cheaper generic alternatives (Shepherd, 2015). Similar provisions have also impacted access to chemical generics in some US states and these are pointers for policy makers to tweak laws to promote competition and access to biosimilars (Shepherd, 2015). Such policies clearly run against the philosophy of the BPCIA's goals of promoting access to affordable healthcare, which would also reduce the government burden on the healthcare budget.

Experts opine that there is great scope for promoting competition to reduce prices as the innovator pharma companies enjoy huge profit margins. Estimates suggest that the average gross margin for these drugs is as much as 98% after the costs for manufacturing the biologics are accounted for (Gal, 2014). That the innovator pharma companies are able to make up to 98% profits indicates the need for erecting barriers to prevent the entry of biosimilar competitors. Analysis by the US Federal Trade Commission on the current competitive climate for biosimilars has concluded that: 'Overall, generic drug competition has substantially reduced many prescription drug prices and total prescription drug expenditures, and increased access to therapeutic drugs for more Americans' (Federal Trade Commission, 2013). In contrast, data from Europe show that biosimilars have stimulated market competition, reducing prices and increasing access to lifesaving drugs (EGMA), which is set to save consumers in the EU between US$15 billion and US$45 billion from 2007 to 2020 (Ramachandra, 2014). These lower prices have also increased the volume of use of biotherapeutics (mostly as biosimilars) by as much as 50% (Ramachandra, 2014). Biosimilar competition for just 11 biologic drugs to US consumers, though significantly less than those in Europe, could reach US$250 billion in the next decade (Miller, 2014).

There are some attempts by the manufacturers of biosimilars and their associations to launch a campaign to counter misconceptions about biosimilars (Pollack, 2013). They have been approaching regulatory agencies and policy makers to support the increasing use of biosimilar drugs that will provide the consumers with safe low-cost alternatives.

19.16.2 Challenges in interchangeability

It is possible to perfectly duplicate traditional chemical drugs, potentially guaranteeing the 'absence of a significant difference' between an innovator's product (an FDA-approved drug) and a proposed equivalent (Roger, 2006; Krishnan *et al.*, 2015; US FDA, 2015). This effective duplication, or bioequivalence, defines the conventional relationship between an innovator 'brand-name' drug and a 'generic' drug which are bioequivalent substitutes for brand-name counterparts. While bioequivalence is possible for a chemically synthesized drug up to a few hundreds of atoms, it is virtually impossible to duplicate exactly the complex biologics with tens of thousands of atoms per molecule (Kerber-Walker, 2013). To compound the problem, biologics could vary slightly across batches from a single, original manufacturer (McCamish, 2013). As a result, companies looking to replicate a biologic must instead use highly similar, but slight variants, of living organisms or processes in creating a biosimilar or biogeneric (Grabowski *et al.*, 2014a, b).

The safety of a chemical generic is typically established through physicochemical identification and demonstration of a similar pharmacokinetic profile (bioequivalence) to the originator product, which is usually adequate to conclude on therapeutic equivalence. In contrast, a biosimilar needs more extensive head-to-head comparison with the reference product, to ensure close resemblance in physicochemical and biologic characteristics, safety and efficacy before approval (Weise *et al.*, 2012). Also, the scientific concepts underlying the comparability exercise for biosimilars are the same as those for changes in the manufacturing process of a given biological, for which sufficient guidance and adequate experience already exist (Weise *et al.*, 2012). In addition, since biosimilars are produced by a different manufacturer, the data requirements for demonstration of biosimilarity are typically more extensive than for demonstration of comparability of a given biological before and after manufacturing changes by the same manufacturer (Weise *et al.*, 2012).

The BPCIA defined 'biosimilarity' as 'that the biological product is highly similar to the reference product notwithstanding minor differences in clinically inactive components' (Mann and Mahinka, 2014). For a conclusion of biosimilarity, however, the BPCIA requires that there are 'no clinically meaningful differences between the biological product and the reference product in terms of safety, purity, and potency' (Mann and Mahinka, 2014). 'Interchangeability' is a higher standard than biosimilarity, requiring a showing that the proposed biosimilar may be substituted for the reference product without the intervention of the healthcare provider who prescribed the reference product (Mann and Mahinka, 2014). In view of the concerns expressed by clinicians about the quality, safety and interchangeability of biosimilars with the reference product, a significant amount of data and evidence on clinical efficacy and safety are required (Weise *et al.*, 2012). These include demonstration of similar physico-chemical characteristics, biological activity, pharmacokinetics and human safety (including immunogenicity) studies and efficacy data for each indication of use of biosimilar with the innovator product, conducted through rigorous comparability studies (Weise *et al.*, 2012).

Unlike chemical generics, there is always a doubt whether the biosimilar can be a true clinical substitute for the innovator product (Krishnan *et al.*, 2015). The major clinical concern is on patient safety due to the repeated switches between the biosimilars and the innovator product causing or increasing immunogenicity with the potential for an adverse event. It is also known that certain inherent differences arising due to a post-approval process or a formulation change could lead to differential immunogenicity that may not be accessed through characterization or during the clinical trials but may also become evident during Phase IV post-marketing pharmaco-vigilance; this may also happen with the innovator product. Tracking such a biologic that has caused an adverse event would be tough if both the innovator product and biosimilar are marketed with the same name. This becomes compounded because the patient population will be large and widely spread. It would therefore be tough for the regulatory agencies to certify that the drug is truly interchangeable in the absence of adequate data. This has become a contentious issue with considerable debate, both in the USA and the EU (Krishnan *et al.*, 2015). However, EMA believes that once a product, viz. biosimilar, is accorded permission for marketing, it means that the biosimilar could be substituted for the innovator product by the treating physician. However, automatic substitution is not possible as per EU law governing similar biological medicinal products. In a later public consultation in 2005, the EMA included in the revised 2005 Guideline a new element with a statement indicating that a biosimilar application when approved for marketing does not certify the interchangeability status (EMA, 2013a, b). Since this is among the provisions of national laws of the EU member states, the EMA has left it to individual members to decide the issue of interchangeability. According to the EU Generics Association more than 12 countries have rules against automatic substitution with only France permitting such switching of biosimilars and innovator product since 1 January 2014. The US FDA has powers for such a declaration in the USA, but so far no such announcement has been made.

Unlike the regulatory pathway in the EU for interchangeability, the BPCI Act of 2009 has a provision that allows biosimilar applicants to pursue a designation of interchangeability for a biosimilar. But it is essential for the biosimilar company to establish that: (i) the biosimilar 'can be expected to produce the same clinical result as the reference product in any given patient'; and (ii) if 'administered more than once to an individual, the risk in terms of safety or diminished efficacy of alternating or switching between use of the biologic product and the reference product is not greater than the risk of using the reference product without such alternation or switch' (Casadevall *et al.*, 2002). According to the relevant US statute, 'interchangeable' means that the biological product could be substituted for the innovator product without the intervention of the treating physician (Casadevall *et al.*, 2002).

Physicians recognize two types of substitution: (i) bioequivalence substitution; and (ii) therapeutic substitution. In the bioequivalent substitution, an innovator drug is replaced with an interchangeable generic drug with the same amount of chemically equivalent active ingredient and comparable activity at a target site. Although there have been some reports that biosimilars being not 'exact' copies may lead to adverse events, these have been refuted by more well-controlled studies that found no evidence that bioequivalent substitution was associated with exacerbations of disease (Sarpatwari *et al.*, 2015). In view of this evidence, the American Medical Association Council on Science and Public Health declared as early as 2007 that evidence of clinical differences between brand-name NTI drugs and their bioequivalent generics 'either does not exist or is extremely weak' (American Medical Association Council on Science and Public Health, 2007). Experts opine that the FDA's bioequivalence requirements are often surpassed by generic drug manufacturers' (Sarpatwari *et al.*, 2015).

19.16.3 Therapeutic substitution

Therapeutic substitution involves products that are not considered bioequivalent. Such therapeutic interchange is possible, as many classes of drugs with multiple chemically related but distinct drugs have a common mechanism of action. For instance, as many as eight types of angiotensin-II-receptor blockers (ARBs) are currently available in the US market. Therapeutic substitution of a brand-name ARB like olmesartan (Benicar) with a generic ARB like telmisartan (Micardis) will save a lot of costs. More so since the innovator product does not offer any significant benefits over the generics in the same class (Sarpatwari *et al.*, 2015). In common with bioequivalent substitution, therapeutic substitution can lead to significant savings for both the patients and the government.

19.16.4 Follow-on biologic substitution

Follow-on biologics authorized by the Biologics Price Competition and Innovations Act (2009) for marketing are required to have the same mechanism of action, route of administration and dosage form as the innovator and not be meaningfully different in respect of safety, purity or potency

(BPCIA, 2009). The FDA has thus far issued four draft guidelines on meeting these requirements (FDA, 2014a, b, c). In the approval process of the FDA, a few biosimilars will be deemed interchangeable, while most will simply be approved as members of the same drug class to treat the same conditions. But interchangeable and non-interchangeable follow-on biologic substitutions are roughly analogous to bioequivalent and therapeutic substitutions, respectively (Sarpatwari *et al.*, 2015). The scientific standards followed by the FDA on biosimilars are identical to the EMA, which has already approved as many as 19 biosimilars in the European market (Megerlin *et al.*, 2013). In spite of the credible European data and experience there are still concerns in the USA about biosimilars, including unfounded fears that small changes in the biosimilar products may lead to unwanted immune responses and therefore are not safe. Such apprehensions have hindered the development of biosimilars with facilitating innovator biological manufacturers successfully lobbying policy makers from allowing automatic substitution in some states in the USA (Pollack, 2013). The IGES analysis also confirms that the biosimilars that are marketed after a process of stringently defined regulatory pathway are quite comparable in terms of quality, safety and efficacy to their reference product (Sandoz, 2011).

19.16.5 Promote generic-friendly policies – sourcing from countries such as India

India is widely considered the pharmacy of the world, primarily due to its ability to supply quality medicines at affordable prices through its strong generic industry and robust competition among the generic manufacturers (MSF; Médicins sans Frontières, 2015). In the area of HIV/AIDS treatment the price of basic HIV treatment was brought down by almost 99% over the course of a decade, from over US$10,000 to around US$100 (MSF, 2015). The MSF sources 80% of its HIV medicines from India to treat over 200,000 people living with HIV all over the world (MSF, 2015). India is the largest provider of generic drugs globally, with Indian generics accounting for over 20% of global exports (India in Business, 2015). The Indian pharma market is the third largest globally in terms of volume with a total market size of US$20 billion with an expected growth of about 20% every year (India in Business, 2015). As of March 2014, 523 Indian pharmaceutical manufacturing facilities were registered with the US FDA, which is the highest number for any country outside the USA (India in Business, 2015). About 40% of all FDA marketing approvals for generic medicines in 2013 were for Indian products (Silverman, 2014). According to the IMS Institute for Healthcare Informatics, 86% of US prescriptions filled in 2014 were for generics and 40% of these were imported from India (Silverman, 2014).

The biotechnology industry of India is growing at an average rate of about 30% with a strong biopharma industry comprising vaccines, therapeutics and diagnostics; diagnostics is the largest sub-sector, contributing nearly 62% of the total revenues of about US$1.90 billion (India in Business, 2015). Some major new facilities are being built in the area of healthcare biotechnology both within and outside India (e.g. Malaysia). Companies are forging alliances with foreign partners such as Stelis Biopharma with Bio-Xcell Biotechnology Park in Malaysia (India in Business, 2015). Cadila Healthcare Ltd is launching a biosimilar for Adalimumab for rheumatoid arthritis and other auto-immune disorders (Exemptia) at one-fifth of the price for the branded version, Humira. Intas Pharmaceuticals is set to launch a biosimilar version of Lucentis (Razumab) (India in Business, 2015).

There are 14 recombinant products available in the market out of the 50 recombinant therapeutic products so far approved. There are about 40 companies, both Indian and multinationals, engaged in R&D, import, manufacturing and marketing of recombinant biotech products (Siddiqui, 2015). The three Indian companies that have initiated plans to make biosimilars for the USA and Europe are all partnered with larger western firms: Biocon has a tie-up with Mylan Inc., USA, Dr Reddy's Laboratories with Merck KGaA, Germany and Intas, with Canada's Apotex Inc. (Siddiqui, 2015).

19.16.6 Investments by biosimilar companies

The biosimilar industry, unlike the generic pharma, is not for small entities as R&D for biologics, even biosimilars, requires strong financial, infrastructure, regulatory and support systems. According to Sandoz, a leader in the EU's biosimilar market and currently in Phase III trials for US approval of filgrastim and pegfilgrastim biosimilars, the cost of biosimilar development can range from US$75 million to US$250 million. This is extremely high compared to the US$2–3 million needed to develop small-molecule bioequivalent drugs. However, even at US$250 million, the costs of this abbreviated pathway remain much less than the cost of going the full BLA route, which is estimated at US$800 million (Sandoz, 2011). This is largely due to the extensive clinical studies and comparability exercise required to demonstrate biosimilarity. In addition, biosimilar manufacturers must invest in clinical trials, manufacturing and post-approval safety monitoring programmes similar to that of the original innovator companies. Because of this investment, cost savings achievable with biosimilars may not be as great as can be experienced with small-molecule generics (Mellstedt *et al.*, 2008; Sandoz, 2011). However, an estimated compound annual growth rate of about 60% from 2010 to 2017 for many companies (both innovator and biosimilars) makes it worthwhile for these companies to venture into the biological market.

19.16.7 International harmonization for biosimilars

Global harmonization of regulatory pathways typically addresses the procedural complexities to help chart a clear global pathway. This should make biosimilars even cheaper and cost-effective as the biosimilar manufacturers will have a single product development cycle for all territories. Such a pathway should be rigorous with highest standards for safety and efficacy. The European Generic Medicines Association and the General Pharmaceutical Association (GPhA) have reportedly put forward such a proposal for regulatory convergence. The EMA has been able to reduce the clinical and non-clinical requirements for approval of biosimilars in Europe, without compromising quality and safety. Yet the limitation is that for a biosimilar to be considered for marketing approval in the EU, the innovator biological also must be licensed in the EU, thus considerably limiting access to products. The biosimilar companies are therefore forced to carry out clinical studies in Europe.

The EMA has been working closely with the FDA to put in place a regulatory framework where, if the clinical trial were to be carried out in the EU or the USA, the clinical data could be sufficient for a biosimilar application in the EU. One of the proposals of the EMA also relates to framing new draft overarching biosimilars guidelines to enable EMA to accept the use of a non-EMA authorized comparator or reference product (GaBI Online, 2013). The EMA has also been working with the WHO towards global harmonization of regulatory guidelines of biosimilars.

The WHO organized a joint workshop with the Korea Food & Drug Administration in 2010 to discuss harmonization of regulatory pathways (Wadhwa *et al.*, 2011). There was representation from national regulatory authorities from ten countries. The workshop recommended that the WHO Guidelines could be used as a tool for harmonizing regulatory guidelines worldwide and that the WHO should take this initiative forward. It appears that there has been no follow-up thereafter.

The EMA has released a long-awaited guidance document with new recommendations on bringing out 'similar biological medicinal products' (biosimilars) into the EU market (EMA, 2014b). These guidelines are an updated version of the Similar Biological Medicinal Products guideline issued earlier. The final guidance was approved by the EMA's Committee for Medicinal Products for Human Use (CHMP) on 23 October 2014, and will be enforceable from 30 April 2015. The guidance states that 'the general principles to be applied [for similar biological medicinal products] are addressed in a guideline taking into account the characteristics of the concerned biological medicinal product published by the Agency'. This guideline also describes the application of the biosimilar approach, the choice of the reference product and the principles of establishing biosimilarity etc. The EMA's guideline attempts to clarify issues relating to a 'similar' biological medicine. The guidance underscores the four core principles, viz. safety, efficacy, quality and biological activity. An important facet of the guidance is the EMA's 'biosimilar approach' in terms of the comparability exercises. Although many of the elements of the guidance are as per the International Conference on Harmonization's (ICH) Q5E document, according to the EMA each biosimilar product will have to be evaluated based on its own merits. Similar to the evaluation of the analytical methods, clinical comparability models and manufacturing processes are used to create and validate the biosimilar product. A major facet of the guidance relates to a product being used for comparison by the generic manufacturer. While the stipulation that the reference product must be one authorized in the EU, the regulator may allow companies to compare a biosimilar 'in certain clinical studies and in vivo non-clinical studies' with a non-authorized comparator to help promote the development of biosimilars as also to avoid the unnecessary repetition of clinical trials. However, the biosimiar applicants need to demonstrate that the reference product approved outside the EU is 'representative of the reference product authorized in the EEA' with proper data and documentation. There is the need for 'bridging data', which will essentially include comparative analytical studies with the biosimilar, the EMA-approved and non-EMA-approved product. The bridging studies would also include pharmacokinetic and pharmacologic data. Many of these steps could be adapted in other countries to work towards harmonization of regulatory systems.

Such a unified International Committee on Harmonization authorized reference standard for biosimilar development will help save huge costs for biosimilar manufacturers, which should also help the users, as combined development costs could be about US$200–$300 million less per product, which could be passed on to the consumer, than when done individually for each geographic region. It is clear that many of the new guidelines made recently are primarily in tune with the EU and US guidelines, with many countries slowly shifting towards a system of bridging studies for approval of biosimilars. Regulatory harmonization will also help those countries with weaker regulations to eventually compete in markets with more regulated markets.

19.17 The World Health Organization Global Strategy and Plan of Action

There are other global initiatives that aim to promote access to healthcare. The Global Strategy and Plan of Action (GSPoA) of the WHO aims to address the twin challenges for medical innovation, viz. financing and incentivizing essential health and biomedical R&D and fostering enhanced access to existing and new medicines especially to people in developing countries (WHO, 2008). The GSPoA was primarily initiated to address medicines for diseases of the poor, the so-called neglected diseases. The World Health Assembly (WHA) adopted in May 2008 Resolution WHA61.21 and Resolution WHA62.16, on a global strategy and plan of action on public health, innovation and IP (WHO, 2008). The WHO has since recognized the epidemiological transition with both rich and poor countries recording significant rise in chronic non-communicable diseases such as cancers for which biotherapeutics offer the only hope. But many remedies are prohibitively expensive, severely limiting access to patients in developing countries. The WHA Assembly (WHO, 2014a) recognized that the new generation of safe, efficacious, quality biotherapeutic medicines that have a positive impact on morbidity and mortality rates for several diseases are not accessible in a timely and adequate fashion due to multiple barriers and high cost. To that end, the WHA recommended promotion of more efficient regulatory frameworks at the national and regional levels that will ensure access to efficacious, quality and safe biotherapeutic products, including similar biotherapeutic products (WHO, 2014b).

19.18 Conclusions

The availability of chemical generic medicines, which now account for over 86% of all prescriptions, have saved US consumers about US$1.5 trillion in the past decade (Generic Pharmaceutical Association, 2011, 2015). This became possible because the US regulatory structure post-Hatch-Waxman permits automatic substitution of innovator chemical drugs with generic medicines, which has become possible under the BPCIA. Policy makers, especially in the USA, must look at how some roadblocks that include recent actions by the FDA and the resultant legal battles before many courts in the states of the USA impact the entry of biosimilars into the market. These barriers to entry not only contravene the spirit of the ACA, which aimed at speedier certification for substitute biologics, but also increase costs and reduce competition, all for no established benefits to patient safety or manufacturer innovation (PPACA, 2010). In actual fact interpretations by the US courts appear to have helped increase or protect exclusivity for biologics that benefited few innovator pharmaceutical companies at the cost of severely restricting access to healthcare, besides the rise in healthcare cost to the government (Shepherd, 2015). There have been some welcome developments that would address the cost issue of medicines, although not directly linked to biologicals. Some US lawmakers concerned about the rise in the prices of generics are planning to help to promote access of affordable healthcare to US citizens. Data released for a Congressional hearing show that the price of about half of generic medicines over the last year went up an average of 10%, some doubled in cost and a few rose by over 500% (Rosenthal, 2014). These include thyroid replacement hormone, the antibiotic doxycycline, the heart pill digoxin and the asthma pill albuterol (Rosenthal, 2014). Senators Amy Klobuchar and John McCain are planning to revise a legislation that will allow prescription drugs from Canada. Since the USA does not have a system of negotiating drug prices unlike many countries, generic medicines are the only means to ensure access to many medicines (Rosenthal, 2014). Such initiatives must be extended to ensure that the ACA Act truly makes healthcare affordable.

Biologics are expected to be at the forefront for the treatment of diseases, with the potential for treatments of hitherto intractable diseases. However, the major concern continues to be the prohibitive cost, putting these drugs beyond the reach of many. Development of newer medicines by the pharma companies must be encouraged through reasonable monopoly through IP protection as companies invest huge sums for R&D. There must be a faster, transparent and robust approval regulatory pathway that will help pharma companies save costs for the introduction of new products into the market. This must be coupled with a system that would facilitate quicker entry of biosimilars into the market.

Notes

1 *Association for Molecular Pathology* v. *Myriad Genetics*, 569 US 12-398 (2013); *Mayo Collaborative Services, DBA Mayo Medical Laboratories, et al.* v. *Prometheus Laboratories, Inc.*, 132 S Ct 1289 (2012).

2 *Association for Molecular Pathology* v. *Myriad Genetics*, 569 US 12-398 (2013).

3 35 U.S.C. § 271(e)(1)

4 *Federal Trade Commission* v. *Actavis, Inc.*, 133 S. Ct. 2223 (2013).

5 Hatch-Waxman Act. Drug Price Competition and Patent Term Restoration Act. Pub. L. No. 98-417, 98 Stat. 1585 (1984). Available at: https://www.gpo.gov/fdsys/pkg/STATUTE-98/pdf/STATUTE-98-Pg1585.pdf (accessed 23 November 2015).

Acknowledgements

The authors would like to thank Dr S.R. Rao, Advisor, Department of Biotechnology, Government of India and the Director-General, Indian Council of Medical Research, New Delhi for encouragement and support.

Disclaimer

The opinions expressed in the paper are the authors' own and do not reflect the views of the Department of Biotechnology or the Indian Council of Medical Research.

The authors have no conflict of interest.

References

American Medical Association Council on Science and Public Health (2007) Generic substitution of narrow therapeutic index drugs. Available at: http://www.ama-assn.org/resources/doc/csaph/x-pub/csaph2a07-fulltext.pdf (accessed 27 January 2015).

Anour, A. (2014) Biosimilars versus 'biobetters' – a regulator's perspective. *Generics and Biosimilars Initiative Journal (GaBI Journal)* 3(4), 166–167.

BCC Research (2012) Biosimilars: Global Markets. BCC Research. Wellesley, Massachusetts. Available at: http://www.bccresearch.com/market-research/biotechnology/biosimilars-global-market-bio090a.html#gsc.tab=0 (accessed 20 November 2015).

Bongers, F. and Carradinha, H. (2009) How to Increase Patient Access to Generic Medicines in European Healthcare Systems: A Report by the EGA Health Economics Committee. Available at: http://www.medicamentsgeneriques.info/wp-content/uploads/2010/10/ega_increase-patient-access_update_072009.pdf (accessed 16 March 2015).

Bouggy, B.R. (2010) Follow-on biologics legislation: Striking a balance between innovation and affordability. *Indiana Health Law Review* 367(7), 368–396.

Bourgoin, A.F. and Nuskey, B. (2013) An outlook on U.S. biosimilar competition. *Drugs Today* 49(6), 399–412.

BPCIA (2009) Biologics Price Competition and Innovation Act. Available at: http://www.fda.gov/downloads/Drugs/GuidanceComplianceRegulatory Information/UCM216146.pdf (accessed 23 November 2015).

Carroll, J. (2014) Biosimilars set to boom as new patent cliff on biologic superstars looms. Available at: http://www.fiercebiotech.com/story/biosimilars-set-boom-new-patent-cliff-biologic-superstars-looms/2014-07-22 (accessed 22 July 2014).

Casadevall, N., Nataf, J., Viron, B, Kolta, A., Kiladjian, J.-J., Martin-Dupont, P., Michaud, P., Papo, T., Ugo, V., Teyssandier, I., Varet, B. and Mayeux, P. (2002) Pure-red cell aplasia and antierythropoietin antibodies in patients treated with recombinant erythropoietin. *New England Journal of Medicine* 346, 469–475.

Center for Drug Evaluation and Research (CDER), US FDA (2012a) *Proprietary Name Review: BLA 125418 – Zaltrap (xxxjaflibercept).* US Food and Drug Administration, Washington, DC.

Center for Drug Evaluation and Research (CDER), US FDA (2012b) *Proprietary Name Review: BLA 125294 – [xxx]-filgrastim*. US Food and Drug Administration, Washington, DC.

Centers for Medicare and Medicaid (2014) *NHE Factsheet.* Available at: http://www.cms.gov/Research-Statistics-Data-and-Systems/Statistics-Trends-and-Reports/%20NationalHealthExpendData/NHE-Fact-Sheet.html (accessed 23 November 2015).

Committee for Medicinal Products for Human Use (CHMP) (2004) Revision 1 Guidelines on similar biological medicinal products. CHMP/437/2004. Available at: http://www.ema.europa.eu/docs/en_GB/document_library/Scientific_guideline/2013/05/WC500142978.pdf (accessed 23 November 2015).

Congress Budget Office (2006) Research and Development in the Pharmaceutical Industry, 7. Available at: https://www.cbo.gov/sites/default/files/cbofiles/ftpdocs/76xx/doc7615/10-02-drugr-d.pdf (accessed 23 November 2015).

Department of Biotechnology (DBT) (2012) *Guidelines on Similar Biologics: Regulatory Requirements for Marketing Authorization in India.* Department of Biotechnology, Government of India, New Delhi.

Dicken, J.E. (2012) The Letter from John E. Dicken, Dir., Health Care, U.S. Government Accountability Office to Sen. Orrin G. Hatch, Committee on Finance (Jan. 31, 2012). Available at: http://www.gao.gov/assets/590/588064.pdf (accessed 23 November 2015).

Dolinar, R.O. and Reilly, M.S. (2013) The future of biological therapy a pathway forward for biosimilars. *Generics and Biosimilars Initiative Journal (GaBI Journal)* 2(1), 36–40.

Don't Trade Our lives Away (2014) Time to Act: Ensure Access to Affordable Biotherapeutic Products. Available at: https://donttradeourlivesaway.wordpress.com/2014/08/26/time-to-act-ensure-access-to-affordable-biotherapeutic-products (accessed 23 November 2015).

EuropaBio (2011) Guide to Biological Medicines – A Focus on Biosimilar Medicines. EuropaBio [Online], 26 October 2011. Available at: http://www.europabio.org/guide-biological-medicines-focus-biosimilar-medicines (accessed 23 November 2015).

European Commission, Enterprise and Industry Directorate-General (ECID) (2012) Consensus Information Paper 2013. What you need to know about Biosimilar Medicinal Products. European Commission. Commission on Implementing Regulation (EU) No 520/2012. *Official Journal of the European Union* L159.

European Medicines Agency (EMA) (2004) Directive 2001/83/EC, as amended by Directive 2003/63/EC and Directive 2004/27/EC European Medicines Agency (2011). Procedural advice for users of the centralised procedure for similar biological medicinal products applications. EMA/940451/2011, March 2013.

European Medicines Agency (EMA) (2005) *Guideline on Similar Biological Medicinal Products. Evaluation of Medicines for Human Use*. EMEA/CHMP/437/04. EMA, London.

European Medicines Agency (2012) Committee for Medicinal Products for Human Use (CHMP) Guideline on similar biological medicinal products containing biotechnology-derived proteins as active substance: quality issues (revision 1). EMA/CHMP/BWP/247713/2012, 24 May 2012.

European Medicines Agency (EMA) (2013a) Draft overarching guidelines on biosimilars: Guideline on Similar Biological Medicinal Products (Reference number: CHMP/437/04 Rev 1). European Medicines Agency. [Online] 22 May 2013. Available at: http://www.ema.europa.eu/docs/en_GB/document_library/Scientific_guideline/2013/05/WC500142978.pdf (accessed 23 November 2015).

European Medicines Agency (EMA) (2013b) Guideline on Similar Biological Medicinal Products (22 May 2013). Available at: http://www.ema.europa.eu/docs/en_GB/document_library/Scientific_guideline/2013/05/WC500142978.pdf (accessed 23 November 2015).

European Medicines Agency (EMA) (2014a) The European Biosimilars Group (EBG). Biosimilar Medicines Increase Access for Patients Across Europe. Available at: http://www.egagenerics.com/index.php/press-room/press-releases/2014/361-biosimilar-medicines-increase-access-for-patients-across-europe (accessed 20 November 2015).

European Medicines Agency (EMA) (2014b) Committee for Medicinal Products for Human Use (CHMP). Guideline on similar biological medicinal products. CHMP/437/04. Rev 1, 23 October 2014. Available at: http://www.ema.europa.eu/docs/en_GB/document_library/Scientific_guideline/2014/10/WC500176768.pdf (accessed 20 November 2015).

European Medicines Agency (EMA) (2015) *European public assessment reports.* Available at: http://www.ema.europa.eu/ema/index.jsp?curl=pages%2Fmedicines%2Flanding%2Fepar_search.jsp&mid=WC0b01ac058001d124&searchTab=searchByAuthType&alreadyLoaded=true&isNewQuery=true&status=Authorised&keyword=Enter+keywords&searchType=name&taxonomyPath=&treeNumber=&searchGenericType=biosimilars&genericsKeywordSearch=Submit (accessed 20 November 2015).

Federal Trade Commission (2009) Emerging Health Care Issues: Follow-On Biologic Drug Competition 69–72. Available at: http://www.ftc.gov/os/2009/06/P083901biologicsreport.pdf (accessed 26 November 2015).

Federal Trade Commission (2013) Public Workshop: Follow-On Biologics: Impact of Recent Legislative and Regulatory Naming Proposals on Competition, 78 Fed. *Reg.* 68840, 68841 (15 November 2013).

Fogel, L.E. and Hanna, P.H. (2014) The Biosimilar Regulatory Pathway and the Patent Dance. December 2014. Available at: https://jenner.com/system/assets/publications/13837/original/The_Biosimilar_Regulatory_Pathway_and_the_Patent_Dance.pdf?1420753075 (accessed 20 November 2015).

Food and Drug Administration, China (2015) *Technical Guideline for the Research, Development and Evaluation of Biosimilars.* Available at: http://www.sfda.gov.cn/WS01/CL0087/115103.html (accessed 20 November 2015).

GaBI Online (2013) Generics and Biosimilars Initiative. EMA issues draft revision of overarching biosimilar guidelines. Mol, Belgium: Pro Pharma Communications International [cited 28 June 2013]. Available at: http://www.gabionline.net/Guidelines/EMA-issues-draft-revision-of-overarching-biosimilar-guidelines (accessed 23 November 2015).

GaBI Online (2014) US$67 billion worth of biosimilar patents expiring before 2020. Available at: http://www.gabionline.net/Biosimilars/General/US-67-billion-worth-of-biosimilar-patents-expiring-before-2020 (accessed 20 January 2014).

GaBI Online (2015) Biosimilar naming conventions around the world. Posted 20/11/2015. Available at: http://gabionline.net/Biosimilars/Research/Biosimilar-naming-conventions-around-the-world (accessed 23 November 2015).

Gal, A. (2014) FTC Presentation. Biosimilars: Commercial Prospective, 5 (4 February 2014). Available at: http://www.ftc.gov/system/files/documents/public_events/Follow-On%20Biologics%20Workshop%3A%20Impact%20of%20Recent%20Legislative%20and%20Regulatory%20Naming%20Proposals%20on%20Competition/gal.pdf (accessed 23 November 2015).

Gal, R. (2011) Biosimilars: Reviewing US Law and US/EU Patents; Bottom up Model Suggests 12 Products and $7-$8B Market by 2020. *Bernstein Research*, 26 May 2011.

Generic Pharmaceutical Association (2011) A Year of Progress, Generic Pharma Association (September 2011). Available at: http://www.gphaonline.org/media/cms/AnnualReport_11.pdf (accessed 23 November 2015).

Generic Pharmaceutical Association (2015) Biosimilars in the EU, Generic Pharmaceutical Association. Available at: http://www.gphaonline.org/gpha-media/gpha-resources/biosimilars-in-the-eu (accessed 23 November 2015).

Generic Pharmaceutical Association & IMS Institute for Healthcare Informatics (GPHA) (2013) Generic drug savings in the U.S. Available at: http://www.gphaonline.org/media/cms/%20GPhA_Generic_Cost_Savings_2014_IMS_presentation.pdf (accessed 27 January 2015).

Generics and Biosimilars Initiative (2011) *US Biosimilars Law May Prove a Barrier to Entry for Biosimilars.* Available at: http://gabionline.net/Biosimilars/Research/US-biosimilars-law-may-prove-a-barrier-to-entry-for-biosimilars (accessed 23 November 2015).

Gershon, S.K., Luksenburg, H., Coté, T.R. and Miles Braun, M. (2002) Pure red-cell aplasia and recombinant erythropoietin. *New England Journal of Medicine* 346, 1584–1586.

Grabowski, H. and DiMasi, J. (2009) Biosimilars, Data Exclusivity, and the Incentives for Innovation: A Critique of Kotlikoff's White Paper 9 (Duke Univ. Dep't of Econ. Working Paper No. 2009-02, Feb. 2009). Available at http://public.econ.duke.edu/Papers//PDF/FinalDraft2_5_09.pdf (accessed 15 November 2015).

Grabowski, H., Long, G. and Mortimer, R. (2011) Implementation of the biosimilar pathway: economic and policy issues. *The Seton Hall Law Review* 41(2), 511–557.

Grabowski, H.G., Guha, R. and Salgado, M. (2014a) Regulatory and cost barriers are likely to limit biosimilar development and expected savings in the near future. *Health Affairs (Millwood)* 33(6), 1048–1057.

Grabowski, H., Guha, R. and Salgado, M. (2014b) Biosimilar competition: lessons from Europe. *Nature Reviews Drug Discovery* 13, 99–100.

Grabowski, H.G., DiMasi, J.A. and Long, G. (2015) The roles of patents and research and development incentives in biopharmaceutical innovation. *Health Affairs* 34, 2302–2310.

Greer, F. (2013) Biosimilar development: the race to market continues. *SGS Life Science Technical Bulletin* 45. Available at: http://www.sgs.com/-/media/global/documents/technical-documents/technical-bulletins/sgs-lss-biosimilars-the-race-to-market-en-13.pdf (accessed 23 November 2015).

Grewal, I.S. (ed.) (2009) *Emerging Protein Biotherapeutics*. CRC Press, Boca Raton, Florida.

Hartung, H.P. (2004) What do we know about the mechanism of action of disease modifying treatments in MS? *Journal of Neurology* 251(Suppl. 5), v12–v29.

Haustein, R., de Millas, C., Hoeer, A. and Haeussler, B. (2011) *Saving Money in the European Healthcare Systems with Biosimilars*. IGES Institute, Berlin.

Health Canada (2010) Guidance For Sponsors: Information and Submission Requirements for Subsequent Entry Biologics (SEBs). Available at: http://www.hc-sc.gc.ca/dhp-mps/brgtherap/applic-demande/guides/seb-pbu/seb-pbu_2010-eng.php (accessed 23 November 2015).

Helfand, C. (2014) AbbVie's Humira competition begins with Cadila's biosimilar launch in India. Available at: http://www.fiercepharma.com/story/abbvies-humira-competition-begins-cadilas-biosimilar-launch-india/2014-12-09 (accessed 23 November 2015).

Hirschler, B. (2013) Analysis: Copycat Biotech Drugs Slow to Take off in Europe, Reuters, 3 December 2013. Available at: http://www.reuters.com/article/2013/12/03/us-biotech-copycatsanalysis-idUSBRE9B205I20131203 (accessed 15 November 2015).
IMS Health (2011) Shaping the biosimilars opportunity: A global perspective on the evolving biosimilars landscape, pp. 1–12. IMS Health, London. Available at: https://docs.340bpvp.com/documents/public/resourcecenter/biosimilars/Shaping_the_biosimilars_opportunity.pdf (accessed 15 November 2015).
India in Business (2015) Pharmaceuticals. Updated September 2015. Available at: http://indiainbusiness.nic.in/newdesign/index.php?param=industry services_landing/347/1 (accessed 20 October 2015).
International Federation of Pharmaceutical Manufacturers & Associations (IFPMA) (2012) Biotherapeutic medicines: Grasping the new generation of treatments. Available at: http://www.ifpma.org/resource-centre/biotherapeutic-medicines-grasping-the-new-generation-of-treatments (accessed 12 January 2016).
International Federation of Pharmaceutical Manufacturers & Associations (IFPMA) (2013) Similar biotherapeutic products: Scientific & regulatory considerations. Available at: http://www.ifpma.org/resource-centre/similar-biotherapeutic-products-scientific-regulatory-considerations (accessed 12 January 2016).
Kanter, J. and Feldman, R. (2012) Understanding and incentivizing biosimilars. *64 Hastings Law Journal* 57, 59.
Kerber-Walker, J. (2013) Small molecules, large biologics, and the biosimilar debate. Arizona Bioindustry Association (18 February 2013). Available at: http://www.azbio.org/small-molecules-large-biologics-and-the-biosimilar-debate (accessed 20 November 2015).
Kotlikoff, L.J. (2008) *Stimulating Innovation in the Biologics Industry: A Balanced Approach to Marketing Exclusivity*. Boston University Working Paper, Boston. Available at: http://people.bu.edu/kotlikof/New%20Kotlikoff%20Web%20Page/Kotlikoff_Innovation_in_Biologics21.pdf (accessed 23 November 2015).
Kotlikoff, L.J. (2009) Biosimilars, Data Exclusivity, and the Incentives for Innovation: Response to Grabowski. Available at: http://www.kotlikoff.net/content/biosimilars-data-exclusivity-and-incentives-innovation-response-grabowski (accessed 1 October 2015).
Krishnan, A., Mody, R. and Malhotra, H. (2015) Global regulatory landscape of biosimilars: emerging and established market perspectives. *Biosimilars* 5, 19–32.
Ledford, H. (2015) First biosimilar drug set to enter US market. *Nature* 517, 253–254.
Mann, J.S. and Mahinka, S.P. (2014) Biosimilars: patent challenges and competitive effects. *LMG Life Sciences Guide 2014,* pp. 33–36.
McCamish, M. (2013) Effect of Naming on Competition and Innovation. Sandoz Biopharmaceuticals 6 (10 December 2013). Available at: http://www.ftc.gov/system/files/documents/public_events/Follow-On%20Biologics%20Workshop%3A%20Impact%20of%20Recent%20Legislative%20and%20Regulatory%20Naming%20Proposals%20on%20Competition/mccamish.pdf (accessed 23 November 2015).
Médicins sans Frontières (2015) PM Modi: Keep India the Pharmacy of the Developing World. 29 September 2015. Available at: http://www.doctorswithoutborders.org/article/pm-modi-keep-india-pharmacy-developing-world (accessed 20 November 2015).
Megerlin, F., Lopert, R., Taymor, K. and Trouvin, J.H. (2013) Biosimilars and the European experience: implications for the United States. *Health Affairs* 32(10), 1803–1810.
Mellstedt, H., Niederwieser, D. and Ludwig, H. (2008) The challenge of biosimilars. *The Annals of Oncology* 19, 411–419.
Miller, S. (2014) Customer Perspective on Biosimilars. Available at: http://www.ftc.gov/system/files/documents/public_events/Follow-On%20Biologics%20Workshop%3A%20Impact%20of%20Recent%20Legislative%20and%20Regulatory%20Naming%20Proposals%20on%20Competition/miller.pdf (accessed 23 November 2015).
Ministry of Food and Drug Safety (MFDS) South Korea (2010) Guideline on Evaluation of Biosimilar Products. Available at: http://www.nifds.go.kr/en/research/bio.jsp (accessed 23 November 2015).
Musselwhite, L. (2009) Biogenerics Op-ed: A cost-cutter that's missing in health bills. News & Observer (NC), 7 October 2009. Available at: http://lp-health@lists.essential.org; http://www.newsobserver.com/opinion/columnists_blogs/story/129059.html (accessed 25 October 2015).
Ohly, C. (2012) The New India Guidelines on Similar Biologics: Regulatory and Market Authorization Requirements. Available at: http://spicyip.com/2012/10/guest-post-new-india-guidelines-on.html (accessed 23 November 2015).
Patient Protection and Affordable Care Act (PPACA) (2010) Text of the Patient Protection and Affordable Care Act, P.L. 111-148, 25 August 2010. Available at: https://www.govtrack.us/congress/bills/111/hr3590/text (accessed 23 November 2015).
Paul, S.M., Mytelka, D.S., Dunwiddie, C.T., Persinger, C.C., Munos, B.H., Lindborg, S.R. and Schacht, A.L. (2010) How to improve R&D productivity: the pharmaceutical industry's grand challenge. *Nature Reviews Drug Discovery* 9, 203–224.
Pharmaceutical Research and Manufacturers of America (PHRMA) (2009) Pharmaceutical Industry Profile 2009. Available at: http://www.phrmajp.org/archives/pdf/profile/PhRMA2009ProfileFINAL.pdf (accessed 23 November 2015).

Pharmaceutical Research and Manufacturers of America (PHRMA) (2011) *Pharmaceutical Industry 2011 Profile.* Available at: http://report.nih.gov/investigators_and_trainees/acd_bwf/pdf/Phrma_Industry_Profile_2011.pdf (accessed 23 November 2015).

Pharmatching Portrays (2012) Pharma & Outsourcing Practice. Biosimilars|Guidelines|Players|Providers. Available at: https://www.pharmatching.com/inforena/pharma-outsourcing-practice-biosimilars-guidelines-players-providers/download/91146/Pharmatching_Portrays_Biosimilars.pdf (accessed 23 November 2015).

Pollack, A. (2013) Biotech Firms, Billions at Risk, Lobby States to Limit Generics. *New York Times*, 29 January 2013. Available at: http://www.nytimes.com/2013/01/29/business/battle-in-states-on-generic-copies-of-biotech-drugs.html (accessed 23 November 2015).

Purvis, L. (2014) Consumer Perspective on Biosimilars, AARP, 8–12. Available at: http://www.ftc.gov/system/files/documents/public_events/Follow-On%20Biologics%20Workshop%3A%20Impact%20of%20Recent%20Legislative%20and%20Regulatory%20Naming%20Proposals%20on%20Competition/purvis.pdf (accessed 21 September 2014).

Ramachandra, R. (2014) *Presentation for FTC Biosimilars* Workshop on Naming Proposals and Impact on Competition, 7 (4 February 2014). Available at: http://www.ftc.gov/system/files/documents/public_events/FollowOn%20Biologics%20Workshop%3A%20Impact%20of%20Recent%20Legislative%20and%20Regulatory%20Naming%20Proposals%20on%20Competition/ramachandra.pdf (accessed 24 November 2015).

Roger, S.D. (2006) Biosimilars: how similar or dissimilar are they? *Nephrology (Carlton)* 11(4), 341–346.

Roox, K., Pike, J., Brown, A. and Becker, S. (2008) *Patent-Related Barriers to Market Entry for Generic Medicines in the European Union*. European Generic Medicines Association, Brussels.

Rosenthal, E. (2014) Lawmakers Look for Ways to Provide Relief for Rising Cost of Generic Drugs. *New York Times*, 24 November 2014. Available at: http://www.nytimes.com/2014/11/25/us/lawmakers-look-for-ways-to-provide-relief-for-rising-cost-of-generic-drugs.html?_r=0 (accessed 15 October 2015).

Roth, V.J. (2013) Will FDA data exclusivity make biologic patents passé? 29 *Santa Clara Computer & High Technology Law Journal* 249, 260–264.

Sandoz (2011) Biosimilar Development. Available at: http://www.sandoz-biosimilars.com/en/biosimilars/development-of-biosimilars-en.shtml (accessed 16 October 2015).

Sarpatwari, A., Choudhry, N.K., Avorn, J. and Kesselheim, A.S. (2015) Paying physicians to prescribe generic drugs and follow-on biologics in the United States. *PLoS Medicine* 12(3), e1001802.

Satyanarayana, K.S., Srivastava, S. and Ganguly, N.K. (2006) Data protection issues in India. *Indian Journal of Medical Research* 123, 723–726.

Schacht, W.H. and Thomas, J.R. (2009) Follow-On Biologics: Intellectual Property and Innovation Issues. Congressional Research Service, 7-5700. Available at: http://www.ipmall.info/hosted_resources/crs/RL33901_090320.pdf (accessed 20 March 2009).

Schiestl, M., Stangler, T., Torella, C., Cepeljnik, T., Toll, H. and Grau, R. (2011) Acceptable changes in quality attributes of glycosylated biopharmaceuticals. *Nature Biotechnology* 29(4), 310–312.

Shankaran, V., Mummya, D., Koepla, L., Bansala, A., Dana, K., Mirick, D.K., Yud, E., Morlockd, R., Ogaled, S. and Ramsey, S.D. (2014) Survival and lifetime costs associated with first-line bevacizumab use in older patients with metastatic colorectal cancer. *The Oncologist* 19(8), 892–899.

Shapiro, R.J. (2008) The Potential American Market for Generic Biological Treatments and the Associated Cost Savings. Available at: http://www.insmed.com (accessed 23 November 2015).

Shepherd, J.M. (2015) *Biologic Drugs, Biosimilars, and Barriers to Entry, 25 Health Matrix 139 (2015)*. Available at: http://scholarlycommons.law.case.edu/healthmatrix/vol25/iss1/8 (accessed 23 November 2015).

Sheppard, A. (2009) Generic Medicines: Essential contributors to the long-term health of society. IMS Health. Available at: http://www.hup.hr/EasyEdit/UserFiles/Granske_udruge/HUP-UPL/IMS.pdf (accessed 16 March 2015).

Siddiqui, Z. (2015) India risks missing boat as biosimilars shake up drug industry. Available at: http://www.reuters.com/article/2015/10/14/us-india-biosimilars-idUSKCN0S82OQ20151014#LA7zpI4I4X6mZfRS.97 (accessed 23 November 2015).

Silverman, E. (2014) Why more Indian generic drugs will make their way to the U.S. *Wall Street Journal* 14 May 2014. Available at: http://blogs.wsj.com/corporate-intelligence/2014/05/14/why-more-indian-generic-drugs-will-make-their-way-to-theu-s/?mg=blogs-wsj&url =http%253A%252F%252Fblogs.wsj.com%252 Fcorporate-intelligence%252F2014%252F05%252F14%252Fwhy-more (accessed 16 March 2015).

So, A.D. and Katz, S.L. (2010) Op-Ed., Biologics Boondoggle. *New York Times*, 7 March.

So, A.D. and Sachs, R. (2012) Making intellectual property work for global health. *Harvard International Law Journal* 53, 107–143.

United States Food and Drug Administration (US FDA) (2012a) *Guidance for Industry: Quality Considerations in Demonstrating Biosimilarity to a Reference Protein Product.* US Food and Drug Administration, Washington, DC.

US FDA (2012c) *Guidance for Industry. Scientific Considerations in Demonstrating Biosimilarity to a Reference Product*. Food and Drug Administration, Rockville, Maryland.

US FDA (2012b) Draft Guidance for Industry: Biosimilars: Questions and Answers Regarding Implementation of the Biologics Price Competition and Innovation Act.

US FDA (2013) Draft Guidance for Industry: Formal Meetings between the FDA and Biosimilar Product Sponsors or Applicants.

US FDA (2014a) Biosimilar guidances. Available at: http://www.fda.gov/Drugs/GuidanceCompliance%20RegulatoryInformation/Guidances/ucm290967.htm (accessed 23 November 2015).

US FDA (2014b) Draft Guidance for Industry: Clinical Pharmacology Data to Support a Demonstration of Biosimilarity to a Reference Product. Available at: http://www.fda.gov/downloads/drugs/guidancecomplianceregulatoryinformation/guidances/ucm397017.pdf (accessed 23 November 2015).

US FDA (2014c) Guidance for Industry. Reference Product Exclusivity for Biological Products Filed Under Section 351(a) of the PHS Act. Available at: http://www.fda.gov/downloads/drugs/guidancecomplianceregulatoryinformation/guidances/ucm407844.pdf (accessed 23 November 2015).

US FDA (2015) Title 21 – Food and Drugs Chapter I – Food and Drug Administration Department of Health and Human Services Subchapter D – Drugs for Human Use. Part 320 – Bioavailability and Bioequivalence Requirements. Available at: http://www.accessdata.fda.gov/scripts/cdrh/cfdocs/cfcfr/CFRSearch.cfm?fr=320.1 (accessed 23 November 2015).

Wadhwa, M., Kang, H., Knezevic, I., Thorpe, R. and Griffiths, E. (2011) WHO/KFDA joint workshop on implementing WHO guidelines on evaluating similar biotherapeutic products, Seoul, Republic of Korea 24–26 August, 2010. *Biologicals* 39, 349–357.

Weise, M., Bielsky, M.C., De Smet, K., Ehmann, F., Ekman, N., Giezen, T.J., Gravanis, I., Heim, H.-K., Heinonen, E., Ho, K., Moreau, A., Narayanan, G., Kruse, N.A., *et al*. (2012) Biosimilars: what clinicians should know. *Blood* 120(26), 5111–5117.

Wiecek, W. and Mikhail, M. (2006) European regulatory guidelines for biosimilars. *Nephrology Dialysis Transplantation* 21, v17–18.

World Health Organization (WHO) (2008) 61st World Health Assembly, May 19–24, 2008, Global Strategy and Plan of Action on Public Health, Innovation and Intellectual Property, at 1, WHA61.21 (24 May 2008). Available at: http://www.who.int/gb/ebwha/pdf_files/A61/A61_R21-en.pdf (accessed 23 November 2015).

World Health Organization (WHO) (2009) Expert Committee on Biological Standardization. Guidelines on Evaluation of Similar Biotherapeutic Products (SBPs). World Health Organization. [Online] October 23, 2009. Available at: http://www.who.int/biologicals/areas/biological_therapeutics/BIOTHERAPEUTICS_FOR_WEB_22APRIL2010.pdf (accessed 23 November 2015).

World Health Organization (WHO) (2014a) Biological Qualifier: An INN Proposal. Programme on International Nonproprietary Names (INN). INN Working Doc. 14.342 Revised draft July. Available at: http://www.who.int/medicines/services/inn/bq_innproposal201407.pdf (accessed 23 November 2015).

World Health Organization (WHO) (2014b) 67th World Health Assembly 24 May. Access to biotherapeutic products including similar biotherapeutic products and ensuring their quality, safety and efficacy. Available at: http://apps.who.int/gb/ebwha/pdf_files/WHA67/A67_R21-en.pdf (accessed 23 November 2015).

World Health Organization (WHO) (2015) The WHO Programme for International Drug Monitoring. [cited: 24 August 2012]. Available at: http://www.who.int/medicines/areas/quality_safety/safety_efficacy/National_PV_Centres_Map/en (accessed 23 November 2015).

20 Changing Paradigm for Intellectual Property Rights Protection in Drug Discovery Research: Where India Stands

Pritom Chowdhury,* Sangeeta Borchetia and Tanoy Bandyopadhyay

Department of Biotechnology, Tocklai Tea Research Institute, Jorhat, Assam, India

20.1 Introduction

The advances in molecular biology and development of recombinant DNA technology during the 1940s and 1970s, respectively, have opened up a new horizon between nature, biology and human health. The invention of the polymerase chain reaction (PCR) technique during the 1980s, which many state as the technology breakthrough of the 21st century, continues to play a pivotal role in biological sciences. This revolution led to the elucidation of genetic code, which underpins the central dogma of molecular biology: DNA makes RNA (specifically messenger or mRNA), which makes proteins. The only exception to this universal theory is RNA tumour viruses, which contain enzymes that use viral RNA as a template for the synthesis of DNA, thereby discovering the principle of reverse transcription (Brown, 2006). These technologies give the biomedical researcher the means to manipulate the gene of their interest, related to a particular disease and identify the genetic codes responsible for expression of a particular disease trait. This advancement in the field of genomics and proteomics has provided information on understanding the mechanism of human diseases. These newer disseminations of knowledge are revolutionizing the field of medical science, which would yield a powerful arsenal of therapies that may offer the opportunities to formulate intervention strategies against pathogens (Collins, 2003; Merrill and Mazza, 2006). The pace of medical invention is accelerating, inspiring better clinical outcomes with less invasive procedures and shorter recovery periods (Merrill and Mazza, 2006).

This paradigm of the rapid advancement in science for the translation of basic scientific research into products for commercialization has accelerated a complexity for intellectual property rights (IPR) protection and generated numerous policy battles. Moreover, it led to increasing number of patents filed and patent grants (Table 20.1). However, unlike in the USA and the EU, where IPR protection is a result of judicial pronouncement, in India the IPR laws have emerged due to the obligations of international conventions. Patent rights were first introduced in India in 1856 and were formalized by the Patent and Design Act in 1911 (PDA, 1911). The passage of the Patent Act of 1970 repealed all previous legislation; however, it excluded pharmaceuticals and agrochemical products from patent eligibility (CGPDTM, 1970). India successfully amended its patent law in 2005 to comply with the World Trade Organization's Agreement on Trade-Related Aspects of Intellectual

*Corresponding author, e-mail: pritomc@gmail.com

Table 20.1. Increasing trend of patent application in major IPR offices, 2004–2014 (State Intellectual patent office (SIPO) of the People's Republic of China; Japan patent office (JPO); United States Patent and Trademark office (USPTO); European Patent office (EPO); Office of the Controller General of Patents, Designs, Trademarks (CGPDTM)).

		2004	2005	2006	2007	2008	2009	2010	2011	2012	2013	2014
China	Patent applications	353,807	476,264	573,178	694,153	828,328	976,686	1,222,286	2,849,906	3,502,683	4,327,819	5,255,996
	Patents granted	NA	NA	268,002	351,782	411,982	581,992	814,825	893,866	1,110,971	1,318,659	1,551,887
Japan	Patent applications	423,081	427,078	408,674	396,291	391,002	348,596	344,598	342,610	342,796	328,436	NA
	Patents granted	124,192	122,944	141,399	164,954	176,950	193,348	222,693	238,323	274,791	277,079	NA
USA	Patent applications	382,139	417,508	452,633	484,955	485,312	482,871	520,277	535,188	576,763	609,052	615,243
	Patents granted	181,319	157,741	196,437	182,928	185,244	191,933	244,341	247,713	276,788	302,948	326,033
EPO	Patent applications	NA	128,665	135,358	141,231	146,244	134,511	151,015	142,822	148,562	148,027	151,981
	Patents granted	NA	53,251	62,777	54,700	59,800	51,952	58,117	62,108	65,655	66,707	64,613
India	Patent applications	17,466	24,505	28,940	35,218	36,812	34,287	39,400	43,197	43,674	42,951	NA
	Patents granted	1,911	4,320	7,539	15,316	16,061	6,168	7,509	4,381	4,126	4,227	NA

Property Rights (TRIPs). However, confusion prevails over product patents in India. Indian patent law instituted in Section 3(d) implies that a mere discovery of a new form of a known substance cannot be patented unless it is significantly more efficacious (CGPDTM, 2005). Protection related to public health is particularly more challenging where there is an uncertainty about the eligibility of materials that may already exist in nature, and the requirements for its novelty and an inventive step. The absence of a marked distinction between patentable invention and discoveries from nature gives rise to great problems; and the requirement of an inventive step under Section 3(d) further exaggerated the hurdle under Indian patent law. However, one should also take note that provisions such as section 3(d) can encourage R&D companies to develop breakthrough drugs and diagnostic modules and not just mere modification of known compounds and protocols. There is a growing disparity among research-driven multinational companies (MNC) over patent issues, which may hinder the progress of R&D activity in India (Chowdhury *et al.*, 2014). Moreover, India is known as the 'pharmacy of the modern world' because of its production of affordable generic drugs (Kapczynski, 2013). Therefore discussion over the changing IPR scenario of drug development worldwide with special context to India is very much warranted. In this chapter we give a perspective, highlighting patent issues over drug research innovation related to technological regulations, public health and harnessing the health benefits of natural compounds for drug development.

20.2 Medical Technological Innovations and the Regulatory System

Regulatory policies dealing with the challenges of medical technology innovation are crucial for effective drug development and the process thereof. This is becoming more complex than at any time previously, and contends with a host of policy, legal and administrative structures at national, regional and international levels. The World Health Organization (WHO) defines health technology as the application of organized knowledge and skills in the form of devices, medicines, vaccines, procedures and systems developed to solve health problems and improve quality of lives (WHO, 2007). The emergence of diverse medical technologies has raised challenges to set an overall standard regulatory system. Regulators face the challenge to balance the benefit of product safety, efficacy and adequate quality with affordability and accessibility. Unjustified regulatory barriers and lengthy marketing authorization processes may delay access to medical technologies to the needed. To overcome this barrier it was felt to have better coherence between different international agencies and wider geographical countries. In the declaration on the TRIPS agreement and Public Health (Doha Declaration), members agreed the need for the World Trade Organization (WTO) agreement on TRIPS to be part of wider national and international bodies (WTO, 2001). During the 49th World Health Assembly (WHA) in May 1996, WHO adopted the mandate to work on the interface between public health and intellectual property (IP) (WHO, 1996). The WHO, WTO and the World Intellectual Property Organization (WIPO) have intensified inter-agency collaboration with each having distinct but complementary mandates to work on issues relating to public health, trade, innovation and IP, respectively (WHO, 2011). They focused on monitoring and assessing the impact of trade agreements, technical and policy support on formulating coherent trade and health policies, and the implementation of TRIPS flexibilities. A major development in this regard was the establishment of the Commission on Intellectual Property Rights, Innovation and Public Health (CIPIH) during 2003. Its main aim is to collect data and proposals from different agencies involved and produce an analysis of IPR, innovation and public health, especially in developing countries (WHO, CIPIH, 2006). In 2009, WIPO established the Global Challenges Program to address interplay between innovation, technology transfer and the dissemination of technology. In 2011, the International Medical Device Regulators Forum (IMDRF) was established to promote an efficient and effective regulatory model for medical devices for regulating safety standards (WTO, 2013). Advancements in the field of medical technology, such as gene therapy, tissue engineering and monoclonal antibody therapy, further necessitate the importance of stringent and uniform regulatory measures for IP protection.

India is a signatory to the Millennium Developmental Goals (MDGs), which is the most comprehensive and specific list of developmental goals ever agreed upon by the world (FICCI, 2013). Medical technology plays a vital role in the delivery of healthcare services in India. Medical professionals rely on medical technology for tests and investigations to aid their clinical decisions. Medical technology also plays a critical role in the treatment of patients, such as in the case of HIV; the dose of the antiretroviral drugs can vary depending on quantification of virus load by real-time PCR (Calmy *et al.*, 2007). Innovation in medical technology can therefore be crucial for a country like India's healthcare system to improve access, enhance quality and reduce costs. The growth in the medical technology sector is parallel with growth in the health sector. It is estimated that the Indian medical technology industry will reach US$14 billion by 2020 (FICCI, 2013). The drivers for it are economic growth, increasing burden of diseases and private investment in health technology, examples include 3M's manufacturing plant in Pune, Becton Dickinson (BD)'s manufacturing facility in Haryana and General Electricals (GE) R&D centre in Bengaluru among many (FICCI, 2013). The government has permitted 100% foreign direct investment (FDI) in the healthcare sector. It has also reduced and rationalized the customs duty on medical technology imports and in addition provides tax incentives to in-house research and development. Despite this, the indigenous private companies are falling behind in medical technology development especially in electronic devices, though innovation in the *in vitro* diagnostics sub-segment is substantial. Public–private partnerships are just starting to gain momentum. India has made vast improvements for creating a favourable environment for technology development, especially with 'Make in India' in place. However, there are still numerous factors that need to be addressed to spur growth in the sector to catch up with the pace of technology development by the USA and China. China accounts for the largest number of patent applications received by any single IP office since 2011 (WIPO, 2014). With its large number of researchers, China is slowly closing the gap in the medical technology patent arena with developed economies such as Germany. Quasi-developed countries, such as Israel, perform favourably well in the medical technology sector in terms of people involved in research. In comparison, India lacks behind, both in the number of researchers and technology output (Fig. 20.1). Indian researchers should focus more on product-oriented research, which will give impetus towards more patents. There is a need for separate regulation for medical technology with clearly articulated, unambiguous processes. The Department of Health and Family Welfare, Government of India have proposed the Drug and Cosmetic (Amendment bill) 2015. The bill introduces separate provision for clinical trials and regulation of medical devices. The bill also introduces a hierarchical structure for regulatory bodies with clear separate regulation for medical devices and legal penalties for any violations (CDSCO, 2014). It is now up to the policy makers to pass the bill to

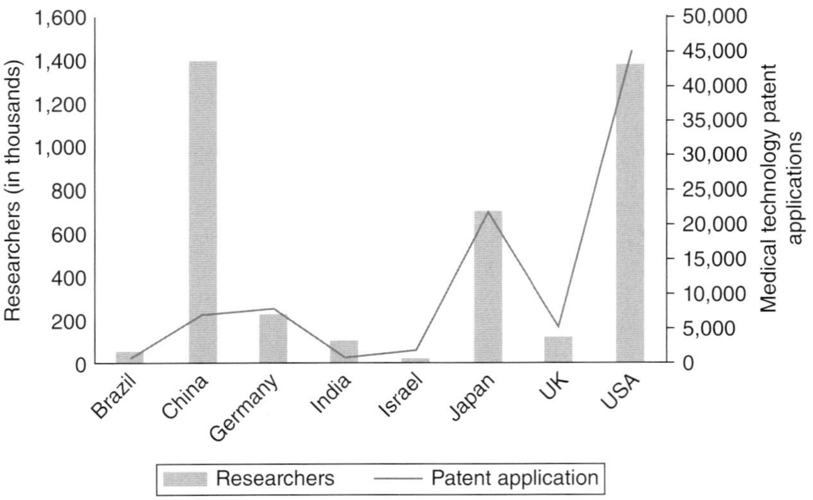

Fig. 20.1. Comparisons of researcher versus medical technology patent application in major patent application countries in 2004–2010 (United Nations Educational, Scientific and Cultural Organization; World Intellectual Property Organization).

make a change in the regulation process. Further, comprehensive training needs to be provided to the regulators at the central and state levels as well as the regulatory affairs departments of medical technology companies. Another important aspect is the need to increase India's GDP expenditure on healthcare. India needs to usher in further reforms in the insurance sector to increase penetration of health insurance while simultaneously providing a safety net for economically weaker sections of the society.

20.3 Biotechnology Patent Scenario

Biotechnology becomes an integral part of drug research, starting from the identification of disease markers, therapeutic development to diagnosis. It helps immensely to the progress of health research but not without complexity. Among the disciplines in science and technology that involve IP protection, it is the patenting of biotechnology inventions, specifically genes, that is a major topic of debate and controversy. Although early patents on medical products of biotechnology, such as insulin, and early genetic engineering techniques have expired, there is an expanding rate of new patent applications for novel gene and protein discoveries, and new processes. The patents include claims for genomic DNA sequences, complementary DNAs, DNA clones, expressed sequence tags (ESTs) and single nucleotide polymorphisms (SNPs) (Kiley, 1992; Knoppers 1999). However, granting patents for these products involves greater complexities to the requirements of successful patents. This complexity is clearly highlighted in the landmark case of *Association of Molecular Pathology* v. *Myriad Genetics* on *BRCA1* and *BRCA2* gene patenting for diagnosis of breast and ovarian cancer.[1] In June 2013 the Supreme Court of America revoked the patents granted and stated that 'A naturally occurring DNA segment is a product of nature and not patent eligible merely because it has been isolated, but cDNA is patent eligible because it is not naturally occurring' (Bakshi, 2014). The court further said that 'Had Myriad created an innovative method of manipulating genes while searching for the BRCA1 and BRCA2 genes, it could possibly have sought a method patent'. Another important case was between *Mayo Collaborative Services* v. *Prometheus Laboratories* over a patent concerning application of threshold level of thiopurine drugs in the treatment of autoimmune diseases, such as Crohn's disease and ulcerative colitis. Prometheus has the exclusive licence of the patents for the diagnostic kits with a method optimized for the therapeutic efficacy for treatment to different patients, which vary in dose-wise response to treatment regime. In 2012, the court reversed the patent on the basis of 'obviousness' to persons skilled in the art (Supreme Court of the United States, 2011).[2] There is a growing disparity among a section of people over this decision. Some believe it will minimize innovation in the field of the emerging personalized medicine sector, whereas other people supported it for social benefit. But the fact is that we are still not very clear on the issue of granting a patent that will stifle or spur innovation. There are hurdles involved with biotechnology product patents, but an analysis of the US and UK case laws show that there is a balance in protecting biotech inventions. One such case example is *Genentech* v. *The Wellcome Foundation* (McGough, 1992), where a patent was granted to Genentech for a tissue plasminogen activator. The patent application was challenged on the basis of obviousness to the persons skilled in the art. The appellate court concluded that the traditional concept of obviousness from the perspective of a hypothetical addressee is unsuitable in the field of biotech, where much technological complexity is involved. Another case that laid a foundation for biotech patents is *Diamond* v. *Chakrabarty* (Robinson and Medlock, 2005), where patent was granted to a genetically engineered bacterium capable of breaking down crude oil.

The guidelines of the Indian Patent Office on biotech patents are not very clear. The word gene or DNA does not appear anywhere in the Indian Patent act, nor is there any special directive issued by the Indian government to accord legal protection to biotech-related inventions. Moreover, distinctions between a gene obtained from nature and a gene modified in the laboratories are not mentioned anywhere, nor there is any important landmark case in India whose verdict may become a pillar for the policy makers. However, it is very important to realize that biotech comprising genomics and proteomics is a research-driven innovative field, where it is imperative to seek patent protection for such inventions. It is a capital intensive industry, where transforming innovation in the lab to the market may take decades. The industry deals with life forms, which in most cases come under the morality issue. Sometimes discovery comprises traditional knowledge, therefore sharing of the benefit is an issue. All these points demarcate a boundary where a revisit is required to assess whether IPR gives adequate protection to biotech products. Many Indian biotech companies are engaged in manufacturing 'biosimilars' i.e. copies of first-generation biologics such as recombinant human insulin and erythropoietin, for which the product patents have expired in developed markets such as in the USA, Europe and Japan. But it leads to the situation where Indian firms are lagging behind in basic research directed towards innovation and improvement of the molecules as such and developing better delivery methods. Unlike the 'synthetic drugs', 'biodrugs' have more complexity in the production process. The work entails selecting suitable cell lines and expression vectors, the purification process and downstream scaling up. Therefore it requires a stringent process and protocol to be followed, which encompasses varied topics of modern biology. Each substance and process is unique and patentable, which makes it hard for generic companies to enter into the field. There is a need for a different policy for biodrugs, considering the complexity involved. The EU has specific guidelines for biodrugs with enhanced property (EMA, 2013). India along with other WTO assigned nations must prioritize a standard set of guidelines for the benefit of mankind.

20.4 Synthetic Drugs and Natural Products Patent

Recent advances in chemistry and computational biology have provided new methods of synthetic drug discoveries. However, this is a challenge as it starts from understanding the patho-mechanism of disease, target identification like gene or protein responsible for particular disease, target validation, lead compound identification, safety screening, lead compound optimization, pre-clinical testing followed by phase 1, 2, 3 clinical trials and then the registration process (Orloff *et al.*, 2009). The whole process may take 10–15 years, during which, among thousands of compounds entered into the R&D pipeline, one may finally receive approval as a drug. The average cost of R&D of each successful drug is estimated to be from US$800 million to US$1 billion (Adams and Brantner, 2006). This is a huge amount required to be pumped into the drug development process. Therefore, it is obvious that R&D-driven pharma companies will thrive under protection from patent regimes to manufacture and sell the approved drug exclusively in order to make a profit and recover the amount invested. However, this situation creates a dilemma where some sections advocate that it will stimulate innovation whereas others raised serious concerns on free access to medicines especially in developing countries due to high pricing (CBO, 2006). In India, recent analysis of court decisions has highlighted this complex situation. The Delhi High Court (HC) has upheld the infringement case filed by Merck Sharp & Dohme (MSD) against Glenmark for violating its patents on diabetes drugs Januvia and Janumet which contain 'sitagliptin phosphate' salt whereas Glenmark Janumet contains 'sitagliptin metmorfin'. The court ruled out the price factor as MSD drug cost in India is not significantly higher than the Glenmark product, therefore it will not hinder access to medicine for common people.[3] However, the Supreme Court put an interim stay on the HC order and the final decision has been put on hold for hearing.[4] This depicts the confusion over the patent row on synthetic molecules in the developing world. In contrast, the developed world has a somewhat clear policy on the issue. Novartis holds patent for the anticancer drug 'Glivec' in the USA, Europe and Japan but the Indian Patent office and subsequently the Indian court did not grant patent by saying it just constitutes a beta-crystalline form of imatinib (Supreme Court of India, 2013).[5] The court ruled that the pharmacological properties of the beta-crystalline form of imatinib mesylate are equally possessed by imatinib in free base form or its salt, thus questioning the enhanced therapeutic efficacy of Glivec, which is mandatory under Section 3(d). Now, if other developing countries start following Section 3(d), which will encompass a larger share of the global market and raises important issues on patenting, will it help or restrict spurring innovations (Kapczynski, 2013)? Surely, it will harm the interests of R&D-driven companies, which will ultimately minimize investment on drug development and will have a negative impact on innovations. Simultaneously, one cannot ignore the fact that it will protect the interests of public health. For example, in the case of patented HIV drugs, which cost almost US$10,000 per person per year, many countries, especially in Africa where HIV spread was rampant, could not afford them. But the same drug from an Indian generic drug maker like Cipla costs US$150 or less (Chien, 2007). Others believe it will encourage undertaking riskier and more extensive research that is required to generate breakthrough drugs (Kapczynski, 2013). To promote necessary trials and health technology assessments for all drugs (novel as well as generic), the UK government has amended its patent law to remove the risk of infringement claims and allow companies to use a patented product when carrying out clinical trials. This change is an addition to the existing 'Bolar Exemption', an EU directive which only protects generic drug makers from the risk of infringement lawsuits if they conducted trials on a patented product (Kupecz *et al.*, 2015).

Natural products have gained popularity in pharmaceutical and healthcare industries for their therapeutic potential. Numerous extracts and compounds isolated from natural products are included in drugs and healthcare supplements (van de Weijer and Barentsen, 2002; Harvey, 2008; Feng *et al.*, 2011). Patents were granted to methods and compositions for preventing or treating cancer by the administration of a combination of therapeutically effective amount of catechins, a group of polyphenols found in green tea along with *Capsicum* extracts.[6] However, conflict with granting patent protection to such compounds is whether they are being deemed as 'invention', a notable criterion for patent, or they are just a mere 'discovery', for which patent cannot be granted. The patent eligibility of natural products isolated or purified from natural materials is less affirmatively defined. An analysis of US court verdicts on natural products showed that an isolated product would not be patentable for its improved purity, unless it results in 'properties and characteristics which were different in kind from those of the known product rather than in degree' (Court of Custom and Patent Appeal, 1938). In one such case, a recrystallized aspirin and a purified form of adrenalin were deemed patentable because the purer forms were therapeutically different from the natural forms (Beuchamp, 2013). On the contrary, purified tungsten from tungsten ore and a naturally occurring bacterial strain were held to be unpatentable because the courts attributed that it is their intrinsic properties and adopted an opposite view that a functional difference was insufficient for transforming a natural product to another 'kind' of substance.[7] The absence of robust standards for determining whether a difference is 'in kind' or 'in degree' or whether it is a 'marked difference', means that the patentability of isolated natural products and the enforceability of patents for these products remain ambiguous. In view of this complicacy and aftermath of some important cases such as Myriad (Bakshi, 2014), Mayo[8] and Chakrabarty (Robinson and Medlock, 2005), the US patent office (USPTO) revisited the patenting policy and issued a comprehensive examination guideline for natural matters in March 2014 (USPTO, 2014). Under this guideline, the patent eligibility of a claim is principally determined by considering whether the claim recites something that is 'significantly' or 'markedly' different from what exists, provided that not all differences rise to the level of 'marked differences' and that 'a marked difference must be a significant difference, i.e., more than an incidental or trivial difference'. Accordingly, a structural difference appears to be important, but not necessarily sufficient, for patent eligibility. When a structural difference is absent or insignificant, a functional difference may also be considered. In the absence of a structural difference, a natural product must exhibit new functions or characteristics preferably with enhanced features that rise to the level of 'marked' differences to be patent-eligible (Wong and Chan, 2014). The European Union Directive (No. 96/9/EC of 11 March 1996) adopted a similar approach. Although the directive did not mention specifically the term 'natural products', it made it clear under Article 3.2 that 'Biological material which is isolated from its natural environment or processed by means of a technical process may be the subject of an invention even if it already occurred in nature' (Gold and Gallochat, 2001). However, many countries including India do not allow the patentability of materials existing in nature. However, despite the possible exclusion of certain natural products from the patent domain, patents may still be obtained for the processes used to produce them in a medicinal form. Sabinsa Corporation obtained patent in India for 'A Method for solubilizing Forskolin and a Composition thereof' (Patent No. IN235492). Forskolins were obtained from the roots of *Coleus forskohlli* and used as eye drops for treating open angle glaucoma (Majeed *et al.*, 2015). India has a vast store of varied natural resources with rich traditional knowledge of medicinal plants and natural remedies; such resources could be used to produce medicines. Countries like India should draw a clear dividing line between products existing in nature, which are not patentable (even if isolated or subject to standard procedures of purification), and products that have been modified or combined in a manner that gives rise to a genuine invention.

20.5 The Way Forward

There is no doubt that research is the key to innovation and IPR protection is the best policy to reward the innovators. Patent protection is given in the fields of medical technology, pharmaceuticals and biotech products in all the major R&D-driven countries (Fig. 20.2). However, with the changing paradigm of drug research wherein high cost is involved, there is a need for a cohesive policy, to strike a balance between the developed world and developing countries. At present, as per TRIPs regulations, countries are allowed to make amendments to protect each public health interest. This is important to promote IPR laws, however, R&D-driven companies should also make provisions to formulate an effective pricing mechanism especially for low income countries. One way to achieve this is through a compulsory licensing model, which needs to be articulated in IPR regimes, wherein the innovator companies receive a royalty or sizable profits while allowing generic companies to provide drugs at an affordable rate. A key factor for IPR policy at this age of 'globalization' is the societal benefit. So what does society consist of? Is it the representations of people from a developed country or developing countries as well? Surely, it comprises of both. They all are eligible for a 'right to health' and this must be taken into consideration by respective governments. The WHO can play a very active role by drafting regulations and guidelines to countries for health insurance policies. In countries like India, where 70% of the population earns less than a US$1 per day and health insurance is not available for the majority of citizens (Satyanarayana, 2008), a stringent cohesive IPR regime needs to develop a policy on 'free access to health' for its citizens. This will help to minimize the burden on personalized medicine cost and at the same time will safeguard the interest of R&D-driven innovative companies. There must be greater cooperation between public–private partnerships. Research for drug discoveries is more targeted towards the diseases of rich societies, where there is the greater possibility of making a profit by the parent company. It should be noted that more than 1 billion people, among whom the overwhelming majority are in the developing world, suffer from neglected tropical diseases (NTD), those for which there are inadequate or nonexistent treatments and a paucity of research and development (Savioli *et al.*, 2006). Incentive must be given to companies that will promote research on developments of drugs for NTD such as tuberculosis, leishmania etc. Since the inception of the IPR regime, significant progress has been made in the direction of implementation. Discussions are continuing on different platforms, policies are being amended, but concentrating solely on IPR will not do the trick for success stories. Different international and national regulatory agencies on social, health and economics should come under the same platform, which may help to formulate more clear

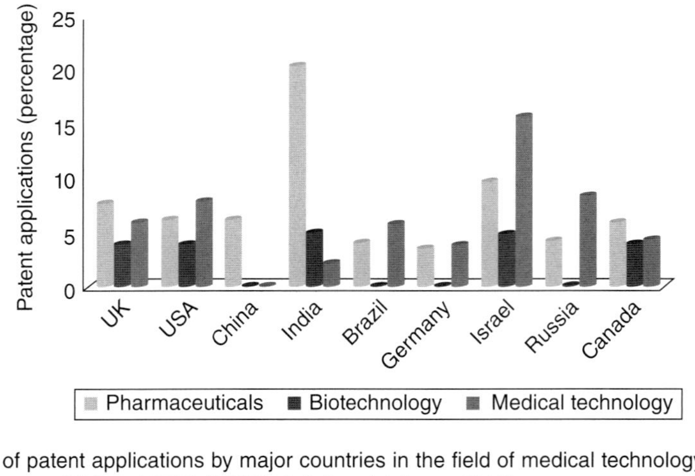

Fig. 20.2. Distribution of patent applications by major countries in the field of medical technology, pharmaceuticals and biotechnology between 1999 and 2013 (WIPO statistics database, 2014).

and stringent policies to avoid economic loss over unwanted cases and confusion over IPR protection at a time of swift progress in drug development and research in general.

Notes

1 US Supreme Court, *Association For Molecular Pathology, et al.*, v. *Myriad Genetics, Inc, et al.*, 569 U.S. (2013), 13 June 2013.

2 US Supreme Court, *Mayo Collaborative Services, et al.* v. *Prometheus Laboratories, Inc.*, 10-1150 US 566. Available at: http://www.supremecourt.gov/opinions/11pdf/10-1150.pdf (accessed 17 June 2015).

3 High Court of Delhi, *Merck Sharp and Dohme Corporation & Anr.* v. *Glenmark Pharmaceuticals Ltd.* Available at: http://lobis.nic.in/ddir/dhc/AKP/judgement/07-10-2015/AKP07102015S5862013.pdf (accessed 1 August 2015).

4 Supreme Court of India, *Glenmark Pharmaceuticals* vs. *Merck Sharp and Dohme Corporation.* Available at: http://supremecourtofindia.nic.in/FileServer/2015-05-14_1431603333.pdf (accessed 6 November 2015).

5 Supreme Court of India, *Novartis AG* V. *Union of India & Others.* Available at: http://supremecourtofindia.nic.in/outtoday/patent.pdf (accessed 1 August 2015).

6 Morré, J.D., 20 March 2007. *Compositions and methods based on synergies between capsicum extracts and tea catechins for prevention and treatment of cancer*. US7192612.

7 US Supreme Court, *Funk Brothers Seed Co.* v. *Kalo Inoculant Co.*, 333 U.S. 127, 16 February 1948. Available at: https://supreme.justia.com/cases/federal/us/333/127/case.html (accessed 1 August 2015).

8 US Supreme Court, *Mayo Collaborative Services, et al.* v. *Prometheus Laboratories, Inc.*, 10-1150 US 566. Available at: http://www.supremecourt.gov/opinions/11pdf/10-1150.pdf (accessed 17 June 2015).

Acknowledgment

Special thanks towards TAF database team of US Patent and Trademark Office (Patent Technology Monitoring team) for sending the link for the biological patent database.

References

Adams, C. and Brantner, V.V. (2006) Estimating the cost of new drug development: is it really $802 million? *Health Affairs* 2, 420–428.

Bakshi, A.M. (2014) Gene patents at the Supreme Court: association for molecular pathology v. myriad genetics. *Journal of Law and the Biosciences* 1, 183–189.

Beuchamp, C. (2013) Patenting nature: a problem of history. *Stanford Technology Law Review* 16, 257–312.

Brown, T.A. (2006) *Genomes: Transcriptomes and Proteomes*, 3rd edn. John Wiley & Sons, New York.

Calmy, A., Ford, N., Hirschel, B., Reynolds, S.J., Lynen, L., Goemaere, E., Garcia de la Vega, F., Perrin, L. and Rodriguez, W. (2007) HIV viral loads monitoring in resource limited regions: optional or necessary? *Clinical Infectious Diseases* 44, 128–134.

CBO (2006) Research and Development in the Pharmaceutical Industry (US Congressional budget office study). Available at: http://www.cbo.gov/publication/18176 (accessed 23 June 2015).

CDSCO (2014) The Draft Drugs and Cosmetics Amendment Bill, 2015. Available at: http://www.cdsco.nic.in/forms/list.aspx?lid=2019&Id=1 (accessed 11 June 2015).

CGPDTM (1970) The Indian Patent Act. Available at: http://www.ipindia.nic.in/IPActs_Rules/updated_Version/sections-index.html (accessed 3 June 2015).

CGPDTM (2005) The Patents (Amendment) Act, 2005. Available at: http://www.ipindia.nic.in/ipr/patent/patent_2005.pdf (accessed 3 June 2015).

Chien, C.V. (2007) HIV/AIDS drugs for sub-Saharan Africa: how do brand and generic supply compare? *PLoS One* 2, e278.

Chowdhury, P., Khan, S.A., Dutta, P. and Mahanta, J. (2014) Pharmaceutical and biotech product patents in India: doldrums or blissfulness? *Nature Biotechnology* 32, 133–134.

Collins, F. (2003) Genomics: the coming revolution in medicine. Available at: https://www.genome.gov/Pages/News/Documents/GlobalAgenda.pdf (accessed 6 November 2015).

Court of Custom and Patent Appeal (1938) *In* re *Merz*, 97 F.2d 599, 600. Available at: https://casetext.com/case/in-re-merz (accessed 6 November 2015).

EMA (2013) Guideline on similar biological medicinal products containing biotechnology derived proteins as active substance: nonclinical and clinical issues. Available at: http://www.ema.europa.eu/docs/en_GB/document_library/Scientific_guidelines/2013/06/WC500144124.pdf (accessed).

Feng, Y., Wang, N., Zhu, M., Feng, Y., Li, H. and Tsao, S. (2011) Recent progress on anticancer candidates in patents of herbal medicinal products. *Recent Patents on Food, Nutrition & Agriculture* 3, 30–48.

FICCI (2013) Enhancing access to healthcare through innovation: Medical technology in India. Available at: http://m.ficci.in/publication-page.asp?spid=20090 (accessed 9 June 2015).

Gold, E.R. and Gallochat, A. (2001) The European Biotech directive: past as prologue. *European Law Journal* 7, 331–366.

Harvey, A.L. (2008) Natural products in drug discovery. *Drug Discovery Today* 13, 19.

Kapczynski, A. (2013) Engineered in India – patent law 2.0. *New England Journal of Medicine* 369, 497–499.

Kiley, T.D. (1992) Patents on random complementary DNA fragments. *Science* 257, 915–918.

Knoppers, B.M. (1999) Status, sale and patenting of human genetic materials: an international survey. *Nature Genetics* 22, 23–25.

Kupecz, A., Roox, K., Dekoninck, C., Schertenleib, D., Stief, M., Sanna, F., Orsingher, M., Miralles, S., Molina, E., Crosse, T., *et al*. (2015) Safe harbors in Europe: an update on the research and Bolar exemptions to patent infringement. *Nature Biotechnology* 33, 710–715.

Majeed, M., Nagabhushanam, K., Natarajan, S., Vaidyanathan, P., Karri, S.K. and Jose, J.A. (2015) Efficacy and safety of 1% forskolin eye drops in open angle glaucoma – an open label study. *Saudi Journal of Ophthalmology* 29, 197–200.

McGough, K.J. (1992) A case for expansive patent protection of biotechnology inventions. *Harvard Journal of Law and Technology* 6, 85–102.

Merrill, S.A. and Mazza, A.M. (2006) *Reaping the Benefits of Genomics and Proteomic Research. Intellectual Property Rights, Innovation and Public Health*, 1st edn. National Academies Press, Washington, DC.

Orloff, J., Douglas, F., Pinheiro, J., Levinson, S., Branson, M., Chaturvedi, P., Ette, E., Gallo, P., Hirsch, G. and Mehta, C. (2009) The future of drug development: advancing clinical trial design. *Nature Review Drug Discovery* 8, 949–957.

PDA (1911) The Indian Patent and Design Act. Available at: http://www.theindianlawyer.in/statutesnbareacts/acts/d42.html (accessed 3 June 2015).

Robinson, D. and Medlock, N. (2007) *Diamond* vs. *Chakrabarty*: a retrospective on 25 years of biotech patents. *Intellectual Property and Technology Law Journal* 17, 12–15.

Satyanarayana, K. (2008) Towards equitable health care: drug prices, and beyond. *Indian Journal of Medical Research* 127, 301–304.

Savioli, L., Engels, D., Daumerie, D., Jannin, J., Alvar, J., Asiedu, K., Gastellu-Etchegorry, M., Simarro, P. and Mariotti, S.P. (2006) Response from World Health Organization [reader response]. *PLoS Medicine* 3, e283.

USPTO (2014) 2014 Procedure for Subject Matter Eligibility Analysis of Claims Reciting or Involving Laws of Nature/Natural Principles, Natural Phenomena, and/or Natural Products. Available at: http://www.uspto.gov/patents/law/exam/myriad-mayo_guidance.pdf (accessed 6 November 2015).

van de Weijer, P.H. and Barentsen, R. (2002) Isoflavones from red clover (Promensil) significantly reduce menopausal hot flush symptoms compared with placebo. *Maturitas* 42, 187–193.

WHO (1996) Revised drug strategy, forty-ninth World Health Assembly (WHA49.14, WHO, 1996). Available at: http://www.who.int/phi/WHA49.14.pdf (accessed 5 June 2015).

WHO (2007) Health technologies, Sixtieth World Health Assembly (WHA60:26). Available at: http://www.who.int/medical_devices/resolution_wha60_29-en1.pdf (accessed 5 June 2015).

WHO (2011) Global strategy and plan of action on public health, innovation and intellectual property, sixty-first World Health Assembly (WHA61.21, WHO, 2008). Available at: http://www.who.int/phi/publications/Global_Strategy_Plan_Action.pdf (accessed 5 June 2015).

WHO, CIPIH (2006) Public health innovation and intellectual property rights (Report of the Commission on Intellectual Property Rights, Innovation and Public Health). Available at: http://www.who.int/intellectualproperty/documents/thereport/ENPublicHealthReport.pdf (accessed 5 June 2015).

WIPO (2014) WIPO IP fact and figures. Economic & statistics series. Available at: http://www.wipo.int/edocs/pubdocs/en/wipo_pub_943_2014.pdf (accessed 5 June 2015).

Wong, A.Y.T. and Chan, A.W.K. (2014) Myriad and its implications for patent protection of isolated natural products in the United States. *Chinese Medicine* 9, 17.

WTO (2001) Declaration on the TRIPs agreement and public health, WTO document WT/MIN(01)/DEC/2 (DOHA WTO Ministerial, 2001). Available at: http://www.wto.org/english/thewto_e/minist_e/min01_e/mindecl_trips_e.htm (accessed 5 June 2015).

WTO (2013) Promoting Access to Medical Technologies and Innovation: Intersection between public health, intellectual property and trade. Available at: http://www.wto.org/english/res_e/booksp_e/pamtiwhowipo-wtoweb13_e.pdf (accessed 5 June 2015).

21 Intellectual Property Rights in Drug Development and Biotechnology

Gerard M. Raj,[1]* and Avinash Arivazhahan[2]

[1]Jawaharlal Institute of Postgraduate Medical Education and Research, Pondicherry; [2]Kasturba Medical College, Manipal, India

21.1 Introduction

Biotechnology is a field of science that brings together natural processes and human interventions. It is the term used to indicate the use of genetic selection and/or engineering to produce commercially useful and/or scientifically fascinating products by living cells. The US Congress Office of Technology Assessment defines biotechnology as 'any technique that uses living organisms (or parts of organisms) to make or modify products, to improve plants or animals, or to develop microorganisms for specific uses' (United States Congress, Office of Technology Assessment, 1989).

Even though the field of biotechnology is gaining wide popularity in recent times, this is a field that has been present and slowly evolving for several centuries. This in-vogue field is commonly grouped under the broad category of life sciences and is concerned with the application of science and technology to living organisms so as to extract new products or knowledge. Biotechnology has come into the limelight recently owing to advances in recombinant DNA techniques, monoclonal antibodies, etc. The extent of resources that goes into developing new technology is varied and vast, making it essential to protect the interests of the scientists, in the form of patents and other intellectual property rights (IPR). Further, the return on investment may be slower as compared to other scientific fields, making it even more important to protect the interests of the scientists and the organization with which they are affiliated.

Biotechnology and genetic engineering techniques are generally applied for developing natural or modified human proteins as drugs (viz. interferons, tissue plasminogen activator, growth hormone and erythropoietin). This has stimulated search for potential drugs among the many natural human proteins. Biotechnology is a wide scientific field, which comprises three major divisions, namely healthcare biotechnology (Red biotechnology), agricultural biotechnology (Green biotechnology) and industrial biotechnology (White biotechnology). There may be a considerable amount of overlap between the three sectors (Spilker, 2009a; AIPLA, 2015).

Intellectual property (IP) refers to the 'creations of the mind', which range from technical inventions to literary and artistic works. Patents, copyrights, trademarks, trade secrets and designs are the different types of IP, all of which are protected by law. The creator of an IP is conferred with exclusive rights over their inventions, i.e. IPR. These IPR for a particular invention are given to the creator for a specified period during which the inventor benefits from their work. IPR need to strike a balance between science and commerce (IPIndia, 2013, 2015; Raj *et al.*, 2015).

*Corresponding author, e-mail: drgmr1111@gmail.com

21.2 History of Intellectual Property Rights

One of the earliest instances that quote this balance between science and commerce is the example of Prince Henry of Portugal. Prince Henry started off a school of navigation and cartography, thereby revealing the location of new territories, especially in the African continent. The secret of cartography was well maintained by the Empire until they had to bank on private merchants for financial reasons. Once the secret was out, the private merchants organized cruises and took a major share of the wealth looted from the new territories.

Similarly in the field of biotechnology, the concept of IPR was not established until the balance between science and commerce was disrupted.

21.3 Forms of Intellectual Property in Biotechnology

Though the most widely applied IP form is that of patents, there are other forms as well, which include trademarks, registered designs and others.

1. Patents are mainly used to protect isolated polynucleic acids, peptides, polypeptides, enzymes, viruses, antibodies, vaccines, seeds, plants, cell lines, transgenic organisms and medical devices. They also cover the methods and techniques used for the preparation of the same.
2. Copyright is a broad term that provides the right to eliminate others from copying creative inventions. The World Intellectual Property Organization (WIPO) helps in global protection of copyrights by administering the WIPO Copyright Treaty (WCT), which was enacted in 2002.
3. Trademarks are more so related to the creative aspects of invention. They cover words/names, icons, graphical designs and multimedia elements.
4. Registered designs are applicable for medical, biophysical, biochemical or bio-electrochemical devices.
5. Trade secrets are generally quoted for laboratory notebooks, design workbooks, customer information, internal process documents and for all data extracted from studies. These trade secrets are also called as 'know-how' in common language.
6. Plant variety protection rights or plant breeders' rights refer to those rights concerning plant varieties and various plant products.
7. Domain name rights are for protection of web addresses online and offline (European IPR Helpdesk, 2014).

21.4 What are Patents?

WIPO defines a patent as 'an exclusive right granted for an invention, which is a product or a process that provides, in general, a new way of doing something, or offers a new technical solution to a problem'(WIPO, 2014).

The Patents Act of India describes a patent as 'a statutory right for an invention granted for a limited period of time to the patentee by the Government, in exchange of full disclosure of his invention for excluding others, from making, using, selling, importing the patented product or process for producing that product for those purposes without his consent' (IPIndia, 2013).

21.5 Prerequisites for Patents

Before applying or considering applying for patent rights, the following criteria should be satisfied.

1. It must be an invention (should not be part of 'prior art' which is the body of existing knowledge).
2. The invention must be new (unobvious).
3. The invention should have involved an 'inventive step', which means that this crucial step in the invention process should not be easily conceivable by experts from the field.
4. It must have an industrial or practical application (Sreenivasulu and Raju, 2008; IPIndia, 2013).

21.6 What Can be Patented and What Cannot?

There is no fixed rule on what can be patented and what cannot.

21.6.1 Non-patentable inventions in India

A. Public order and morality exception – Section 3(b) – As per this section an invention would not be patentable if it is immoral or against public order, harmful to human, animal or plant life or harmful to the environment.
B. Discovery of living things or non-living substances in nature – Section 3(c).

C. Plants and animals in whole or any parts thereof other than micro-organisms but including seeds, varieties and species – Section 3(j).
D. Essentially biological processes for the production or propagation of plants and animals – Section 3(j).
E. Any process for the medicinal, surgical, curative, prophylactic, diagnostic or therapeutic or other treatment of human beings or animals to render them free of disease or to increase their economic value or that of their products – Section 3(i).
F. New use or new property of known substance – Section 3(d).
G. Methods of agriculture or horticulture – Section 3(h).
H. Traditional knowledge – Section 3(p). This section excludes from patentability 'an invention which, in effect, is traditional knowledge or which is an aggregation or duplication of known properties of traditionally known component or components.' This exclusion is but one of several provisions inserted into the new act in an effort to prevent the exercise of proprietary rights in India's genetic resources and indigenous knowledge.

21.6.2 Patentable inventions in India

A. The Manual of Patent Practice and Procedure (MPPP) regards claims to genetically modified Gene Sequence/Amino Acid Sequence, a method of expressing the sequence, an antibody against the protein/sequence, a kit containing such antibody/sequence as having a single inventive concept and capable of being granted a patent.
B. Gene sequences, DNA sequences without having disclosed their functions are not patentable for lack of inventive step and industrial application.
C. Living entities of artificial origin such as micro-organism vaccines are patentable.
D. Biological material such as rDNA, plasmids are patentable provided they are produced by substantive human intervention.
E. Processes for producing chemical and biological substances using microorganisms including lower plants and animals are patentable.
F. Modified microorganism and process thereof.
G. Process for modification/isolation of microbes.
H. Isolated nucleic acids encoding gene, first time isolation of a molecule; novel peptides, novel peptide analogues, proteins, vaccines, antibodies, recombinant: DNA, RNA, amino acids, antibodies, primers, recombinant oligonucleotides and primers, genes and process thereof; DNA-related inventions such as preparing plasmids, vectors etc.; composition/formulation thereof.
I. Cell lines – a cell line is patentable if artificially produced.
J. Hybridoma technology: patents are also allowed on hybridoma technology, but not on protoplast fusion.
K. rDNA, cDNA, r-RNA, r-antibodies. Expressed sequence tag's, or expressed sequence tags (ESTs), are small fragments of genetic material obtained by reverse transcriptions of messenger RNA (mRNA) from expressed genes. The gene sequence, or EST, can be patented if it has a use, such as if it works as a probe.
L. Bioprobes, biosensors.
M. Diagnostic/equipment kits, research tools.
N. Methods of enzyme purification.
O. Environment cleaning using biological materials such as solid or liquid waste (industrial/domestic) treatment, water treatment, mitigation of air pollution, bio-remediation.
P. Treatment of plants (AIPLA, 2015).

21.7 How to Find Patents

- Free-of-charge databases: organizations such as the European Patent Office (EPO), the WIPO as well as some National Patent Offices provide free access to patent information on their websites. For example, Google as a private initiative offers access to the full US Patents database and similarly, WIPO provides data on previous patents through Patent Information Services (PIS).
- Commercial patent databases, managed by private companies, provide access to databases covering the whole patent literature or sectoral patent information.
- Patent alerts of branch organizations. For example, the German association of plant breeders (BDP) publishes a monthly overview of new (published/granted) patents in green biotechnology (Rang, 2006; Freedman, 2009).

21.8 Indian Situation

In a country like India, strict enforcement of patent laws is very hard. If lifesaving drugs are put under stringent patent laws, the accessibility to these drugs can be very difficult owing to the high cost of acquiring rights to produce or procure such drugs. Since the inception of the Indian Patent Act in 1970, 'process patent' was given more importance than 'product patent'. Hence, the same drug could be manufactured using an alternative process ('reverse engineering') resulting in acquiring patent and selling of drugs at a cheaper price than the original innovator company. Most of the Indian pharmaceutical companies implemented this ploy and flourished.

Enforcement of the TRIPS (Trade-Related Aspects of Intellectual Property Rights) agreement under the WTO (World Trade Organization) brought about two significant changes to the Indian patenting system. The duration of patent protection was extended from a period of 7 years to 20 years and the other major modification is that patents have to be granted to the composition of the product and not just the process in making them. The shift from the process to product patent may result in a loss of around US$650 million within the next few years (Malhotra *et al.*, 2008).

The Indian pharmaceutical industry, particularly the smaller companies, sustained a massive blow as they now have to focus more on their R&D activities to compete with the international companies. Moreover, new drug development is a laborious task considering the money invested (around US$900 million). The drug prices in India can also be adversely affected by the TRIPS agreement. In pharmaceutical parlance, the first patent lawsuit in the country was between Roche, a Swiss company, and Cipla, an Indian generic company. In 2009, the Supreme Court of India upheld the decision of the Delhi High Court of not providing injunction to sales of the generic version of Tarceva (erlotinib) manufactured by Cipla. Earlier, Roche acquired the patent from the Indian Patent Office (IPO) for the anticancer drug erlotinib (Tarceva) and appealed to the Delhi High Court for prohibiting sales from Cipla. The price of Roche's patented version was Rs4800 (US$100) whereas that of the Cipla (generic version) was Rs1600 (US$33).

The patent right for Glivec (imatinib mesylate), an anticancer drug, was acquired by Novartis (a Switzerland-based pharmaceutical company) in almost 40 countries all over the world except in India. Novartis offered Glivec free of charge to 95% of patients prescribed the drug in India through public funding programmes. Even then, the Supreme Court of India denied patent rights to Novartis based on Sections 3(b) and 3(d) of the Patent Act (Amendment in 2005). Though this is a welcome move for our generic Indian companies which sell the drugs at nominal rates, it may severely jeopardize the state of multinational pharmaceutical companies in India. This may even reflect in such a way that interference in import of patented life-saving drugs from these pharmaceutical companies can occur in the future (Wall Street Journal, 2013).

Another revolutionary decision taken by the India Patent Office is to provide 'compulsory licence' for production of sorafenib (Nexavar) to Natco, a Hyderabad-based generic company. As of 2008, the Indian patent right for sorafenib was with Bayer, a German company. However, Bayer denied Natco's proposal for a voluntary licence to synthesize and sell the drug at a subsidized price. Hence, now with the acquisition of a compulsory licence, Natco can sell sorafenib at Rs8800 (US$140) compared to Bayer's Rs2.8 lakhs (US$4500) per month for those patients with hepatic or metastatic renal cell cancers.

The manufacturers of tenofovir (Viread), Gilead Sciences, were opposed against filing a patent application in India. The cost of Viread was US$5700 compared to that of the generic version, which was priced at US$800. Hence, the majority of those people living with AIDS/HIV (PLHA) around the globe consumed the generic forms of tenofovir produced in India.

Though recently the Controller General of Patents, Designs, and Trademarks of India issued a revision to the Draft Guidelines for Examination of Patent Applications in the Field of Pharmaceuticals (the 'Revised Draft Guidelines' (August, 2014)), there are certain grey areas that continue to exist in the guidelines, making India's obligations under the WTO-TRIPS agreement inconsistent.

Only if India creates a congenial environment for sales of innovative drugs, the much-hyped tag of India transforming into a 'pharmerging market' may not be possible in the near future. In summary, the advantageous patent system at present has supported the growth of the Indian pharmaceutical industry. However, now with the advent of TRIPS, it is high time that Indian companies should undertake high-quality R&D in producing novel drugs of international standard. A more comprehensive and transparent IPR policy is the need of the hour to strengthen the patent laws in India, thereby removing the obstacles for investments in India's healthcare market (Raj *et al.*, 2015).

21.9 International Treaties

Patents provide only territorial (regional) protection and are not universal rights. For instance, a patent right for a drug in the USA does not confer the same to India. Hence, patent rights for drugs have to be sought on a country-by-country basis. But owing to many inherent impediments, applying for patents in all countries at the same time is not feasible. To overcome this setback, a pharmaceutical company can initially file for a Patent Cooperation Treaty (PCT) International Application, which gives allowance for the time delay in applying to the individual countries (contracting countries) (Raj *et al.*, 2015).

Enforced in 1970, PCT is an international agreement that supports applicants (drug companies) in seeking patent protection internationally for their inventions. India signed the PCT on 7 September 1998, and now it is one of the 188 contracting countries.

Most of the countries, except the USA, follow the 'first-to-file' (FTF) system in which the patent rights are vested with the person who first files a patent application. In the USA, the 'first-to-invent' (FTI) system was followed until 16 March 2013, when the 'first-inventor-to-file' (FITF) system was introduced in accordance with the America Invents Act, and is the one now practised (USPTO, 1994; Spilker, 2009b).

Research universities and other academia are involved in rigorous research activities generating scientific evidence, so that all their efforts are translated into clinical tests and therapies in due course of time. However, time after time the academic setup has failed to extend support for the researchers in commercializing their valuable products such as various diagnostic kits. Hence, academicians or academic institutions as a whole have to team up with pharmaceutical companies in converting their scientific discoveries into a commercial product which can be utilized at the bedside. In this process, the institution has to pass on a patented invention to a third-party private company though with a licensing fee. Many ethical and policy issues may develop due to these transitions in patenting rights.

Universities also tend to practise a bitter system wherein they obtain a transfer agreement ('reach through') from subsequent researchers who work in the same area as that of the original patent on a new invention possessed by the university. The term 'reach through' is a diabolic statement in the context of patent rights as it is a practice of conditioned licensing where the successive researchers have to share the IPR of any additional discoveries they make by using the original patented invention (research tool). Hence, it can become a great hindering force in further research activities in the same field (Rang, 2006; Lo, 2010).

21.10 Patent Law and Genetics

Anything under the sun is patentable, if and only if it has a human intervention. This principle is in accordance with the 'product-of-nature' doctrine, which prohibits the patenting of products originating in the natural world. The 'product-of-nature' doctrine forbids patenting of products originating in the natural world; hence, human intervention is a *sine qua non* element. Therefore, human intervention is the critical demarcation between the genes and proteins found inside the body and the DNA and protein sequences that have been patented. Hence, only genes and proteins that have been purified, isolated or modified can be patented. Even genetically engineered plants and animals can be patented.

Unlike patents in other fields, other inventors cannot 'invent around' the discovery of a DNA sequence or a functional gene, as doing so would infringe upon the original patent.

Patenting human genes is controversial as the impact of such patenting practices may endanger future biomedical research and genetic testing of gene-related diseases. Myriad Genetics Inc. of Utah, USA had acquired the patent for *BRCA1* and *BRCA2* gene mutation testing and was offering the test at an exorbitant rate of around US$3000. None the less, in 2013 the US Supreme Court revoked the patent rights from Myriad Genetics on the grounds that 'a naturally occurring DNA segment is a product of nature and not patent eligible merely because it has been isolated'. Hence, obtaining or providing patents for genes or gene-related products/processes is indeed a complex state (Human Genomic Variation, 2007; Lo, 2010).

21.11 Conclusion

The capacity to obtain a patent to a known protein derived by biotechnology techniques has been debatable and precisely what can actually be patented varies in different parts of the world. Patents for biotechnology products range from acquiring patents for procedures involved in the production, purification or procedures used for expressing the gene. This has meant that having a patent on a biotechnology product is sometimes insufficient to simply allow the patent holder to make their patented product because of other blocking patents that exist (e.g. for a purification step). However, the technology has also advanced so rapidly that the methods used to develop and manufacture the first generation of biotech products may well be irrelevant to those developed subsequently.

Hence, patents are indeed the sole custodians of any creation – including biotechnological and pharmaceutical inventions.

References

AIPLA (2015) Biotechnology patenting in India and related issues. Available at: http://www.aipla.org (accessed 31 October 2015).

European IPR Helpdesk (2014) Intellectual property in Biotechnology – Fact sheet. Available at: http://www.iprhelpdesk.eu. (updated June 2014; accessed 31 October 2015).

Freedman, T. (2009) *Career Opportunities in Biotechnology and Drug Development*, 1st edn. Cold Spring Harbor Laboratory Press, New York.

Human Genomic Variation (2007) In: Campbell, M.A. and Heyer, L.J. (eds) *Discovering Genomics, Proteomics, & Bioinformatics*, 2nd edn. Dorling Kindersley Pvt. Ltd, Noida, Uttar Pradesh, India.

Intellectual Property India (IPIndia) (2013) The Patents Act, 1970. Available at: http:// ipindia.nic.in/ipr/patent/patent_Act_1970_28012013_book.pdf (updated 26 January 2013; accessed 31 October 2015).

Intellectual Property India (IPIndia) (2015) Manual of Patent Office Practice and Procedure. Available at: http://www.ipindia.nic.in (updated 22 March 2011; accessed 31 October 2015).

Lo, B. (2010) Patenting and Commercialization of Discoveries. In: *Ethical Issues in Clinical Research: A Practical Guide*, 1st edn. Lippincott Williams & Wilkins, Philadelphia, Pennsylvania.

Malhotra, S., Shafiq, N. and Pandhi, P. (2008) *A Comprehensive Clinical Research Manual.* Jaypee Brothers Medical Publishers (P) Ltd, New Delhi.

Raj, G.M., Priyadarshini, R. and Mathaiyan, J. (2015) Drug patents and intellectual property rights. *European Journal of Clinical Pharmacology* 71(4), 403–409.

Rang, H.P. (2006) *Drug Discovery and Development – Technology in Transition*, 1st edn. Elsevier Limited, Philadelphia, Pennsylvania.

Spilker, B. (2009a) Biotechnology. In: Spilker, B. (ed.) *Guide to Drug Development – A Comprehensive Review and Assessment.* Lippincott Williams & Wilkins, Philadelphia, Pennsylvania, pp. 119–131.

Spilker, B. (2009b) Patent Activities and Issues. In: Spilker, B. (ed.) *Guide to Drug Development – A Comprehensive Review and Assessment.* Lippincott Williams & Wilkins, Philadelphia, Pennsylvania, pp. 1144–1158.

Sreenivasulu, N.S. and Raju, C.B. (2008) *Biotechnology and Patent Law: Patenting Living Beings*, 1st edn. Manupatra Information Solutions P Limited, Noida, Uttar Pradesh, India.

United States Congress, Office of Technology Assessment (1989) *New Developments in Biotechnology: PatentingLife.* Special Report, OTA-BA-370. US Government Printing Office, Washington, DC.

United States Patent and Trademark Office (USPTO) (1994) Available at: http://www.uspto.gov/aia_implementation/ faqs_first_inventor.jsp (updated 28 October 2015; accessed 31 October 2015).

Wall Street Journal (2013) Novartis loses Glivec patent battle in India. Available from: http://www.wsj.com/articles/SB10001424127887323296504578395672582 (updated 1 April 2013; accessed 31 October 2015).

World Intellectual Property Organization (WIPO) (2014) *What is Intellectual Property?* Available at: http://www.wipo.int/about-ip/en/index.html#ip (accessed 31 October 2015).

22 Leishmaniasis: Drug Development and Intellectual Property Issues

Ashim Borah,* Nitai Basak and Asim Kumar Jana

National Institute of Technology, Jalandhar, Punjab, India

22.1 Introduction

Leishmaniasis is a vector-borne systemic infection caused by members of the dimorphic genus of parasitic intracellular protozoa *Leishmania*. The parasites causing the disease are transmitted to the mammalian host by 20 different types of female sandflies (Pearson and Sousa, 1996). Initially, the protozoa exists as an extracellular promastigote form, which after internalization into mononuclear cells of a mammalian host, reproduces and finally transforms into the intracellular, aflagellate amastigote form (Sacks and Kamhawi, 2001). Leishmaniasis complexity varies ranging from a simple self-healing, asymptomatic localized skin infection or lesion to development into a dreadful life-threatening visceral form of the disease in severe cases. The most common symptoms in patients with leishmaniasis include cough, abdominal pain, diarrhoea, fever, epistaxis, splenomegaly and hepatomegaly.

Leishmaniasis, being amongst the world's most neglected illnesses, primarily affects the unprivileged or indigent individuals of the developing nations. It is predominant across the tropical and subtropical districts of Asia, Africa, America, the Mediterranean and southern Europe. The disease is endemic in around 90 countries, of which 75 are developing countries (Alvar *et al*., 2012). The transmission of *Leishmania* parasites usually occurs through two distinct modes: (i) anthroponotic, in which the infectious protozoan parasite carried by humans is transferred to other animals through a female sandfly vector; and (ii) zoonotic, where dogs and rodents transmit the infection to humans via phlebotomine sandflies vector (Alvar *et al*., 2004). The former mode of transmission is found to be common in Asia, while the latter mode is prevalent in Africa, Europe and the Americas. Visceral leishmaniasis is the most dreadful parasitic infection, which is fatal in nature. A recent study shows that about 92% of all the cases of visceral leishmaniasis (VL) reported worldwide occur in Asian countries including India, Bangladesh and Nepal. Brazil and Sudan are also major contributors to VL cases (Postigo, 2010). VL is transmitted through different varieties of sandflies in different regions. For example, *Leishmania donovani* and *Leishmania infantum* are the main causal agents of VL in the Indian subcontinent and the Mediterranean basin, respectively (Reithinger *et al*., 2007). The internalized protozoan parasite does not affect the mammalian host immediately. It undergoes an initial incubation period of a few months, wherein differentiation of inactive promastigotes into active amastigotes takes place. Thereafter, the active amastigotes start affecting the mammalian host, which results in development of different symptoms. Starting with the typical symptoms of intermittent fever, weight loss and massive hepatosplenomegaly, it advances to haemorrhages, epistaxis, cachexia and pancytopenia. Due to the growing prevalence and severity of leishmaniasis infection, the World Health Organization (WHO, 2007) has targeted to eliminate leishmaniasis by 2015. Due to the absence of antileishmanial vaccines in the current market, our prime dependence on chemotherapy for treating VL in the current scenario has increased significantly. Chemotherapy poses severe side-effects to patients post-treatment.

*Corresponding author, e-mail: ashim.borah087@gmail.com

For several decades, pentavalent antimonials have been used as the primary antileishmanial compound for effective treatment of leishmaniasis. However, recently, a few cases of leishmaniasis have been reported in Bihar (India), where the protozoan parasites have shown a developed resistance to the pentavalent antimonials, thereby jeopardizing their clinical efficacy (Sundar *et al.*, 2007).

A new phenomenon which is now posing significant challenges to the efficient control of VL is the development of human immunodeficiency virus (HIV)-VL co-infection during treatment process. This has now provoked us to further advance our knowledge of new metabolic pathways and drug targets in the parasite and thereafter develop more stringent new alternatives to treat leishmaniasis. Therefore the identification and selective screening of novel chemical compounds which have effective antileishmanial properties as well as low cytotoxicity has become imperative.

22.2 *Leishmania* spp. Life cycle and Transmission

Leishmania transmission primarily occurs via the bite of infected female sandflies belonging to one of two genuses: *Phlebotomus* and *Lutzomyia*. Out of around 500 known species of phlebotomine sandfly, there are only 25 species that are considered as competent vectors (Sharma and Singh, 2008). The dimorphic life cycle of *Leishmania* protozoan parasites alternates between the phlebotomine sandfly vector and the mammalian host. Infected sandflies can inoculate a host during a blood meal. Within the vector, the protozoan parasites exist as highly motile flagellated promastigotes. Promastigotes are long and slender (approx. 23 μm × 3 μm), enveloped in a thick glycocalyx composed of a variety of extensively branched, glycosylphosphatidylinositol-anchored glyco-conjugates, including lipophosphoglycan and the metalloprotease glycoprotein (Dedet *et al.*, 1999). Once the promastigotes are introduced into a mammalian host, they are further internalized by different phagocytic cells such as macrophages or dendritic cells (Sacks and Sher, 2002). After phagocytosis, maturation of promastigotes from early phagosome phase to late phagosomes takes place inside membrane-bound organelles. Thereafter, fusion of late phagosomes with lysosomes takes place to form phagolysosomes (Korner *et al.*, 2006). Inside the host cells, under appropriate environmental stimulation promastigotes differentiate into the amastigotes. The time period for this overall process is around 24–72 h. Amastigotes (2–6 μm size) have a vestigial flagellum covered by a densely packed glycocalyx. The glycocalyx is primarily composed of two types of lipids, namely glycoinositol phospholipids and glycosphingolipids. Amastigotes survive and multiply within the harsh, highly acidic (pH 4.0) environment of the phagolysosomes by maintaining a neutral internal pH. When a sandfly bites an infected host, the amastigotes-containing macrophages are ingested. The internalized amastigotes are released in the sandfly midgut during digestion after which the process of differentiation into promastigotes begins.

22.3 Diagnosis of Leishmaniasis

Demonstration of the protozoan parasite in smears or biopsies is the primary step in the effective diagnosis of leishmaniasis. The primary sources of tissue specimen for diagnosis of *Leishmania* parasite includes spleen, bone marrow or lymph nodes (for VL), and skin lesions of the infected person (for cutaneous leishmaniasis; CL) (Singh *et al.*, 2006). The obtained tissue smear is fixed with absolute alcohol and stained with Giesma or Leishman stain before examination under a light microscope. The light microscopic view of a stained tissue smear containing amastigotes shows the cytoplasm in pale blue and the nucleus in red colour. It also shows the presence of a rod-shaped structure adjacent to the nucleus, which is known as kinetoplast. The visualization of kinetoplast under the microscope is the final confirmation of the patient contacting the leishmaniasis parasite (Singh *et al.*, 2006).

Serology tests, such as enzyme-linked immunosorbent assay (ELISA) (Mittal *et al.*, 1991), indirect immunofluorescence (IFAT) (Edrissian and Darabian, 1979) and direct agglutination test (DAT) (El-Safi and Evans, 1989), are the alternatives to conventional diagnosis based on the clinical presentation and visualization of the parasite (Jensen *et al.*, 1996). Parasite DNA can also be detected by use of the polymerase chain reaction (Andresen *et al.*, 1997).

22.4 Treatment

The geographical location and different species of *Leishmania* involved in the infection are two major factors taken into consideration while treating leishmaniasis. Targeting amastigotes with specific antileishmanial compounds is the main concern

Table 22.1. Summary of characteristics of antileishmanial drugs in current use.

Drug	Trade name	Route of administration	Regimen	Drug price (US$)	Average cure rate (%)	Issues
Pentavalent antimonials	Pentostam, Sodium stibogluconate, Glucantime	Intramuscular	20 mg/kg daily for 30 days	15–150	50	Resistance, toxicity in HIV-co-infection
Amphotericin B	Fungizone	Intravenous	7–20 mg/kg total dose for up to 20 days	60–150	97	Toxicity
Amphotericin B lipid complex	Ambisome	Intravenous	10–20 mg/kg total dose in 5–10 doses over 10 days	1000–2500	98	Expensive, retreatment of antimonial failures/ Intolerance and HIV-leishmania co-infection
Miltefosine	Miltex	Oral	100 mg/kg for 28 days	300	94	Resistance

while treating leishmaniasis in humans. The pentavalent antimonials sodium stibogluconate and meglumine antimoniate have been used as standard drugs for treating leishmaniasis across different geographical locations for more than seven decades. As there was no drug for leishmaniasis until 1996, an alternative treatment was Pentostam™. Intravenous liposomal amphotericin B was approved by the US FDA for the treatment of VL in 1997. ThermoMed, a device that heats the skin with radio waves, was registered in 2003 for CL. Most of the available drugs have limitations, such as cytotoxicity and problems with liver function. Antimonials have severe side-effects, including cardiac arrhythmia and acute pancreatitis. Amphotericin B leads to adverse side-effects, such as infusion-related fever, chills, hypokalaemia, nephrotoxicity and anaphylaxis. Researchers have been working to find a better antileishmanial agent with reduced side-effects. Recently, in 2014, the US FDA approved miltefosine as the drug for treating all forms of leishmaniasis. Commercially available drugs used for leishmaniasis along with their specifications are tabulated in Table 22.1.

22.5 Drugs Used for Treatment of Leishmaniasis

In the current scenario, there are only about 30 potential compounds with antileishmanial effects, awaiting US FDA approval. A major portion of initial potential drug candidates becomes screened out in the clinical trial stage. Pentavalent antimonial compounds, glucantime (meglumine antimoniate) and pentostam (sodium stibogluconate) are still the first-line treatment drugs (Dedet *et al.*, 1999). As second-line treatments, amphotericin B, pentamidine, aminosidine and miltefosine are recommended. Currently, the use of antimonial compounds has been restricted owing to shortcomings they produced. Its effectiveness against CL and VL has shown variable results as reported in various research articles. Moreover, the emergence of significant resistance by parasites against these antimonials has increased with time (Herwaldt, 1999).

To reduce the high cost of the treatment involved, several new generic versions of pentostam are produced (Gasser *et al.*, 1994). Intra-lesional administration may be a therapeutic choice but each lesion has to be injected individually, moreover, the potential dispersion of infection is still not guaranteed (Santos *et al.*, 2008). Amphotericin B, because of its high leishmanicidal activity, constitutes a choice in patients who indicated imperviousness to treatment with antimonials (Berman, 2010). The real constraining variable for the utilization of this medication is its cytotoxicity. Currently, toxic impacts of amphotericin B have been to a great extent improved with the approach of new lipid preparations. In these preparations, deoxycholate has been superseded by different lipids that veil amphotericin B from sensitive tissues. This way the reticulo-endothelial cells

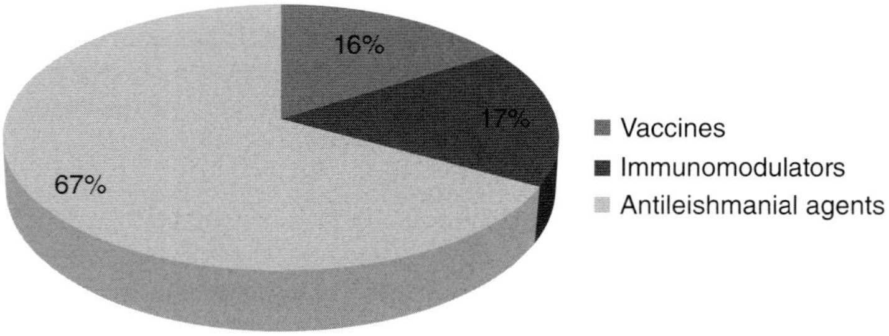

Fig. 22.1. Percentage distribution of major developed products in the current drug market.

can preferentially uptake the amphotericin B drug. Thus, this method of drug delivery serves the dual role of increased efficacy and reduced cytotoxicity. Three lipid-related preparations of amphotericin which are economically accessible include AmBisome, Abelcet and Amphocil (Desjeux, 2004). AmBisome has undergone various experimentations by different research groups. Some studies even demonstrated its success in patients with CL and VL, particularly in areas where antimonial resistance has been detected (Mishra *et al.*, 2010). Pentamidine is one of the medications for clinical utilization in all types of leishmaniasis. However, distinct scientific tests regarding pentamidine usefulness throughout Colombia and Peru exhibited inconsistent results. In general, the use of this drug has declined due to its low efficacy and toxicity.

Treatment of VL through a novel oral medication of miltefosine in immunocompromised patients has been considered as a standout amongst the most significant recent advances in antileishmanial therapy. However, few studies have exhibited the obliviousness of *Leishmania* spp. from the new world, including: *L. braziliensis*, *L. guyanensis* and *L. mexicana* (Pearson and Sousa, 1996). The results were disappointing because it was expected that miltefosine would provide a better alternative to drug therapy in America. Other drugs have been clinically used such as azoles (fluconazole, ketoconazole, itrakonazole and posaconazole), allopurinol and sitamaquine. In addition, the use of products that stimulate the immune system is another alternative in leishmaniasis treatments.

22.6 *Leishmania*-Related Patent Applications of Various Potential Antileishmanial Agents

The literature review revealed that there were around 8515 patents granted between 1974 and 2010 having the keyword 'leishmania'. Of these, 867 patents were classified according to the developed product: vaccines (139 patents), immunomodulators (148 patents) and drugs (580 patents). Figure 22.1 shows the percentage distribution of these three developed products in the current patent market. The major onus (67%) corresponds to antileishmanial agents, which demonstrates the high interest of studies to improve the alternative treatment.

Further, the distribution of patents filed on treatment of leishmaniasis in the last two decades is graphically represented in Fig. 22.2. The chart shows a gradual increase in each year, with a sudden increase of articles on patent filing in the last decade. There are basically three important explanations that justify this event. First, in recent years sophisticated spectroscopic techniques and novel methods of synthesizing potent compounds have been developed by chemical innovation. Second, the concept of 'virtual drug screening' is now possible with the knowledge of parasite genome sequences obtained from various bioinformatics databases. This has offered the opportunity for drug discovery based on the identification and validation of new molecular drug targets. Finally, a great deal of networking and research collaborations among scientific communities have been taken into account specifically to pursue tropical diseases research. The Drug for Neglected Diseases Initiative (DNDi) is a project producing significant results in the study of tropical diseases.

Different strategies have been developed to search new antileishmanial drugs: (i) optimization of therapy with existing antileishmanial drugs; (ii) development of analogues from current antiprotozoal drugs and identification of agents to be used against other indications; (iii) combination therapy with conventional drugs; (iv) search for new rational drug design; and (v) the exploration of natural products as a source of antileishmanial

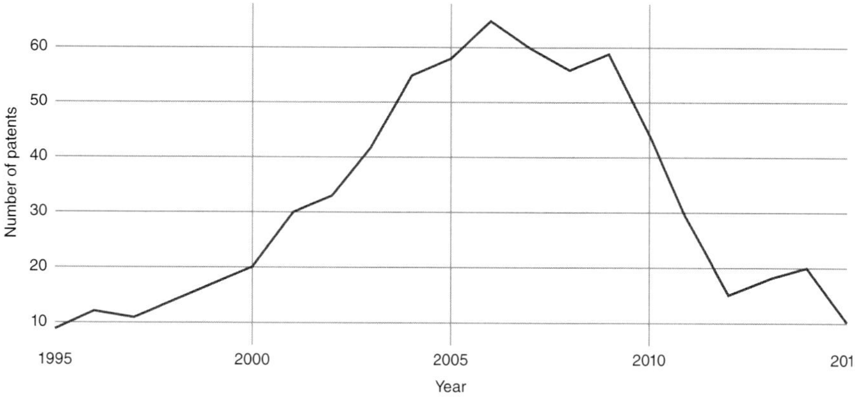

Fig. 22.2. Graph showing the distribution of various patents filed on treatment of leishmaniasis from 1995 to 2015 (Google patents with key word 'leishmaniasis' as accessed on 15 October 2015).

Table 22.2. Antileishmanial compounds with their respective antileishmanial activity (Rama *et al.*, 2013).

Antileishmanial compounds	Antileishmanial activity – IC_{50} (µM)
Rhodacyanin derivatives	0.09
Aryl isoquinoline derivatives	2.02
Semisynthetic almiramide analogue	1.6
Nitrobenzylphosphoramide mustard analogue	5.13
Aza-analogues of linear triaryldications	0.72
Chalcone derivatives	4.6
Simple phenyl derivatives	0.87
Sodium stibogluconate	2.7

agents (Na-Bangchang and Karbwang, 2009). In this sense, 580 patents have been compiled, assessed and categorized agreeing the innovator's methodology in the extension of antileishmanial drug discovery (Table 22.2).

An *in vitro* evaluation of amastigotes and promastigotes has been reported by Rama *et al.* (2013) in a patent on oxoisoaporphines derivatives against *L. amazonensis*, *L. infantum*, *L. braziliensis* and *L. guyanensis*. Diaryltriazines are very potent HIV-1 non-nucleoside reverse transcriptase inhibitors. Based on this application, Augustyns disclosed a patent on di-substituted triazine dimers as anti-infective agents (Rama *et al.*, 2013). They found that the diaryltriazine monomers not only exhibit antileishmanial activity but also antimalarial and anti-HIV properties. Nitroimidazole drugs are mainly used to treat anaerobic microbial infections. The Global Alliance for TB Drug Development patented the nitroimidazo oxazine and the oxazole analogues as useful compounds for the treatment of both *Mycobacterium tuberculosis* and *Leishmania donovani*.

The Aoyama Gakuin Educational Foundation patented the extract of the brown algae *Sargassum yamadae*. The invention was based on *in vivo* and *in vitro* analysis of antileishmanial activity (Rama *et al.*, 2013). This extract can be given as a dietary supplement or in any other dosage form to treat leishmaniasis patients. The compound showed antileishmanial activity that is at par to amphotericin B.

A few researchers from Hoshi University have carried out *in vitro* studies on the pharmaceutical composition of rhodacyanine and benzo[a]phenoxanthin derivative against *L. donovani* (Rama *et al.*, 2013). In their patent, it was disclosed that both of the compounds can be used for the prevention or treatment of leishmaniasis. It was found that the introduction of fluorine atoms plays a major role in antileishmanial activity. All 15 benzo[a]phenoxanthin derivatives showed antileishmanial activity, but it was not as effective as miltefosine. Monzonte (2011) filed a patent on hybrid compounds of 4- and 8-aminoquinolines for the treatment of malaria and other parasitic diseases, including leishmaniasis (Rama *et al.*, 2013). The use of aryl isoquinoline derivatives as anti-infective agents, such as anticandida, antileishmania, antitrypanosoma and antiplasmodium, as well as for the prevention of biofilm formation by human pathogenic bacteria, was invented by Bringmann.

Sereno filed a patent on the methods for treating leishmaniasis and trypanosomiasis with α-hydoxyacetophenone and N-hydroxybenzamide (Rama *et al.*, 2013). It was claimed that the metronidazole derivatives are active specifically against *L. donovani* and *L. tropica* promastigotes. The *in vitro* antileishmanial activity of these compounds against *L. donovani* and *L. tropica* showed better activity than metronidazole, but were not as effective as amphotericin B.

Spinosyns are the fermented products derived from *Saccharopolyspora spinosa* bacteria. Established uses of spinosyns as insecticides, pesticides and for wound healing in mammals have been summarized in the patent disclosed by Kritikou (Rama *et al.*, 2013). The spinosad (~85% spinosyns A and 15% spinosyns D) and spinetoram (semisynthetic spinosyns) are the main compounds investigated in the patent for their use against cancer and viral and protozoal infections.

The activity against *Leishmania* parasites was determined by *in vitro* and *in vivo* tests conducted on *L. donovani* promastigote- and amastigote-infected mice macrophages. From this, it was concluded that the spinosad is active against parasites, with IC50 values of 9.5 ± 3.8 μM for amastigotes and 9.0 ± 3.0 μM for promastigotes.

A group of inventors from Ohio State University conducted *in vitro* and *in vivo* studies for leishmaniasis on the root extracts of *Pentalinon andrieuxii* separated from the hemisynthetic compounds for antileishmanial activity (Rama *et al.*, 2013). Out of 20 natural compounds analysed, 14 were sterols, three were coumarins, two were pentalinonosides and one each belonged to pentalinonsterols and triterpenes. Among these, one of the sterol compounds (6,7-dihydroneridienone) was potent against *L. mexicana* promastigotes. Sodium stibogluconate was used as a control for comparing the antileishmanial activity of the patented active compounds.

In parallel, the design of specific inhibitors as possible means of controlling the parasites' growth without damaging the host was explored. It was possible due to the knowledge of potential target in *Leishmania* parasite. Several synthetic compounds have been patented and it demonstrated their antileishmanial potency by inhibition of the enzymatic activities, such as cysteine proteases, histone deacetylase, lipoxygenase etc. Black and Beaulieu filed a patent on cysteine protease inhibitors.[1]

Acea Biotech filed a patent on novel water-soluble polyene macrolides. They have also provided methods and novel drug formulations for the prevention, treatment and inhibition of both fungal and parasitic diseases with improved bioavailability and reduced toxicity. The *in vitro* efficacy of Corifungin™ (Acea Biotech) was evaluated against *L. donovani* and its IC_{50} was 0.09 μM, which was comparable to the activity of amphotericin B (IC_{50} 0.05 μM).

Many patent applications featuring novel methods for preventing, treating and detecting leishmaniasis have been filed. Table 22.3 features a few novel patents that have been successful in their proposed applications and are now available to diagnose or cure leishmaniasis.

22.7 Natural Products with Antileishmanial Properties

Natural sources, either as pure compounds or as standardized extracts, provide unlimited opportunities for new drug leads. In particular, the use of plants as natural extracts has a long history of efficacy and safety in human populations. A few studies have been embraced to assess the inhibitory impacts of natural extracts on *Leishmania* parasites, together with patented options. The *in vitro* and *in vivo* activity of triterpene saponins isolated from the plants *Maesa balansae* and *M. lanceolata* has demonstrated both prophylactic and therapeutic activity against visceral and cutaneous species of *Leishmania* (Rama *et al.*, 2013).

On the other hand, the use of phospholipases from *Crotalus durissus terrificus* venom as a biocide component for the treatment of viral, bacterial and parasitic infections, such as infections caused by *Plasmodium* and *Leishmania*, were evaluated. Other natural products that have been explored include syringomycinlipodepsipeptide isolated from a plant bacterium *Pseudomonas syringae*,[2] alkaloids from *Picralima nitida*,[3] quassinoids from plant extracts of Simaroubaceae family and monoterpene compounds such as limonene and pinene extracted from essential oil of citrus.[4]

The development of natural products as a treatment strategy should be complemented with pharmaceutical technology, with the aim to design herbal standardized products or conventional medicaments, which could be more available to the endemic population. The validation of traditional natural preparations used by natives could reduce the cost and studies.

Table 22.3. Novel patented compounds with their proposed applications.

Patent title	Publication number	Novel applications
Amine containing flavonoids	US 20150239870 A1	The present invention provides novel compounds with highly potently anti-leishmaniasis activity, whilst displaying low toxicity. The invention provides compounds which are water-soluble and orally bioavailable, and suitable for formulation into anti-leishmaniasis compositions suitable for oral administration, whilst achieving the desired efficacy.
Dronedarone	US 20150150841 A1	Several advantages such as decreasing the dose of administrated drugs to avoid side effects and avoid apparition of resistances to the chosen treatment with time. One additional feature of such formulation is to avoid/reduce toxicity reactions of the skin on contact with the formulation.
Polypeptides for the diagnosis and therapy of leishmaniasis	US8865180 B2	Useful for the detection of anti-leishmanial antibodies in individuals where the immune responses are elicited against *Leishmania* species of Indian strains and similar or closely related to Indian *Leishmania* strains, the compounds are also useful as a vaccine and therapeutic agent to prevent and treat leishmaniasis.
Leishmania antigens for use in the therapy and diagnosis of leishmaniasis	US6638517 B2	The present invention provides compositions and methods for preventing, treating and detecting leishmaniasis, as well as for stimulating immune responses in patients.
Liposome carriers in chemotherapy of leishmaniasis	US05891257	The present invention relates to a novel technique in treatment of leishmaniasis, consisting in the incorporation of an anti-leishmanial drug into liposomes, and introduction of the 'encapsulated' agent into the body of an infected mammal. By this procedure, the effectiveness and duration of action are prolonged, and also drug toxicity is decreased.
Recombinant polyprotein vaccines for the treatment and diagnosis of leishmaniasis	US 20130177584 A1	The present invention is generally directed to compositions and methods for preventing, treating and detecting leishmaniasis.
Leishmaniasis antigen detection assays and vaccines	US 20130236484 A1	The present invention relates to isolated visceral leishmaniasis (VL) antigens that are useful in therapeutic and vaccine compositions for stimulating a VL specific immunological response.
Leishmania vaccine using sand fly salivary immunogen	US 20140294875 A1	The present invention provides vectors that contain and express *in vivo* or *in vitro* sand fly *L. longipalpis* salivary antigens that elicit an immune response in animal or human against Leishmania.

22.8 Conclusion

It has been observed that diverse chemical scaffolds with antileishmanial activity have been disclosed by different inventors, from simple molecules (e.g. 2,5-dihydroxybenzene derivatives) to more complex ones (e.g. spinosad). Some of the inventors have verified the activity of their inventions *in vitro*, while others have demonstrated their activity *in vivo*. However, none of these compounds was screened against all *Leishmania* species. In addition to their complexity, the number of compounds in a particular patent varied from 5 to 100. These antileishmanial compounds may also act as anticancer, antimalarial and antitrypanosomal agents.

With the passage of time, more researchers have been working on antileishmanial compounds, which is evident from the increasing number of research papers and patents. However, mode of action was not addressed in most of the patents. It is very difficult to compare or rank the compounds unless the mode of action is known. Moreover, these compounds were not screened against all *Leishmania* species. In parallel, an increment of holistic comprehensions of new metabolic pathways in the parasite could offer better treatments for the fate of antileishmanial medications. The periodic monitoring of the delivery mechanism, clinical response and resistance of available antileishmanials is crucial for maintaining their potency and lifespan.

Overall, the development of antileishmanials has been generally slow, which calls for development of new drugs. Miltefosine is the only oral drug that has shown significant efficacy against VL. As disclosed in a recent report, in Nepal, relapse with miltefosine was observed in one-fifth of the patients.[2] Hence, there is scope for further research into antileishmanial agents.

Notes

1 Cameron Black, Christian Beaulieu, 26 April 2012. *Cathepsin cysteine protease inhibitors for the treatment of various diseases*. US 2012/0101053 A1.

2 Takemoto, J.Y., Forney, J.R., Healey, M.C., Yang, S. and Werbovetz, K.A., 30 October 2001. *Method of controlling protozoan infections using syringomycin-family lipodepsipeptide*. US6310037 B1.

3 Iwu, M.M., Klayman, D.L., Jackson, J.E., Tally, J.D. and Andersen, S.L., 1 March 1994. *Alkaloids of Picralima nitida used for treatment of protozoal diseases*. US5290553 A.

4 Chritian Diehl and Chami De Diehl Silvia, 17 April 2008. *Topical and intravaginal microbicidal and antiparasitic compositions comprising quassinoids or quassinoids containing plant extracts*. US20080089958 A1.

Acknowledgement

AB is grateful to NIT Jalandhar for financial support.

References

Alvar, J., Carmen, C., Ricardo, M., Javier, M. and Javier, N. (2004) Canine leishmaniasis. *Advances in Parasitology* 57, 1–88.

Alvar, J., Velez, I.D., Bern, C., Herrero, M., Desjeux, P. and Cano, J. (2012) Leishmaniasis worldwide and global estimates of its incidence. *PLoS One* 7(5), e35671.

Andresen, K., Gasim, S., Elhassan, A.M. and Kharazmi, A. (1997) Diagnosis of visceral leishmaniasis by the polymerase chain reaction using blood, bone marrow and lymph node samples from patients from the Sudan. *Tropical Medicine & International Health* 2, 440–444.

Berman, J. (2010) Recent developments in leishmaniasis: epidemiology, diagnosis and treatment. *Current Infectious Disease Reports* 7, 33–38.

Dedet, J.P., Pratlong, F., Lanotte, G. and Ravel, C. (1999) Cutaneous leishmaniasis: the parasite. *Clinics in Dermatology* 17, 261–268.

Desjeux, P. (2004) Leishmaniasis: current situation and new perspectives. *Comparative Immunology, Microbiology and Infectious Diseases* 27, 305–318.

Edrissian, G.H. and Darabian, P. (1979) A comparison of enzyme-linked immunosorbent assay and indirect fluorescent antibody test in the serodiagnosis of cutaneous and visceral leishmaniasis in Iran. *Transactions of the Royal Society of Tropical Medicine and Hygiene* 73, 289–292.

El-Safi, S.H. and Evans, D.A. (1989) A comparison of the direct agglutination test and enzyme-linked immunosorbent assay in the serodiagnosis of leishmaniasis in Sudan. *Transactions of the Royal Society of Tropical Medicine and Hygiene* 83, 334–337.

Gasser, R.A., Magill, A.J., Oster, C.N., Franke, E.D., Grogl, M. and Berman, J.D. (1994) Pancreatitis induced by pentavalent antimonial agents during treatment of leishmaniasis. *Clinical Infectious Disease* 18, 83–90.

Herwaldt, B.L. (1999) Leishmaniasis. *Lancet* 345, 1191–1199.

Jensen, A.T., Gaafar, A., Ismail, A. and Theander, T.G. (1996) Serodiagnosis of cutaneous leishmaniasis: assessment of an enzyme-linked immunosorbent assay using a peptide sequence from Gene B Protein. *The American Journal of Tropical Medicine and Hygiene Use* 55, 490–495.

Korner, U., Fuss, V., Steigerwald, J. and Moll, H. (2006) Biogenesis of leishmania major-harboring vacuoles in murine dendritic cells. *Infection and Immunity* 74, 1305–1312.

Mishra, J., Saxena, A. and Singh, S. (2010) Chemotherapy of leishmaniasis: past, present and future. *Current Medicinal Chemistry* 14, 1153–1169.

Mittal, V., Bhatia, R. and Sehgal, S. (1991) Serodiagnosis of Indian kala-azar: evaluation of IFA, ELISA, and CIEP tests. *Journal of Communicable Disease* 23, 131–134.

Monzonte, L. (2011) Antileishmanial current drugs and relevant patents. *Recent Patents on Anti-infective Drug Discovery* 6, 1–26.

Na-Bangchang, K. and Karbwang, J. (2009) Current status of malaria chemotherapy and the role of pharmacology in antimalarial drug research and development. *Fundamental and Clinical Pharmacology* 23, 387–409.

Pearson, R.D. and Sousa, A.Q. (1996) Clinical spectrum of leishmaniasis. *Clinical Infectious Diseases* 22, 1–13.

Postigo, J.A. (2010) Leishmaniasis in the World Health Organization Eastern Mediterranean region. *International Journal of Antimicrobial Agents* 36, 62–65.

Rama, M., Kumar, N.V. and Balaji, S. (2013) A comprehensive review of patented antileishmanial agents. *Future Science* 4, 37–56.

Reithinger, R., Dujardin, J.C., Louzir, H., Primez, C., Alexander, B. and Brooker, S. (2007) Cutaneous leishmaniasis. *Lancet Infectious Disease* 7, 581–596.

Sacks, D. and Kamhawi, S. (2001) Molecular aspects of parasite–vector and vector–host interactions in leishmaniasis. *Annual Review of Microbiology* 55, 453–483.

Sacks, D. and Sher, A. (2002) Evasion of innate immunity by parasitic protozoa. *Nature Immunology* 3, 1041–1047.

Santos, D.O., Coutinho, C.E., Madeira, M.F., Bottino, C.G., Vieira, R.T. and Castor, H.G. (2008) Leishmaniasis treatment – a challenge that remains: a review. *Parasitology Research* 103, 1–10.

Sharma, U. and Singh, S. (2008) Insect vectors of leishmania: distribution, physiology and their control. *Journal of Vector Borne Diseases* 45, 255–272.

Singh, R.K., Pandey, H.P. and Sundar, S. (2006) Visceral leishmaniasis (kala-azar): challenges ahead. *Indian Journal of Medical Science* 123(3), 331–344.

Sundar, S., Chakravarty, J. and Rai, V.K. (2007) Amphotericin B treatment for Indian visceral leishmaniasis: response to 15 daily versus alternate-day infusions. *Clinical Infectious Diseases* 45, 556–561.

World Health Organization (2007) *Global plan to combat neglected tropical diseases 2008–2015.* Available at: http://apps.who.int/iris/bitstream/10665/69708/1/WHO_CDS_NTD_2007.3_eng.pdf (accessed December 2015).

Index